T0380550

Electrostatics and Magnetostatics

Reiner M. Dreizler • Cora S. Lüdde

Electrostatics and Magnetostatics

Springer

Reiner M. Dreizler
Institut f. Theoretische Physik
Goethe Universität Frankfurt
Frankfurt/Main, Hessen, Germany

Cora S. Lüdde
Institut f. Theoretische Physik
Goethe Universität Frankfurt
Frankfurt/Main, Hessen, Germany

ISBN 978-3-662-69932-4 ISBN 978-3-662-69933-1 (eBook)
https://doi.org/10.1007/978-3-662-69933-1

Translation from the German language edition: "Theoretische Physik 2" by Reiner M. Dreizler and Cora S. Lüdde, © Springer-Verlag Berlin Heidelberg 2005. Published by Springer Berlin Heidelberg. All Rights Reserved.

© The Editor(s) (if applicable) and The Author(s), under exclusive license to Springer-Verlag GmbH, DE, part of Springer Nature 2024

This work is subject to copyright. All rights are solely and exclusively licensed by the Publisher, whether the whole or part of the material is concerned, specifically the rights of translation, reprinting, reuse of illustrations, recitation, broadcasting, reproduction on microfilms or in any other physical way, and transmission or information storage and retrieval, electronic adaptation, computer software, or by similar or dissimilar methodology now known or hereafter developed.
The use of general descriptive names, registered names, trademarks, service marks, etc. in this publication does not imply, even in the absence of a specific statement, that such names are exempt from the relevant protective laws and regulations and therefore free for general use.
The publisher, the authors and the editors are safe to assume that the advice and information in this book are believed to be true and accurate at the date of publication. Neither the publisher nor the authors or the editors give a warranty, expressed or implied, with respect to the material contained herein or for any errors or omissions that may have been made. The publisher remains neutral with regard to jurisdictional claims in published maps and institutional affiliations.

This Springer imprint is published by the registered company Springer-Verlag GmbH, DE, part of Springer Nature.
The registered company address is: Heidelberger Platz 3, 14197 Berlin, Germany

If disposing of this product, please recycle the paper.

Preface

The historical development of electrodynamics and of the special theory of relativity is interconnected. The formulation of the theory of relativity is a consequence of the fact that electromagnetism is not compatible with the principle of relativity formulated on the basis of classical mechanics. The central equations of electrodynamics, the Maxwell equations, are not form invariant under Galilei transformations. This would imply that different results should be observed for electromagnetic phenomena in different inertial systems. Experiments, in particular those by Michelson and Morley, show that this is not the case.

The material presented here follows the historical development of the theory of electromagnetism from its beginnings until its completion with the theory of relativity. Each of the chapters contains two parts. The first part retraces the development of the theory with the topics

- Basic concepts and equations of electrostatics (Chaps. 1 and 2)
- Calculation of stationary electric fields for given distributions of charges and materials. Practical aspects: electrical circuits and response of materials to stationary electric fields (Chaps. 3 and 4)
- Generation of stationary magnetic fields by stationary electric currents, response of materials to magnetic fields: para- and ferro-magnetism—on the basis of classical models (Chap. 5)
- Law of induction, Maxwell equations, the wave solutions of the inhomogeneous equations (transmitters), the wave solutions of the homogeneous equations (propagation), energy and momentum of electromagnetic fields (Dreizler and Lüdde, 2024, Electrodynamics and Special Theory of Relativity (Springer Berlin Heidelberg), Chap. 1)
- Additional applications: transformers, crystal optics, metal optics, wave guides, diffraction, electromagnetic radiation by antennas, bremsstrahlung, Čerenkov radiation (Dreizler and Lüdde, 2024, Electrodynamics and Special Theory of Relativity (Springer Berlin Heidelberg), Chap. 2)
- Michelson-Morley experiments, simple form of Lorentz transformation, consequences of the Lorentz transformation: addition theorems (velocity and accelerations), length contraction, time dilatation, equal time problem, linear spaces with non-Euclidean metric, Minkowski space, relativistic formulation of mechanics

and electrodynamics, historical remarks (Dreizler and Lüdde, 2024, Electrodynamics and Special Theory of Relativity (Springer Berlin Heidelberg), Chap. 3)

The second part tries to expand and amplify the coverage of this material by additional questions and suggestions under the label *Details*. In order to cope with the length of the material, we have split this book into two separate parts

- Part 1. Stationary electric and magnetic Problems in: Dreizler and Lüdde, 2024, Electrostatics and Magnetostatics (Springer Berlin Heidelberg), ISBN: 978-3-662-69932-4
- Part 2. Dynamic Problems and the special Theory of Relativity in: Dreizler and Lüdde, 2024, Electrodynamics and Special Theory of Relativity (Springer Berlin Heidelberg), ISBN: 978-3-662-69941-6

It is necessary to add some comments concerning the different systems of units, which are used for electric and magnetic quantities. The two commonly used units in electrodynamics are extensions of two systems of mechanics, namely the SI-system and the CGS-system. These extensions are:

- the rationalised MKSA-system, also called the SI-system. The letter A stands for the unit of electrical currents 'Ampère'. This system is usually preferred for technical applications.
- The Gauss CGS-system, which is usually just called the Gauss system. The unit of the electric current in this system, the 'statamp', can be directly expressed in terms of the mechanical CGS units. This system is the usual choice for theoretical considerations.
- Other systems of units, that are used, are:
  The electrostatic system of units (esu).
  The electromagnetic system of units (emu).
  The Heaviside-Lorentz system of units.

Five constant factors are needed to handle the equations involved in electrodynamics to and accommodate the different systems of units. These constants,

$$k_\mu \longrightarrow k_e, k_d, k_m, k_f,$$

are associated with specific physical quantities. They are discussed and justified in Appendix B. Additional lists of the units of various physical quantities are reproduced in Appendix B.2. In the main text all basic equations are given in a form employing these constants, so that they are valid for all systems of units. The exception is Chapter 2 in Dreizler and Lüdde, 2024, Electrodynamics and Special Theory of Relativity (Springer Berlin Heidelberg), where only the units of the Gauss system are used.

Mathematical topics, e.g. special functions of mathematical physics, the theory of distributions, Green's functions, etc., are interspersed in the main text. A short compilation of mathematical material for the subjects

- Equations of vector analysis
- Angular functions
- Radial functions
- Linear spaces with non-Euclidean metric

is also presented in the appendix of both parts as a support for working with the relations of electrodynamics.

Frankfurt/Main, Germany
November 2023

Reiner M. Dreizler
Cora S. Lüdde

It is with a heavy heart that I must add an addendum to this foreword. Unfortunately, Prof. Dr Dreizler passed away shortly before this work was completed. He was a very patient teacher who always endeavoured to impart knowledge without placing himself at the centre. The collaboration with him was always peaceful, but also characterised by many fruitful discussions. It was an honour to work with this generous and tolerant person and I thank him posthumously for this.

I would also like to thank my family and friends for their patience, understanding and support during this project. Last but not least, I would also like to express my thanks to the team of Springer Verlag, Heidelberg, for the friendly cooperation and technical support.

Frankfurt/Main, Germany
April 2024

Cora S. Lüdde

Contents

1 Basic Elements of Electrostatics 1
1.1 Preliminary Remarks 1
1.2 The Coulomb Law 7
1.3 The Electric Field 11
1.4 The Integral Theorem of Gauss 25
1.4.1 Simple Applications of the Gauss' Theorem 28
1.5 Details 34
1.5.1 Millikan's Experiment for the Determination of the Charge of the Electron 34
1.5.2 Calculation of the Asymptotic Electric Dipole Field 39
1.5.3 The Electric Field of a Circular Disk and of a Sphere (Both Uniformly Charged) 42

2 Basic Equations of Electrostatics 49
2.1 Basic Equations of Electrostatics 49
2.1.1 Integral Form 50
2.1.2 Differential Form 51
2.2 Additional Discussion of Electrostatics in Terms of Integrals 55
2.2.1 Electric Fields in Conductors 58
2.3 Distributions 62
2.3.1 Pragmatic Approach 63
2.3.2 Properties of the δ-Function 66
2.3.3 The Derivatives of the δ-Function 69
2.3.4 Generalised Functions 70
2.3.5 Some Additional Properties of the δ-Function 73
2.4 Representation of Charge Densities by Distributions 74
2.5 The Electric Potential 78
2.6 The Electric Field and the Storage of Energy 90
2.7 Details 93
2.7.1 Calculation of the Electric Potential of a Sphere and a Spherical Ring, Both Uniformly Charged 93

3 Solution of the Poisson Equation: Simple Boundary Conditions 97
3.1 Problems with Spherical Symmetry 97
3.2 Examples with Azimuthal Symmetry 104

3.3 General Problems: Solution with the Full Set of Spherical Coordinates ... 113
3.4 Multipole Moments ... 121
3.5 Details ... 132
3.5.1 The Potential of a Uniformly Charged Sphere in a Uniformly Charged Sphere ... 132
3.5.2 The Electric Potential of a Rotational Ellipsoid ... 135

4 Solution of the Poisson Equation: General Boundary Conditions ... 143
4.1 Remarks on the Classification of Boundary Conditions ... 144
4.2 Solution of Dirichlet Problems ... 147
4.3 Green's Functions ... 162
4.4 Capacitors ... 174
4.5 Polarisation of Dielectric Materials ... 180
4.6 The Technique of Complex Potentials ... 197
4.7 Details ... 210
4.7.1 Surface Charge of a Grounded Metal Sphere Induced by a Point Charge ... 210
4.7.2 The Potential of Two Point Charges Outside of a Grounded Metal Sphere ... 213
4.7.3 Comments on the Green's Function of Stationary Potential Problems ... 216
4.7.4 Evaluation of the General Formula for a Potential in the Case of Spherical Symmetry ... 224
4.7.5 The Capacity of a Spherical and a Cylindrical Capacitor ... 227
4.7.6 Potential of a Straight Wire with a Dielectric Cylinder ... 230

5 Magnetostatics ... 239
5.1 The Electric Current ... 240
5.2 Stationary Magnetic Fields ... 246
5.2.1 Experimental Basis ... 246
5.2.2 The Law of Ampère ... 249
5.2.3 The Formula of Biot-Savart ... 253
5.3 The Magnetic Vector Potential ... 258
5.4 Matter in a Magnetic Field ... 273
5.4.1 Global Models of Magnetisation ... 274
5.4.2 The Three Magnetic Fields ... 280
5.4.3 The Magnetic Material Equation ... 282
5.4.4 The Behaviour of B and H on Interfaces ... 284
5.4.5 Explicit Survey of Matter in Magnetic Fields ... 285
5.5 Forces on Charges in Magnetic Fields ... 293
5.6 Details ... 300
5.6.1 Mathematical Properties of the Vector Potential ... 300
5.6.2 The Magnetic Field of a Circular Current ... 303
5.6.3 The Current Density of a Circulating Point Charge ... 315

5.6.4 The Gyromagnetic Factor of a Uniformly Rotating Homogeneous Charge Distribution 317
5.6.5 The Magnetic Field of a Uniformly Magnetised Sphere..... 320
5.6.6 Definition of the Unit Ampère 323

A Literature 327
A.1 Books and Literature Quoted in the Text 327
A.2 Introductory Texts 327
A.3 Electrodynamics 328
A.4 Mathematics 328
A.5 Special Functions and Handbooks 329

B Systems of Units in Electrodynamics 331
B.1 The Systems 331
B.2 Tables 337

C Additional Mathematical Topics 343
C.1 Equations of Vector Analysis 343
C.2 Multiple Products of Vectors 343
C.3 Product Rules for the Application of the ∇-Operator 344
C.4 Double Application of ∇ 344
C.5 Differential Operators in Spherical and Cylinder Coordinates 345
C.5.1 Spherical Coordinates 345
C.5.2 Cylinder Coordinates 345
C.6 Angular Functions 346
C.6.1 Legendre Polynomials $P_l(x)$ 346
C.6.2 The Functions $Q_l(x)$ 348
C.6.3 Associated Legendre Functions 349
C.6.4 Spherical Harmonics 350
C.7 Radial Functions 352
C.7.1 The Hypergeometric Functions $F(a, b, c\,;\, x)$ 352
C.7.2 The Confluent Hypergeometric Functions $F(a, c\,;\, x)$ 354

Index 357

Basic Elements of Electrostatics 1

Classical electromagnetic phenomena can be understood completely in terms of Maxwell's equations (a set of partial differential equations for electric and magnetic fields) and a set of supplementary equations, which address the response of different materials to these fields. For this reason one could begin with the specification of these equations and concentrate on the discussion of methods for their solution and on their interpretation in terms of physical concepts. It is, however, more useful to follow the historical development of the theory and assemble the relevant equations piece by piece. If one follows this path, the first topic one has to address, is *electrostatics*, which involves qualitative statements on electric charges and the structure of matter. The more quantitative aspects are concerned with the fundamental electric forces between stationary charged objects, as expressed in *Coulomb's law*. The basic form of this law characterises the forces between pairs of ideal point charges. An extended charged body can be imagined to consist of a superposition of point charges. The superposition principle allows the consideration of arbitrary charge distributions and the characterisation of stationary electric fields in terms of the *theorem of Gauss*. This theorem can be used for the direct calculation of electric fields of simpler charge distributions and for establishing the basic differential equation of electrostatics, the *Poisson equation*. With the aid of these equations the treatment of more advanced aspects, as the response of materials to external stationary electric fields, is possible.

1.1 Preliminary Remarks

The discussion of electric and magnetic phenomena is directly tied to the discussion of the structure of matter. One of the examples is ferromagnetism. A complete understanding of the ferromagnetic properties of materials requires the knowledge of the quantum structure of solids. The historical development of electrodynamics

© The Author(s), under exclusive license to Springer-Verlag GmbH, DE, part of Springer Nature 2024
R. M. Dreizler, C. S. Lüdde, *Electrostatics and Magnetostatics*,
https://doi.org/10.1007/978-3-662-69933-1_1

shows, however, that it is possible to deal with electric and magnetic properties of matter, without the consideration of its complete microscopic structure.

The basic understanding of electrodynamics has been completed within a period of 80 years, that is from 1785 until 1865. The beginning were the observations of Charles Augustin de Coulomb concerning the electrostatic forces between charged objects. The second basic observation followed in the year 1820. Hans Christian Øerstedt found that electric currents produce magnetic fields. Michael Faraday studied the interplay of time dependent electric and magnetic fields only a few years later (starting in 1831). In 1864 James Clerk Maxwell published the equations, which constitute the basis of the field of electrodynamics.

By contrast, our knowledge of the microscopic aspects of matter has been gathered in the twentieth century, starting with the discovery of the atomic nucleus by Ernest Rutherford, Hans Geiger and Ernest Marsden in the year 1906, the first quantum theory of the atom by Niels Bohr and the understanding of the ferromagnetic properties of matter, which was initiated by Werner Heisenberg in 1928.

A cursory look at the development of the topic shows, that the formulation of electrodynamics proceeded without the necessary knowledge of the details of the structure of matter. In order to understand the bridge between the point of view of the nineteenth century and the present one, it is useful to consider some (really simple) statements on electric charges.

The first statement is: *electric charges* are quantised. This means, that there exists in nature a smallest unit of charge, which can not be divided further by ordinary means. The value of this elementary charge is approximately[1]

$$\begin{aligned} e_0 &= 1.602\,18 \cdot 10^{-19}\,\text{Coulomb} \\ &= 4.803\,27 \cdot 10^{-10}\,\text{statcoul} \end{aligned}$$

(Details of the units involved will be discussed below). The elementary charge is carried by the elementary particles

$$\begin{aligned} &\text{the electron} \qquad q_{\mathrm{e}} = -e_0 \\ &\text{the proton} \qquad q_{\mathrm{p}} = +e_0 \,. \end{aligned}$$

The sign of the charge is a question of convention. The convention, which is used today, is due to Benjamin Franklin, even though he had no knowledge of these

[1] A value of $1.602\,176\,634 \cdot 10^{-19}$Coulomb has been agreed upon in 2021 by international consent. This value should be referred to as the basic unit of charge, as the fractional charge of quarks has not been found to be free.

elementary particles. Each charge Q, found in nature at the level of and above the level of chemical investigations, has the value

$$Q = \pm n\, e_0 \qquad (n = 1, 2, \ldots)\,.$$

The first experimental proof of this statement is provided by the 'oil drop experiment' of Robert Millikan in 1910 (Detail 1.5.1). The discrete structure of the charges can not be recognised in most macroscopic experiments. For instance, in a light bulb of 100 watts operated at 220 volts about $n \approx 3 \cdot 10^{18}$ elementary charges per second pass through the filament. This number is too large for an experimental observation of the discrete structure of the charges.

The statements concerning the smallest charge unit have to be modified. It has been found, starting in 1970, that there exists a substructure of the proton, the quark structure. According to the quark model of strongly interacting particles (as the proton) these particles are composed of several quarks. The model suggests that the proton p consists of two up-quarks and one down-quark

$$p \longrightarrow (u\, u\, d)$$

with the respective charges

$$q_u = \frac{2}{3}\, e_0 \qquad q_d = -\frac{1}{3}\, e_0\,,$$

so that charge of the proton is

$$q_{\mathrm{p}} = 2q_u + q_d = \frac{4}{3}\, e_0 - \frac{1}{3}\, e_0 = e_0\,.$$

Even though free quarks have not yet been observed, there is sufficient evidence for the correctness of this model. One might even be tempted to speculate that quarks themselves may be composite particles and thus a further definition of the elementary charge would be possible or necessary.

Independent of the question whether one uses e_0 or $1/3\, e_0$ as a fundamental unit of charge, the statement is: Charges, which occur in nature, are quantised. In macroscopic experiments this quantisation can not be easily observed. The quantity e_0 will be used as the elementary charge in the following text.

A second introductory statement is a brief remark on the composition of atoms. If one divides, in the manner of the school of sophists, one cubic centimetre of a material into two halves, divides one of the halves again and continues this process, one arrives after about 10^{25} divisions at the chemical building blocks of matter, the atoms. Individual atoms are electrically neutral, if viewed from the outside. Inside the atom one finds a spatial charge distribution. An atom contains a nearly point-like, positively charged atomic nucleus and a negative 'charge cloud' of electrons. The positively charged part can be decomposed into neutrons (neutron number N) and protons (proton number Z). The standard nomenclature for nuclei with $A = N + Z$ nucleons is

$$_{\mathrm{Z}}\mathrm{Element}^{\mathrm{A}}_{\mathrm{N}} \qquad \text{with} \quad \mathrm{A} = \mathrm{N} + \mathrm{Z}\,,$$

as e.g. for the normal hydrogen nucleus

$$_{1}\mathrm{H}^{1}_{0} \longrightarrow 1 \text{ proton}$$

or for the deuteron, (the heavy hydrogen isotope) with the nucleus

$$_{1}\mathrm{H}^{2}_{1} \longrightarrow 1 \text{ proton},\ 1 \text{ neutron}\,.$$

A lead nucleus contains always 82 protons, but there exist about 30 different isotopes. The lightest isotope is

$$_{82}\mathrm{Pb}^{194}_{112}$$

(with a half life of $\tau \approx 11$ min), the most abundant

$$_{82}\mathrm{Pb}^{208}_{126}$$

is stable, the heaviest lead isotope

$$_{82}\mathrm{Pb}^{214}_{132}$$

has a half life of $\tau \approx 27$ min.

The characterisation of the electrons in an atom as a charge cloud is an attempt to express the statistical nature of quantum mechanics in a simple, verbal form. The electrons in an atom can not be localised. One can only state a measure, which describes the probability to find an electron at a given point of the space around the nucleus.

In order to characterise atoms further one should recall the following numbers: The medium diameter of the negative charge cloud is approximately

$$d_{\mathrm{atom}} \approx 1 \cdot 10^{-8}\,\mathrm{cm}\,.$$

This number is, with minimal deviation, valid for all atoms. The medium diameter of the atomic nuclei is of the order 10^{-13} cm and varies with the nucleon number as $A^{1/3}$. In particular one finds the diameters

$$\begin{array}{ll} \mathrm{H} & d_{\mathrm{H}} \approx 1.4 \cdot 10^{-13}\,\mathrm{cm} \\ \mathrm{Pb} & d_{\mathrm{Pb}} \approx 14 \cdot 10^{-13}\,\mathrm{cm}\,. \end{array}$$

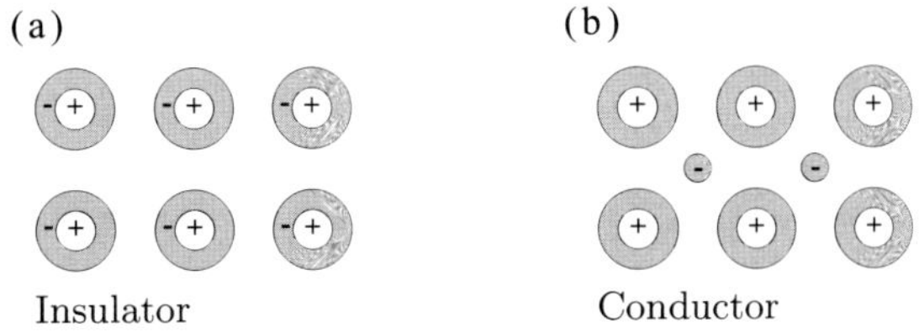

Fig. 1.1 A simple model of matter

The radius of the charge cloud is roughly larger by a factor of 10^5 than the radius of the atomic nucleus.

With respect to the three standard building blocks of matter one can quote the numbers:

	Mass m	Charge q	Magn. moment μ
Proton p	$1.6726 \cdot 10^{-24}$ g	$+e_0$	$1.5210 \cdot 10^{-3}$
Neutron n	$1.6749 \cdot 10^{-24}$ g	0	$-1.0419 \cdot 10^{-3}$
Electron e^-	$9.109 \cdot 10^{-28}$ g	$-e_0$	1.0012

Proton and neutron have approximately the same mass, the ratio of a proton mass with respect to an electron mass is $m_p/m_e \approx 1840$. The total mass of an atom is practically concentrated in the atomic nucleus. The number of electrons N_e in a neutral atom is equal to the number of protons Z. This is, as will be discussed later, the reason why an atom can be considered to be neutral, if viewed from the outside. The magnetic moments μ of the three elementary particles (in the units of Bohr magnetons) will also be discussed at a later stage in detail (Chap. 5.3). For the moment one should only note, that a neutron possesses magnetic properties if viewed from the outside, even though it is electrically neutral.

The third set of remarks concerns some direct electrical properties of macroscopic matter. With respect to the electrical properties one distinguishes two types of materials[2]

► Insulators $\Longleftrightarrow$ Conductors.

Examples are glass, rubber and plastic on one side and metals as well as silicates on the other. A simple model of the structure of the different solids is the following. In each solid one finds a more or less regular and a more or less rigid lattice of atomic nuclei. The difference between insulators and conductors is due the different behaviour of the electrons in the solids. The electrons of each atom (or molecule) of an insulator are bound strictly to the nuclei. There exist no freely movable charges throughout the bulk material (Fig. 1.1a). On the other hand, the outer electrons in

[2] A different name for insulator is isolator, dielectric or dielectric material.

an atom loose their electric binding to the individual nuclei in a metallic solid (Fig. 1.1b). They can move relatively freely within the material.

The fact, that electrons are responsible for the electric conductivity, was first demonstrated[3] by Edwin Hall in 1879.

Simple electrical experiments have been known since early times. A piece of amber acquires a negative charge, if it is rubbed with a piece of fur. The piece of fur carries a positive charge. An explanation on the atomic level of this simple experiment is the following: The binding energies of the outer electrons in the fur atoms and the amber atoms are different. It is larger in the amber. According to the classical energy principle it is energetically more advantageous to have the electrons in the amber. Rubbing of the materials provides the contact, which is necessary for the transfer of the electrons to the material with the lower binding energy.

In general one can state (independent of the mechanism responsible for the charging of objects) a

$$\begin{aligned} &\text{positive charging} \quad \longrightarrow \quad \text{electrons are removed from the atoms} \\ &\text{negative charging} \quad \longrightarrow \quad \text{electrons are added to the atoms}\,. \end{aligned}$$

It is always the electrons, which are moving between the materials, as the protons in the atomic nuclei are bound approximately 10^6 times more strongly than the electrons in the atom or molecule.

Such experiments demonstrate charge conservation. In the fur-amber experiment the atoms of the two materials involved are initially electrically neutral

$$Q_{\text{fur atom}} = 0 \qquad Q_{\text{amber atom}} = 0\,.$$

After the experiment one finds

$$Q_{\text{fur ion}} = +|Q| \qquad Q_{\text{amber ion}} = -|Q|\,.$$

The total charge of the fur-amber system is still zero. The number of electrons, which have left the fur is equal to the number of electrons, which are now in the amber.

Charge conservation is also observed in nuclear or in subnuclear processes. Nuclear reactions are only possible, if charge is conserved. For example, if carbon isotopes C^{13} are bombarded with protons, one of the possible reactions is the production of nitrogen isotopes N^{13} and neutrons

$${}_6C_7^{13} + {}_1H_0^1 \longrightarrow {}_7N_6^{13} + {}_0n_1^1\,.$$

[3] E. Hall, Amer. Journ. Math. **2** (1879), p. 287.

If neutrons are observed, a specific nitrogen isotope is produced. The sum of the charge on both sides of the equation describing the nuclear reaction is $7e_0$.

The fact, that charge is conserved but not mass, can be observed in the reaction of an electron with its antiparticle, the positron,

$$e^- + e^+ \longrightarrow 2\gamma \,.$$

In this process two massive particles with the total mass $m = 2m_e$ are annihilated and two γ-particles, a particular form of mass-less electromagnetic radiation is created. The charge number on both sides of this reaction equation is zero.

Such simple experiments can form the basis of a qualitative view of electric (and magnetic) phenomena. The quantitative approach begins in the 18th century with the basic experiment of electrostatics, which established the Coulomb law, a statement on the forces between stationary electric charges.

1.2 The Coulomb Law

The tool, which was used by C.A. Coulomb for the determination of the electrostatic force law, is a torsion balance: Two spheres of metal with equal mass and equal diameter are attached to a horizontal rod, which can rotate. One of the spheres on the rod is charged (e.g. using a charge q_1, obtained by rubbing the metal). A third sphere with the charge q_2 is placed at a point, which is a distance r_{12} away from q_1 (Fig. 1.2). One can use the resulting rotation of the apparatus for the determination of the force between the two charges. Gravitational forces can, as explained below, be neglected. By variation of the magnitude and the sign of the charges as well as their distance, the electric force between the spheres can be determined. The ensuing *Coulomb force law* can be expressed in the form

$$\boldsymbol{F}_{12} = k_e \, q_1 \, q_2 \, \frac{\boldsymbol{r}_{12}}{r_{12}^3} = -\boldsymbol{F}_{21} \,. \tag{1.1}$$

The vector $\boldsymbol{F}_{12}$ is a vector describing the force, which the charge q_1 exerts on the charge q_2. The following remarks can be offered concerning this law:

(1) It has the same form as the corresponding law of gravitation. In gravitation only an attractive force between masses is known. In the electrostatic world exists

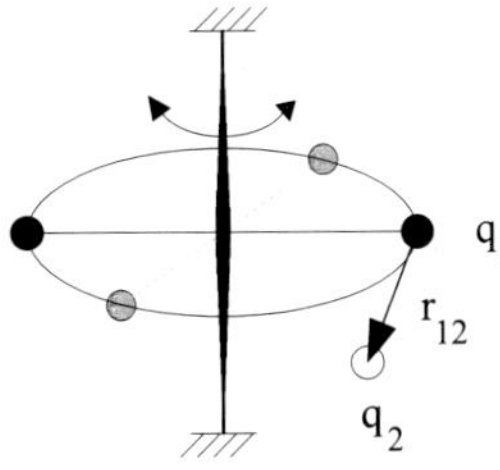

Fig. 1.2 Coulomb's torsion balance (Schematic representation)

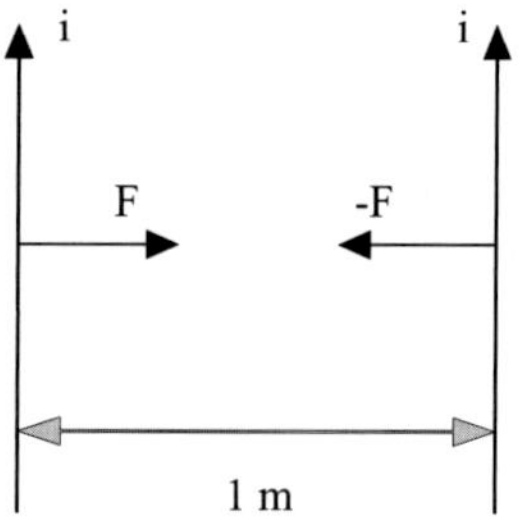

Fig. 1.3 Experimental definition of the unit 1 Ampère

an attractive force (for charges of a different sign) and a repulsive force (for charges of the same sign). Electrostatic forces obey, as do gravitational forces, the third axiom of Newton.

(2) Coulomb's law is, as indicated by the notation, only valid for point charges (the electric equivalent of the concept of a mass point).

(3) The choice of the constant k_e determines the units, by which the charges are measured. From the possible systems of units in electromagnetism only two (the SI-system and the Gauss system) will be used in the following:[4]

(a) The *international system of units* SI (System International). In this system (which is also referred to as the MKSA-system) the three mechanical units (meter, kilogram, second) and an additional unit for the charge q

$$[q] = \text{Coulomb} = \text{C}$$

are employed. The unit of the charge (C) is defined via the electric current. For stationary currents one has

$$\text{strength of the current} = \frac{\text{amount of charge}}{\text{amount of time}} \quad \longrightarrow \quad i = \frac{\Delta q}{\Delta t} \,.$$

The definition of the unit of charge implies two statements:

(i) 1 Coulomb is the amount of charge, which flows through a cross section of a conducting wire, if the strength of the current is 1 Ampère (1 A for short).

(ii) If a current of the same strength flows in two parallel wires, which are 1 meter apart (situated in a vacuum, see Fig. 1.3), and if the two wires attract each other with a force of $2 \cdot 10^{-7}$ N per m, then the strength of the current in each wire is 1 Ampère (see Detail 5.6.6 for a full explanation).

[4] A more complete compilation of electromagnetic systems of units including the factors necessary for the conversion of the units is found in Appendix B.1.

With this definition one can (via a series of equations, which will be discussed later) determine the constant k_e. It has the value

$$k_{e,\mathrm{SI}} = 8.9875 \cdot 10^9 \, \frac{\mathrm{N\,m^2}}{\mathrm{C^2}} \approx 9 \cdot 10^9 \, \frac{\mathrm{N\,m^2}}{\mathrm{C^2}} \,.$$

The constant is also given in the form

$$k_{e,\mathrm{SI}} = \frac{1}{4\pi\,\varepsilon_0 \,.}$$

The alternative constant ε_0 is referred to as the 'electric field constant' or as the 'electric constant of the vacuum'. Its value is

$$\varepsilon_0 = \frac{1}{4\pi\,k_{e,\mathrm{SI}}} = 8.8542 \cdot 10^{-12} \frac{\mathrm{C^2}}{\mathrm{N\,m^2}} \approx 8.9 \cdot 10^{-12} \frac{\mathrm{C^2}}{\mathrm{N\,m^2}} \,.$$

The SI-system is, according to international convention, the official system of electric (or with suitable extensions electromagnetic) units.

(b) The *Gauss or CGS-system*, which is used more extensively in the theoretical literature, as the resulting equations are more transparent.

  (i) The unit of the charge is defined by the choice $k_{e,\mathrm{CGS}} = 1$. It can, as a consequence, be related directly to mechanical quantities

$$[q] = (\mathrm{dyn\,cm^2})^{1/2} = \frac{\mathrm{g^{1/2}cm^{3/2}}}{\mathrm{s}} = 1 \text{ statcoul} = 1 \text{ esu} \,.$$

  The Coulomb law states: Two point charges, of which each carries the charge $q = 1$ statcoul (or alternatively 1 esu = 1 electrostatic unit), experience a force of 1 dyn if the distance between them is 1 cm.

  (ii) The unit of the current strength in the CGS-system is

$$[i] = \frac{\text{statcoul}}{s} = \frac{\mathrm{g^{1/2}cm^{3/2}}}{\mathrm{s^2}} = 1 \text{ statamp} \,.$$

(c) In order to relate the units of the current strength of the two systems, one first needs to note the statement: Two charges of 1 Coulomb with a separation of 1 m experience of force of

$$F = 9 \cdot 10^9 \, \frac{\mathrm{N\,m^2}}{\mathrm{C^2}} \cdot 1^2\,\mathrm{C^2} \cdot \frac{1}{1^2\,\mathrm{m^2}} = 9 \cdot 10^9\,\mathrm{N} \,.$$

  (i) One then has to answer the question: Which units of the charges have to be used in the CGS-system so that they experience the same force at the same separation? The answer is

$$\frac{q^2}{10^4\,\mathrm{cm}^2} = 9 \cdot 10^9 \cdot 10^5\,\mathrm{dyn}$$

or

$$q = \left[9 \cdot 10^{18}\right]^{1/2} \mathrm{statcoul}\,.$$

The statement on the basis of the more exact numerical values is

$$1\,\mathrm{C} = 2.997925 \cdot 10^9\,\mathrm{statcoul} \approx 3 \cdot 10^9\,\mathrm{statcoul}\,.$$

(ii) It is possible to use this definition of the unit of the electric charge in order to compare the strength of the electrical forces directly with the strength of gravitation. The comparison can be established by considering the hydrogen atom. The electric force of a proton on an electron (using the absolute value in the CGS-system) is

$$F_{\mathrm{el}} = \frac{e_0^2}{r^2} \qquad F_{\mathrm{grav}} = \frac{\gamma\, m_{\mathrm{e}} m_{\mathrm{p}}}{r^2}$$

or

$$\frac{F_{\mathrm{el}}}{F_{\mathrm{grav}}} = \frac{e_0^2}{\gamma\, m_{\mathrm{e}} m_{\mathrm{p}}} \approx 2.3 \cdot 10^{39}\,.$$

The electric force is much stronger. For the discussion of the structure of microscopic matter (atomic or solid state physics), gravitation plays no role.

(iii) Finally one should ask the question: Is the $1/r^2$-behaviour of the Coulomb law really correct? An accurate answer has been given in an experiment in 1936 by Plimpton and Lawton.[5] The result of this experiment, which will be discussed further down, is: The exponent α in Coulomb's law with the form $1/r^{\alpha}$ is

$$1.999\,999\,998 \le \alpha \le 2.000\,000\,002\,.$$

On the basis of these statements one may, in analogy to gravitation, discuss the concepts of electric potential and electric field.

[5] S.J. Plimpton and W.E. Lawton, Phys. Rev. **50** (1936), p. 1066. The history of attempts to verify the $1/r^2$-law is discussed by L.P. Fulcher, Phys. Rev. **A33** (1986) p. 759.

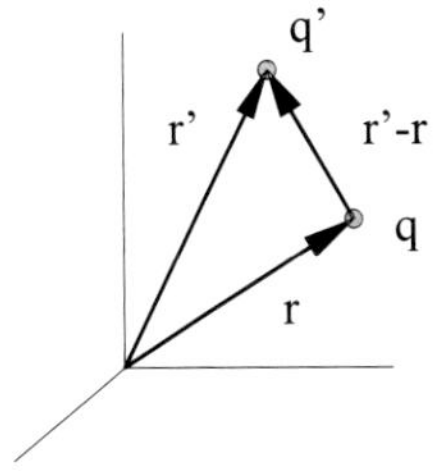

Fig. 1.4 Geometry for the definition of the electric field

1.3 The Electric Field

One may consider the following situation (similar to gravitation) in order to define the concept of an electric field: A point charge q is at the position $\boldsymbol{r}$. If one places a second point charge q' at the position $\boldsymbol{r}'$ (Fig. 1.4), where the magnitude of the first charge $|q|$ is significantly larger than $|q'|$. The operational definition of the electric field $\boldsymbol{E}$ created by the charge q, which is located at the position $\boldsymbol{r}$, at the position $\boldsymbol{r}'$ is then

$$\boldsymbol{E}_{\mathrm{q}}(\boldsymbol{r}', \boldsymbol{r}) = \lim_{q' \to 0} \frac{\boldsymbol{F}_{q \,\mathrm{on}\, q'}}{q'} = k_e \, q \, \frac{(\boldsymbol{r}' - \boldsymbol{r})}{|\boldsymbol{r}' - \boldsymbol{r}|^3} \,. \tag{1.2}$$

This equation indicates: In order to measure the electric field produced by the charge q at the position $\boldsymbol{r}$, one has to use a sequence of smaller and smaller charges q' at the position $\boldsymbol{r}'$ and determine the limiting value of the force divided by the charge q'. This limiting process can be executed in a simple manner on the basis of Coulomb's law. For this reason one may assume that it does not make any difference if one discusses the situation in terms of forces or in terms of fields. This is only correct for a stationary situation (with fixed charges), for which the introduction of the concept of a field can be viewed as a model of a force (or vice versa). The interpretation of the formulation in terms of the field concept is: A charge positioned at a point of space (as at the origin of a coordinate system in Fig. 1.5a) generates an

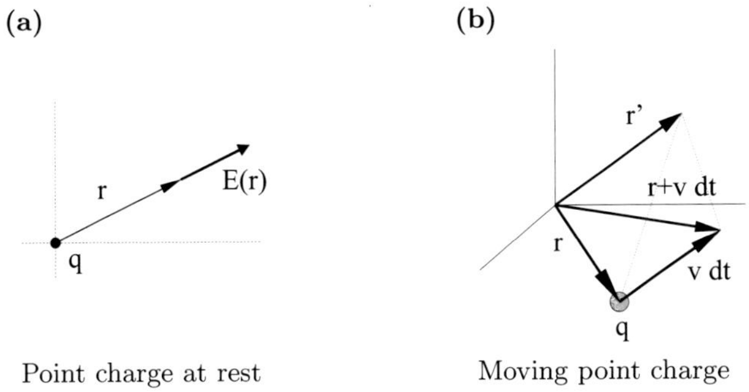

Fig. 1.5 Concerning the concept of fields

electric field $\boldsymbol{E}_{\mathrm{q}}(\boldsymbol{r})$ in a point $\boldsymbol{r}$ independent of the presence of another charge. The force on a charge q' at the position $\boldsymbol{r}$ is then

$$\boldsymbol{F}_{\mathrm{q\ on\ q'}}(\boldsymbol{r}) = q'\boldsymbol{E}_q(\boldsymbol{r})\,.$$

The situation is different, if one deals with moving charges (the dynamical situation). If one considers a point charge q moving at a uniform velocity $\boldsymbol{v} = \text{const.}$, one has to answer the question, whether the information, that the point charge has moved by $\boldsymbol{v}\,\mathrm{d}t$, can be observed at the same instant in another point $\boldsymbol{r}'$ or whether this information reaches this point with some retardation (Fig. 1.5b). Such questions have led to the development of the theory of relativity. The theory of relativity implies that any information concerning the change of the position of a point charge can be transferred only with a velocity up to the velocity of light. It therefore implies retardation. Effects of retardation can be discussed in a simpler fashion, if one bases the discussion on the transformation properties of electric *and* magnetic fields rather than on the concept of forces. The discussion of the retardation is more involved, if it is based on the concept of forces.

For a visualisation of electric fields one uses field lines. The field lines of a point charge are characterised by the statements

(a) The field vectors $\boldsymbol{E}(\boldsymbol{r})$ are tangents of the field lines in each point of space.
(b) The orientation associated with the field lines corresponds to the direction of the field vectors.

The fields of point charges (positive, negative), illustrated in Fig. 1.6a and b, indicate the following semi-quantitative properties expressed by the field lines: The 'density of the lines' is larger in the vicinity of the charge. A larger density of the lines expresses a stronger field. The decrease of the density with the distance from the charge corresponds to the $1/r^2$ law. If one is interested in a semi-quantitative comparison of pictures of field lines of charges of different size, one has to

(c) choose the number of field lines proportional to the quantity of the charge, as e.g. for $2q$ in Fig. 1.6c in comparison with q $(q > 0)$ in Fig. 1.6a.

A main task of electrostatics is the calculation of electric fields for a given, stationary distribution of charges. To do this one uses the **superposition principle**. A distribution of point charges q_i $(i = 1 \ldots N)$ at the positions $\boldsymbol{r}_i$ generates the electric field

$$\boldsymbol{E}(\boldsymbol{r}) = \sum_{i=1}^{N} \boldsymbol{E}_i(\boldsymbol{r}) = \sum_{i=1}^{N} \frac{k_e\, q_i}{|\boldsymbol{r} - \boldsymbol{r}_i|^3}(\boldsymbol{r} - \boldsymbol{r}_i) \qquad (1.3)$$

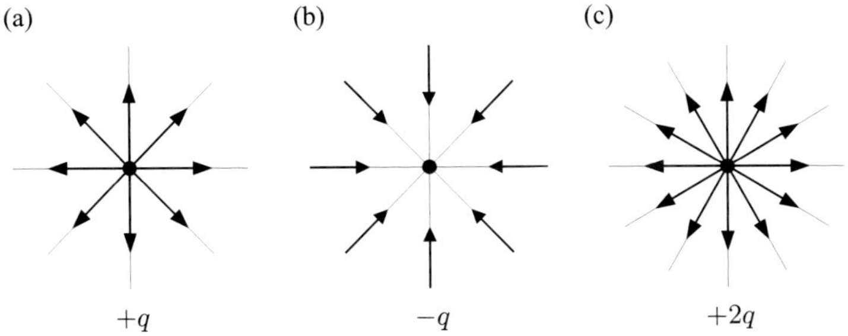

Fig. 1.6 Illustration of field lines for different point charges

at the position $\boldsymbol{r}$ (Fig. 1.7). The superposition principle is (like the Coulomb law) a result summarising our experience. It can be applied if the equations for the calculation of electric fields are linear in the components of the field $\boldsymbol{E}$.

An explicit, though simple example for a direct calculation of fields with (1.3) is the field of an electric dipole. A **dipole** consists of two point charges $+q$ and $-q$ $(q > 0)$, which are separated by a distance $2a$. It is useful to place the dipole along one of the coordinate directions (say, the z-axis). The origin of the coordinate system is the centre of the dipole (Fig. 1.8a). In view of the rotational symmetry (here with respect to the z-axis) one can restrict the discussion to one coordinate plane (as the x-z plane). For the distances $r_\pm$ of an arbitrary point in this plane from the two charges one finds (Fig. 1.8b)

$$r_+ = \left[x^2 + (z-a)^2\right]^{1/2}$$

$$r_- = \left[x^2 + (z+a)^2\right]^{1/2} .$$

The field components in the corresponding directions of the coordinates are

$$E_{+x} = E_+ \cos\theta_+ = k_e q \,\frac{x}{\left[x^2 + (z-a)^2\right]^{3/2}}$$

Fig. 1.7 Concerning the superposition principle

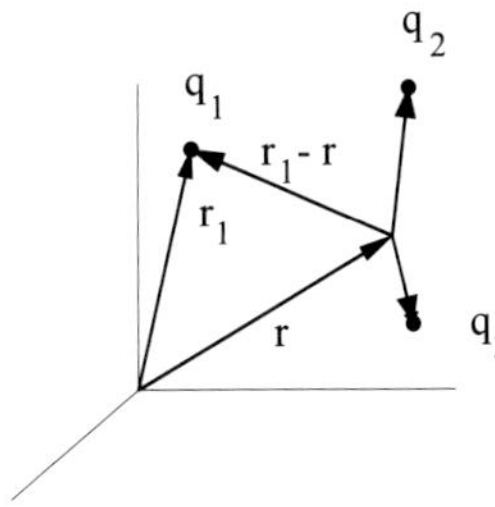

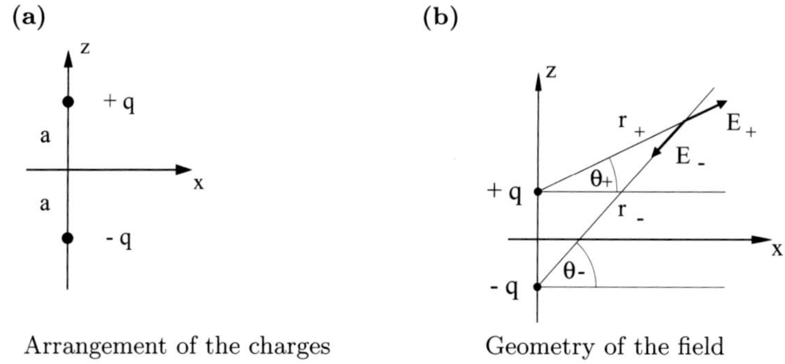

Fig. 1.8 The electric field of a dipole

$$E_{+z} = E_+ \sin\theta_+ = k_e q \frac{z-a}{\left[x^2+(z-a)^2\right]^{3/2}}$$

$$E_{-x} = E_- \cos\theta_- = -k_e q \frac{x}{\left[x^2+(z+a)^2\right]^{3/2}}$$

$$E_{-z} = E_- \sin\theta_- = -k_e q \frac{z+a}{\left[x^2+(z+a)^2\right]^{3/2}} .$$

The vector of the dipole field is obtained by superposition of the contribution of the two charges

$$\begin{aligned} \boldsymbol{E}(x,z) &= (E_x,\ E_z) \\ &= \left(k_e q x \left(\frac{1}{r_+^3} - \frac{1}{r_-^3}\right),\ k_e q \left(\frac{(z-a)}{r_+^3} - \frac{(z+a)}{r_-^3}\right)\right) . \end{aligned}$$

The expression for the field vector in an arbitrary point of 3-dimensional space can be obtained by appealing to the symmetry: Replace the x-coordinate by the distance $\rho = \left[x^2+y^2\right]^{1/2}$ with respect to the z-axis and project the ρ-component of the field vector $\boldsymbol{E}$ onto the x - z respectively y- z planes (Fig. 1.9). The result is

$$\begin{aligned} \boldsymbol{E}(\boldsymbol{r}) &= \left(E_x,\ E_y,\ E_z\right) \\ &= \left(k_e q \rho \cos\varphi \left(\frac{1}{r_+^3} - \frac{1}{r_-^3}\right),\ k_e q \rho \sin\varphi \left(\frac{1}{r_+^3} - \frac{1}{r_-^3}\right),\right. \\ &\qquad \left. k_e q \left(\frac{(z-a)}{r_+^3} - \frac{(z+a)}{r_-^3}\right)\right) . \end{aligned}$$

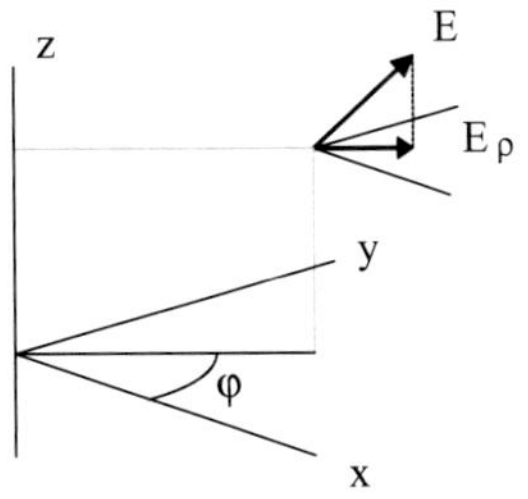

Fig. 1.9 Dipole field: Calculation of the field in three-dimensional space

The quantities r_+ and r_- are now

$$r_+ = \left[x^2 + y^2 + (z-a)^2\right]^{1/2} \qquad r_- = \left[x^2 + y^2 + (z+a)^2\right]^{1/2} .$$

In order to obtain a quantitative idea of this field, one looks e.g. at the field for points on the x-axis respectively the z-axis.

For the x-axis one has $z = 0$ and $y = 0$ and therefore

$$r_+ = r_- = \left[x^2 + a^2\right]^{1/2}$$

as well as (compare Fig. 1.10)

$$\boldsymbol{E}(x, 0, 0) = \left(0,\ 0,\ -\frac{2k_e q\, a}{\left[x^2 + a^2\right]^{3/2}}\right) .$$

The resulting vector has only a component in the z-direction. Figure 1.10a shows the function $E_z(x)$ for the values of the parameters $a = 0.5$ cm and $q = 1$ statcoul

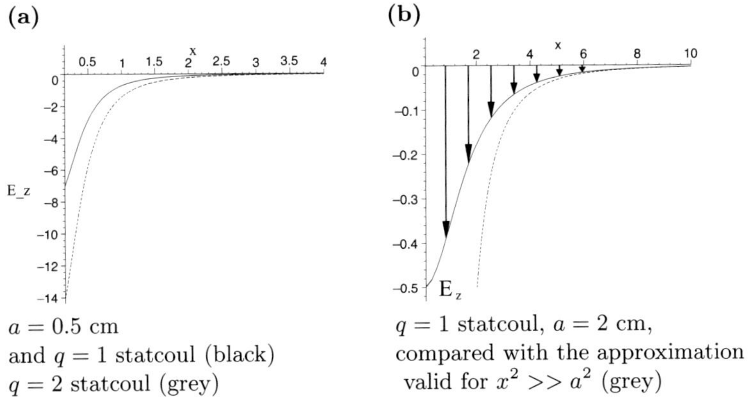

Fig. 1.10 Dipole field along the x-axis ($k_e = 1$)

(black) as well $q = 2$ statcoul (grey). The function E_z takes the form

$$E_z(x, 0, 0) \xrightarrow{x^2 \gg a^2} -\frac{2k_e q\, a}{x^3}$$

for larger separation from the centre of the dipole $\left(x^2 \gg a^2\right)$. Figure 1.10b shows the exact function (black) in comparison with an approximation for $x \gg a$ (grey). The approximation differs considerably for points along the axis with $x < 4a$.

The product of the value of the charge and the separation of the charges is called the **moment of the dipole** p

$$p = 2a\, q\,. \tag{1.4}$$

If one is far away from the dipole one observes only the dipole moment and not its individual features. A dipole with twice the charges $q' = 2q$ and half the separation $a' = a/2$ generates the same field in the x-direction for $x \to \infty$. For the z-axis with $x = y = 0$ one finds

$$\boldsymbol{E}(0, 0, z) = \left(0,\ 0,\ k_e q \left(\frac{(z-a)}{\left[(z-a)^2\right]^{3/2}} - \frac{(z+a)}{\left[(z+a)^2\right]^{3/2}}\right)\right).$$

For the resolution of the roots one has to take care of the sign of the radicant. For this reason one finds for points in the region outside of the dipole $|z| > a$ the field

$$\boldsymbol{E}(0, 0, z) = \left(0, 0, \pm\frac{2\, k_e\, p\, z}{(z^2 - a^2)^2}\right) \quad \begin{Bmatrix} z > a \\ z < -a \end{Bmatrix}$$

and

$$\boldsymbol{E}(0, 0, z) = \left(0,\ 0,\ -\frac{2k_e q(z^2 + a^2)}{(z^2 - a^2)^2}\right) \quad \begin{Bmatrix} z < a \\ z > -a \end{Bmatrix}$$

for points with ($|z| < a$) (that is between the two charges).

The result for the direction of the field on the dipole axis is shown in Fig. 1.11a. The direction is away from the positive and towards the negative charge. It is singular at the position of the charges $z = \pm a$. For large separations from the dipole one finds, that the field decreases as $1/r^3$

$$\boldsymbol{E} \xrightarrow{z \to \infty} \frac{2k_e p}{z^3}\,.$$

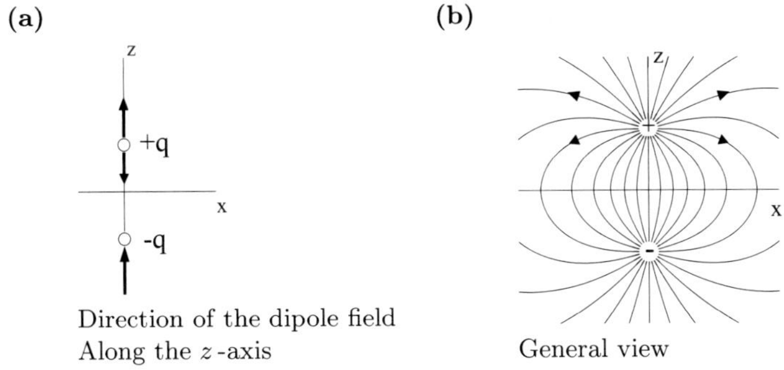

Fig. 1.11 The field lines of a dipole

The $1/r^3$-behaviour does not only hold for points on the dipole axis, but for all distant points. According to Detail 1.5.2 the general result for a dipole field in a direction, that is characterized by an angle θ with respect to the z-axis, has the form

$$|\boldsymbol{E}| = \frac{k_e p}{r^3}\left[1 + 3\cos^2\theta\right]^{1/2} .$$

The dipole field decreases more rapidly than the field of a point charge for all distant points.

A more quantitative illustration of the complete dipole field is shown in Fig. 1.11b. All field lines begin at the positive charge and end at the negative charge. Shown is the situation in a plane through the x-z axes, however the lines are rotationally symmetric with respect to the axis of the dipole. The field lines go over into the field lines of the point charges $\pm q$ in the vicinity of these charges.

In the similar fashion one can discuss the field for a larger set of point charges. The calculation might perhaps be slightly more tedious, but presents no difficulties in principle. Figure 1.12 shows the field lines of a stretched quadrupole with the charges q, $-2q$, q at a separation a along the x-axis.

An equation for the field lines can be obtained by remembering that the direction of the field lines and the field itself coincide in each point. This feature can be expressed by the vector product $\boldsymbol{E}(\boldsymbol{r}) \times \mathbf{d}\boldsymbol{r} = \mathbf{0}$ where $\mathbf{d}\boldsymbol{r}$ is an infinitesimal vector

Fig. 1.12 Field lines of a stretched quadrupole

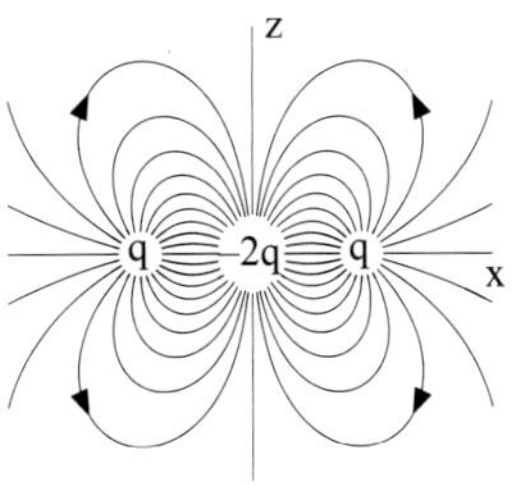

tangential to the field line in the point $\boldsymbol{r}$. For a situation in a coordinate plane (e.g. in the x- z plane as in the example above) this leads to

$$\boldsymbol{E}(\boldsymbol{r}) \times \mathbf{d}\boldsymbol{r} = \begin{vmatrix} \boldsymbol{e}_x & \boldsymbol{e}_y & \boldsymbol{e}_z \\ \mathrm{d}x & 0 & \mathrm{d}z \\ E_x & 0 & E_z \end{vmatrix} = \Big(\boldsymbol{E}_x(x, z)\mathrm{d}z - \boldsymbol{E}_z(x, z)\mathrm{d}x \Big) \boldsymbol{e}_y = \mathbf{0} \,.$$

The expression in the bracket is a differential equation, which can be used to calculate equations for the field lines in the form $z = z(x)$ or $x = x(z)$ from the knowledge of the field components.

From a practical point of view it is, however, also necessary to determine the field of a macroscopic charged body rather than a finite assembly of a few point charges. As the number of point charges is rather large in this case (of the order 10^{20}), it is useful to work with the concept of **charge densities**, that is space charge density, surface charge density and (the possibly more academic variant) linear charge density.

The concept of a space charge density can be introduced in the following fashion: Electrons, which are attached to or missing from the atoms, can globally be characterized as a space charge density ρ (Fig. 1.13a), which can be defined as

$$\rho(\boldsymbol{r}') = \frac{\mathrm{d}q'(\boldsymbol{r}')}{\mathrm{d}V'} \,,$$

where $\mathrm{d}q'$ is an infinitesimal charge in an infinitesimal volume element $\mathrm{d}V'$ at the position $\boldsymbol{r}'$ divided by the volume element. This definition has to be regarded with some caution. A volume element of atomic or subatomic dimension (needed for the limiting process envisaged) would yield a function $\rho(\boldsymbol{r}')$ with strong variations from point to point depending on the fact, whether elementary charges are included in the element or not. On the other hand macroscopic experiments do not require such fine resolution. This allows one to work with the picture that the true charge distribution (whatever it is) can be averaged over sufficiently large regions of space (Fig. 1.13b), so that one can consider the averaged charge elements

$$\mathrm{d}q' = \overline{\rho(\boldsymbol{r}')}\, \mathrm{d}V' \equiv: \rho(\boldsymbol{r}')\, \mathrm{d}V'$$

as a point charge element, leading to a smooth function $\rho(\boldsymbol{r}')$. Each infinitesimal point charge $\mathrm{d}q'$ at a point $\boldsymbol{r}'$ then contributes to the infinitesimal field $\mathrm{d}\,\boldsymbol{E}$ at the position $\boldsymbol{r}$ (Fig. 1.14)

$$\mathrm{d}\,\boldsymbol{E} = k_e\, \mathrm{d}q' \,\frac{(\boldsymbol{r} - \boldsymbol{r}')}{|\boldsymbol{r} - \boldsymbol{r}'|^3} = k_e \rho(\boldsymbol{r}')\, \frac{(\boldsymbol{r} - \boldsymbol{r}')}{|\boldsymbol{r}' - \boldsymbol{r}|^3}\, \mathrm{d}\,V' \,.$$

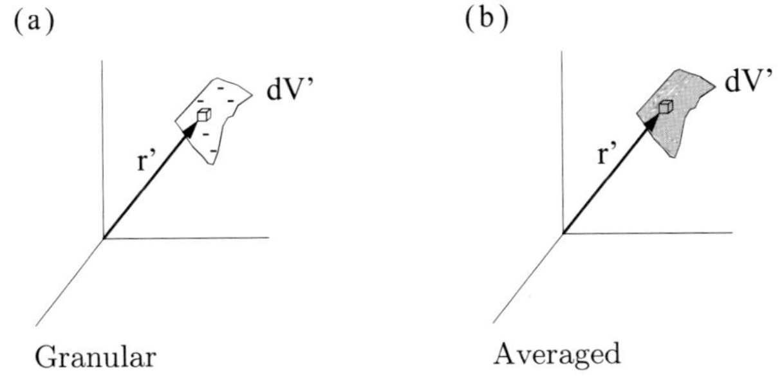

Fig. 1.13 The space charge density

The field due to the charge distribution of a three-dimensional charged body is then obtained by the addition of such vector contributions, that is integration over its volume

$$\boldsymbol{E}(\boldsymbol{r}) = \iiint_{\text{body}} \mathrm{d}\,\boldsymbol{E} = k_e \iiint_{\text{body}} \rho(\boldsymbol{r}')\,\frac{(\boldsymbol{r} - \boldsymbol{r}')}{|\boldsymbol{r} - \boldsymbol{r}'|^3}\,\mathrm{d}\,V'\,. \tag{1.5}$$

One integrates over the primed coordinates. Application of this expression implies triple integration, e.g. in Cartesian coordinates for the x-component of the field

$$E_x(\boldsymbol{r}) = k_e \iiint_{\text{body}} \rho(x', y', z')\frac{(x - x')\,\mathrm{d}x'\,\mathrm{d}y'\,\mathrm{d}z'}{\left[(x - x')^2 + (y - y')^2 + (z - z')^2\right]^{3/2}}\,.$$

The evaluation of these integrals can be done analytically in the simplest cases, but can be cumbersome in general. One has to find more practical methods for the calculation of the fields generated by spatial charge contributions.

A surface charge distribution is defined in a similar fashion. In many cases (for instance for a charged conductor) the total charge is confined to a layer of atomic dimensions at the surface of the body. In this case it is useful to work with a (suitably averaged) surface charge density

$$\mathrm{d}q' = \sigma(\boldsymbol{r}')\,\mathrm{d}f'\,,$$

Fig. 1.14 Calculation of fields: Geometry

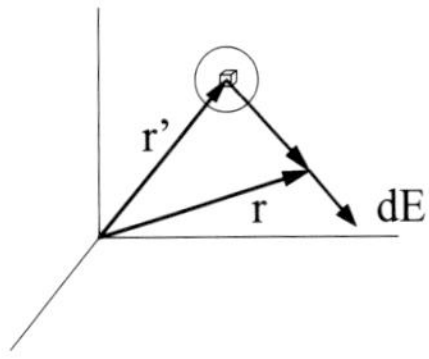

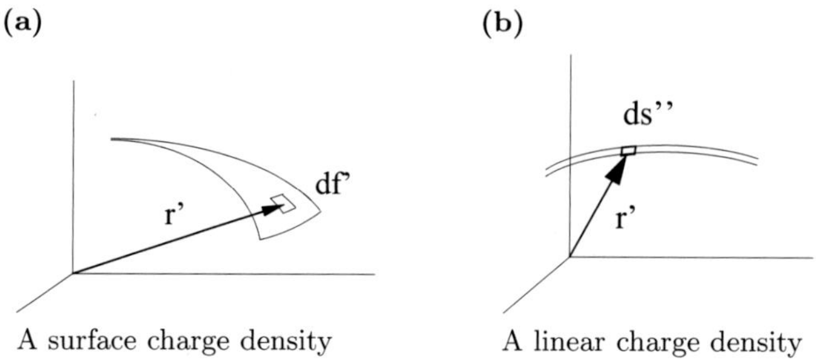

Fig. 1.15 More charge densities

where $\mathrm{d}f'$ is the area of an infinitesimal surface element (Fig. 1.15a). The field of the total charge distribution has to be calculated by a double integral over the specified surface in this case by

$$\boldsymbol{E}(\boldsymbol{r}) = k_e \iint_{\text{surface}} \sigma(\boldsymbol{r}') \frac{(\boldsymbol{r} - \boldsymbol{r}')}{|\boldsymbol{r} - \boldsymbol{r}'|^3} \mathrm{d}\, f' . \tag{1.6}$$

Equation (1.6) stands for a vectorial summary of three integrals.

The linear charge density is the one-dimensional equivalent of these charge densities. One may image a charged tube of atomic cross section and define the (averaged) linear charge density

$$\mathrm{d}q' = \lambda(\boldsymbol{r}')\mathrm{d}s' .$$

The quantity $\mathrm{d}s'$ is the infinitesimal line element along the tube (Fig. 1.15b). For the electric field of this charge distribution one writes

$$\boldsymbol{E}(\boldsymbol{r}) = k_e \int_{\text{curve}} \lambda(\boldsymbol{r}') \frac{(\boldsymbol{r} - \boldsymbol{r}')}{|\boldsymbol{r} - \boldsymbol{r}'|^3} \mathrm{d}\, s' . \tag{1.7}$$

The components of the electric field are obtained by integration along a specified curve in space.

Some direct examples can be used to illustrate the application of the possibilities described above (see also Detail 1.5.3).

The first problem is: A charge Q ($Q > 0$) is distributed uniformly along a circle of radius R, which is placed around the origin in the z-y plane (Fig. 1.16a). The task is the calculation of the electric field for a point' on the x-axis through the centre of

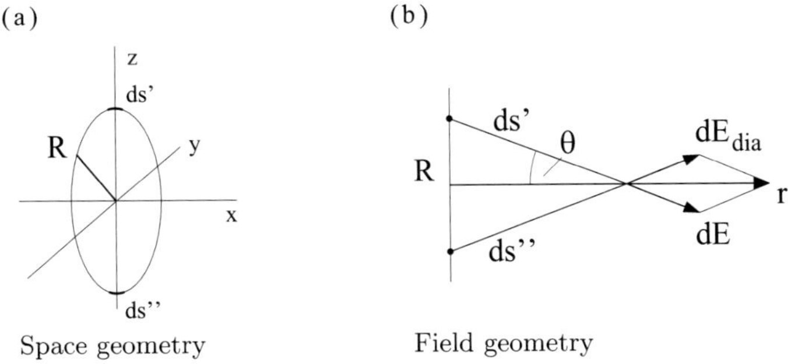

Fig. 1.16 Electrical field of a uniformly charged ring for points on the axis through the ring

the ring.[6] As the charge density is constant

$$\lambda = \frac{Q}{2\pi R} = \text{const.},$$

one obtains a contribution $\mathrm{d}\boldsymbol{E}$ of a line element $\mathrm{d}s' = R\,d\phi'$ of the circle in a point of the axis, which is at a distance r from the centre of the circle. For an equivalent diametrical element $\mathrm{d}s''$ the contribution on the axis is $\mathrm{d}\boldsymbol{E}_{\text{dia}}$ (Fig. 1.16b). The two infinitesimal field vectors include the same angle θ with respect to the x-axis. The components of the two infinitesimal fields, perpendicular to the axis, cancel. The two contributions in the direction of the axis have the same length, so that they add up to

$$\begin{aligned} \mathrm{d}\,E_{axis} &= 2(\mathrm{d}\,E)\,\cos\theta = 2\frac{r}{\sqrt{r^2+R^2}}\,\mathrm{d}\,E \\ &= 2\,\frac{r}{\sqrt{r^2+R^2}}\,k_e\,\frac{Q}{2\pi R}\,\frac{1}{(r^2+R^2)} = \frac{k_e Q}{\pi R}\,\frac{r\,\mathrm{d}\,s'}{(r^2+R^2)^{3/2}}\,\mathrm{d}\,s'\,. \end{aligned}$$

Integration over half the circle gives

$$\begin{aligned} E &= \int_{\text{half circle}} \mathrm{d}E_{axis} = k_e\,\frac{Q}{\pi R}\frac{r}{(r^2+R^2)^{3/2}}\,R\int_0^{\pi}\mathrm{d}\varphi' \\ &= k_e Q\,\frac{r}{(r^2+R^2)^{3/2}}\,. \end{aligned}$$

[6] The calculation of the field for other points of space is, as can be found in Chap. 2.4, more difficult.

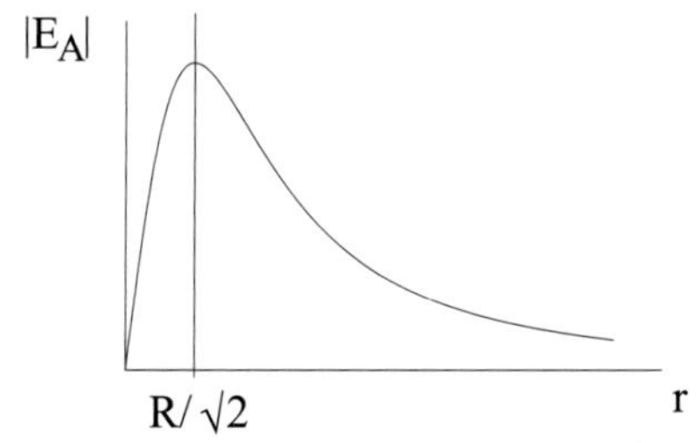

Fig. 1.17 Electrical field on the axis of a charged circular ring

The variation of the field with the distance r from the centre of the circle is shown in Fig. 1.17. The field has the value zero at the centre of the circle, it grows first linearly with r, reaches a maximum at $r = R/\sqrt{2}$ and decreases for large values of r as

$$E \xrightarrow{r\to\infty} k_e \frac{Q}{r^2} .$$

The ring looks like a point charge Q if viewed from a large distance.

The second problem is the calculation of the electric field of a circular disk (radius R) with a uniform distribution of surface charges

$$\sigma = \frac{Q}{\pi R^2}$$

for points on a perpendicular axis through the centre of the disk. The results of the previous problem can be used in this case by discretising the disk into infinitesimal rings (Fig. 1.18a), by calculating the contributions of each individual ring and adding these contributions (see Detail 1.5.3). The resulting field on the axis as a function of the distance r from the centre of the disk is given by (Fig. 1.18b)

$$E_{disk}(r) = \frac{2k_e Q}{R^2}\left[1 - \frac{r}{\left[r^2 + R^2\right]^{1/2}}\right] .$$

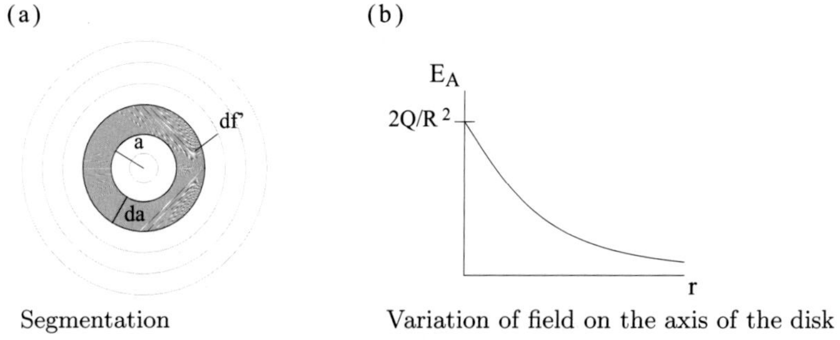

Fig. 1.18 The electric field of a uniformly charged circular disk

For $r = 0$ it has the value $2k_e Q/R^2$ and decreases steadily towards the value zero. For large r-values one finds, because of

$$\frac{r}{\left[r^2 + R^2\right]^{1/2}} = \frac{1}{\left[1 + (R/r)^2\right]^{1/2}} = \left(1 - \frac{1}{2}\left(\frac{R}{r}\right)^2 + \ldots\right),$$

once more the result

$$E \xrightarrow{r \to \infty} k_e \frac{Q}{r^2} + O\left(\frac{1}{r^4}\right).$$

If viewed from a large distance the disk also looks like a point charge.

Two additional limits are of interest:

- The disk is reduced to a point

$$R \to 0, \quad Q = \text{const.}$$

  In this case one finds again

$$E_A \xrightarrow{R \to 0} k_e \frac{Q}{r^2}$$

  the field of a point charge.
- The circular disk is extended to a uniformly charged plane with a constant surface charge density

$$R \to \infty, \quad \sigma = \frac{Q}{\pi R^2} = \text{const.}$$

  The total charge of the plane is infinite

$$Q_{\text{plane}} = \sigma \int \mathrm{d}f \to \infty\,.$$

  The limiting value of the field is in this case

$$\lim_{R \to \infty} E = 2k_e \pi \sigma\,.$$

  The position of the axis is not important in this limit. The result is an electric field that has the same constant value for each point in space and is orthogonal to the plane with the charge density (Fig. 1.19a and b). This limit is often used as an approximation for a field of a flat plate with finite extend, neglecting the effects of the edge. This is acceptable if one is not too close to the edge and not to far from the plate. The field is, with these restrictions, nearly uniform.

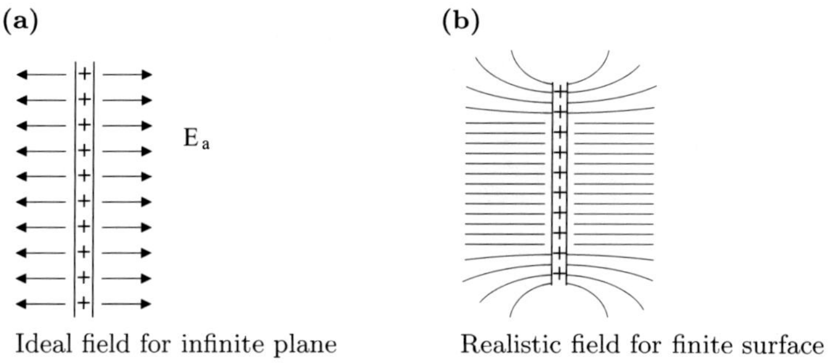

Fig. 1.19 The electric field of a charged plate

The last problem will only be indicated at this point, as the result will be obtained with simpler methods at a later stage. The task is to find the electric field of a uniformly charged sphere (radius R) with uniform charge density

$$\rho = \frac{Q}{\frac{4}{3}\pi R^3} = \text{const.}$$

The problem can be solved by dissecting the sphere into circular disks with the thickness ds and add the contribution of all infinitesimal disks. The radius a of a disk at position s (Fig. 1.20) is $a^2 = R^2 - s^2$. The charge of the disk is therefore

$$\mathrm{d}q' = \rho\pi a^2 \mathrm{d}s\,.$$

The elementary calculation has to be performed separately for points inside the sphere $r < R$ and outside the sphere $r > R$. The position of the axis does not play any role because of the symmetry. The calculation is also presented in Detail 1.5.3. The result for the radial field is

$$E(r) = \begin{cases} k_e \dfrac{Qr}{R^3} & r \leq R \quad \text{(interior)} \\ k_e \dfrac{Q}{r^2} & r > R \quad \text{(exterior)}\,. \end{cases} \tag{1.8}$$

Fig. 1.20 The calculation of the field of a uniformly charged sphere

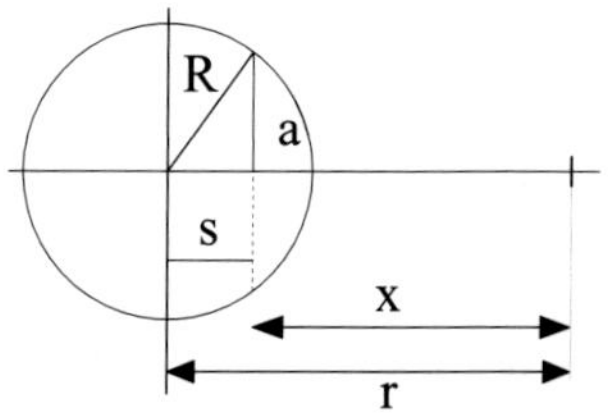

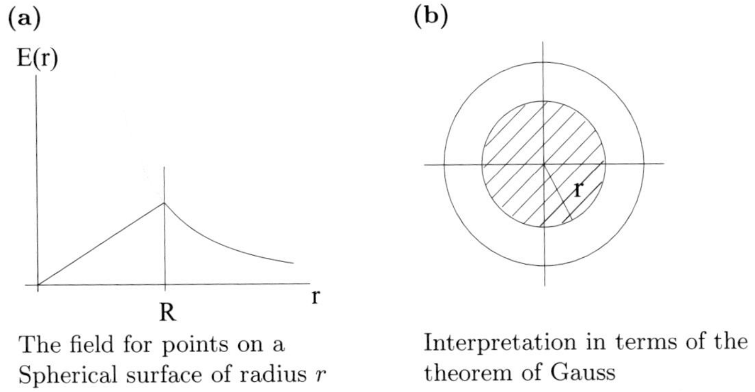
(a)
E(r)
R
r
The field for points on a
Spherical surface of radius r
(b)
r
Interpretation in terms of the
theorem of Gauss

Fig. 1.21 Electric field $E(r)$ of a uniformly charged sphere

The field rises linearly up to the surface of the sphere and decreases in the same fashion as the field of a point charge in the exterior region (Fig. 1.21a). The field is continuous at the surface $r = R$, but the derivative of the field function with respect to r is not.

The result for the interior can be interpreted in a simple fashion. It has the form

$$E(r) = k_e \left(Q \frac{r^3}{R^3} \right) \frac{1}{r^2} = k_e \frac{Q_{\text{in}}(r)}{r^2} \, .$$

Q_{in} is the charge, which is enclosed by a sphere of radius r (Fig. 1.21b). This result is a special case of the theorem of Gauss, which will be discussed in the next section.

1.4 The Integral Theorem of Gauss

The integral theorem of Gauss (see Vol. 1, Math. Chap. 5.3.3) is a useful tool for the further development of electrostatics, but it can also be used for the direct calculation of (electric) fields. In the simplest application one encloses a point charge q within a spherical surface of radius r. The surface is then divided into infinitesimal surface elements $\mathbf{d}\boldsymbol{f}$. The direction of the vector characterising each of these elements is the vector normal to the surface (Fig. 1.22). The surface integral with the field of a point charge can be calculated directly

$$\begin{aligned} \oiint_{\text{sphere}} \boldsymbol{E} \cdot \mathbf{d}\boldsymbol{f} &= k_e \iint \left(\frac{q}{r^2}\, \boldsymbol{e}_{\text{r}} \right) \cdot \left(r^2\, \mathrm{d}\Omega\, \boldsymbol{e}_{\text{r}} \right) \\ &= k_e q \iint \mathrm{d}\Omega = 4\pi k_e q \, , \end{aligned} \tag{1.9}$$

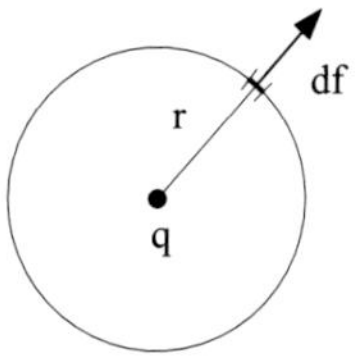

Fig. 1.22 Theorem of Gauss: Simple application

or explicitly in the standard systems of units

$$\oiint_{\text{sphere}} \boldsymbol{E} \cdot \mathbf{d}\boldsymbol{f} = 4\pi\, q \qquad \text{in the CGS-system}$$

$$\oiint_{\text{sphere}} \boldsymbol{E} \cdot \mathbf{d}\boldsymbol{f} = \frac{q}{\varepsilon_0} \qquad \text{in the SI-system}\,.$$

As the result is independent of the radius of the surface, it is valid for any spherical surface around the point charge q. The discussion can be extended directly to a situation with an arbitrary closed surface around the point charge.

For a point charge, which is enclosed within an arbitrary surface F, the statement

$$\oiint_{\text{F}} \boldsymbol{E} \cdot \mathbf{d}\boldsymbol{f} = \oiint_{\text{sphere}} \boldsymbol{E} \cdot \mathbf{d}\boldsymbol{f}' = 4\pi k_e q$$

is also valid, for instance also for surfaces with an indention (Fig. 1.23a). If there is no charge within the surface F (Fig. 1.23b) one has

$$\oiint_{\text{F}} \boldsymbol{E} \cdot \mathbf{d}\boldsymbol{f} = 0\,.$$

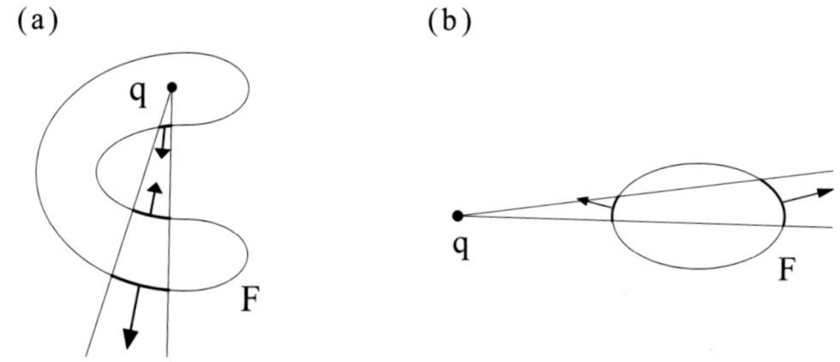

Fig. 1.23 Theorem of Gauss: More possibilities

The surface integral of an electric field over an arbitrary surface

$$\iint_{\mathrm{F}} \boldsymbol{E} \cdot \mathbf{d}\boldsymbol{f} = \Phi_{\mathrm{F}}$$

is called **electric flux** Φ_{F} or, more precisely, the flux of the electric field $\boldsymbol{E}$ through the surface F. This terminology is based on the interpretation of the field lines as the flow lines of a hypothetical fluid. The surface integral measures the quantity of fluid, which traverses the specified surfaces. In this spirit one may call the charges as the sources ($q = +$) or the sinks ($q = -$) of the electric field. The statement

$$\Phi_{\mathrm{F}} = \oint\!\!\!\oint_{\mathrm{F}} \boldsymbol{E} \cdot \mathbf{d}\boldsymbol{f} = \begin{cases} 4\pi k_e q & q \text{ within F} \\ 0 & q \text{ not in F} \end{cases} \tag{1.10}$$

which is true for a point charge in any closed surface is cited in the form:

The flux of the electric field of a point charge through a closed surface around the charge equals $4\pi k_e$ times the point charge enclosed.[7]

The simple form of the **Gauss' theorem** (or Gauss' law) applied in electrostatics can be extended:

(i) If the point charges $q_1, q_2, \ldots q_N$ are enclosed in a closed surface F, the statement

$$\oint\!\!\!\oint_{\mathrm{F}} \boldsymbol{E}_i \cdot \mathbf{d}\boldsymbol{f} = 4\pi k_e q_i \qquad (i = 1, 2, \ldots N)$$

is valid for each of the charges. Addition of the individual contributions with the superposition principle yields for the total field

$$\boldsymbol{E} = \sum_{i=1}^{N} E_i$$

the result

$$\oint\!\!\!\oint_{\mathrm{F}} \boldsymbol{E} \cdot \mathbf{d}\boldsymbol{f} = 4\pi k_e \sum_{i=1}^{N} q_i \,.$$

The flux of the total field equals $4\pi k_e$ times the total enclosed charges.

[7] This includes the case, that the point charge is zero.

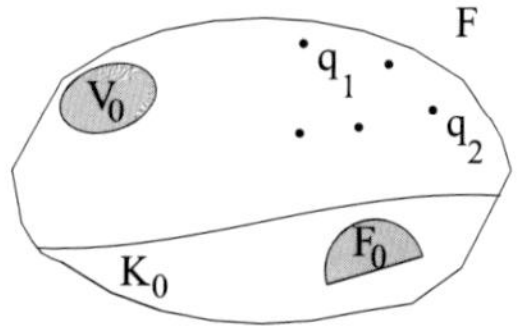

Fig. 1.24 Charge distributions

(ii) If there exist space, surface or linear charges beside the point charges, one also finds

$$\oiint_F \boldsymbol{E} \cdot \mathbf{d}\boldsymbol{f} = 4\pi k_e\, Q_{\text{in}}\,. \tag{1.11}$$

Q_{in} represents the total charge enclosed in the surface, which is calculated as

$$Q_{\text{in}} = \sum_i q_i + \iiint_{V_0} \rho(\boldsymbol{r}')\,\mathrm{d}V' + \iint_{F_0} \sigma(\boldsymbol{r}')\,\mathrm{d}f' + \int_{K_0} \lambda(\boldsymbol{r}')\,\mathrm{d}s'\,.$$

The volume V_0 is the part of the volume enclosed by F which contains the space charges, F_0 the part of the enclosed surface which is covered with surface charges, and K_0 are the lines containing the line charges (Fig. 1.24). This assumes a decomposition of all real charges into point charges and a limiting process as used in (i).

(iii) The general form of the theorem of Gauss is a summary of Coulomb's law and the superposition principle.

1.4.1 Simple Applications of the Gauss' Theorem

The theorem of Gauss contains the central differential equation of electrostatics, which will be discussed in the next section. It can, however, also be used for the direct calculation of electric fields.

If it is possible to identify closed surfaces perpendicular to the field, on which the magnitude of the field is constant (Fig. 1.25), the evaluation of the surface integral is rather trivial. This is indicated in the following examples.

- The first example is a uniformly charged sphere (radius R) with a uniform spherical charge density

$$\rho = \frac{Q}{(4/3)\pi R^3} = \text{const.} \qquad \text{for} \quad r \le R\,.$$

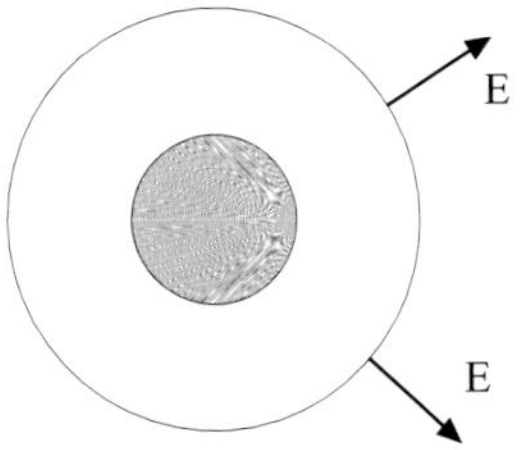

Fig. 1.25 Gauss' theorem: Charge distribution with spherical symmetry

It demonstrates, that one can obtain results with minimal effort by application of the Gauss' theorem (compare the comments on p. 24). The symmetry surfaces are concentric spheres. This allows to evaluate the flux integral

$$\oiint_{\text{sphere}} \boldsymbol{E} \cdot \mathbf{d}\boldsymbol{f} = \iint E(r)\, r^2 \mathrm{d}\Omega = E(r)\, r^2 4\pi \,.$$

The following cases have to be distinguished concerning the enclosed charge: The enclosed charge for $r \le R$ is

$$Q_{\text{in}} = \rho \iiint_{\text{sphere}} \mathrm{d}V' = \frac{4}{3}\pi r^3 \rho = Q\frac{r^3}{R^3}\,,$$

so that

$$4\pi r^2 E(r) = 4\pi k_e \left(Q\frac{r^3}{R^3} \right)$$

or after resolution

$$E(r) = k_e Q \frac{r}{R^3}\,.$$

For $r \ge R$ the total charge Q is enclosed, so that

$$Q_{\text{in}} = \frac{4}{3}\pi R^3 \rho = Q$$

and

$$E(r) = k_e \frac{Q}{r^2}\,.$$

If the spherical charge distribution is not uniform

$$\rho(\boldsymbol{r}) = \rho(r)$$

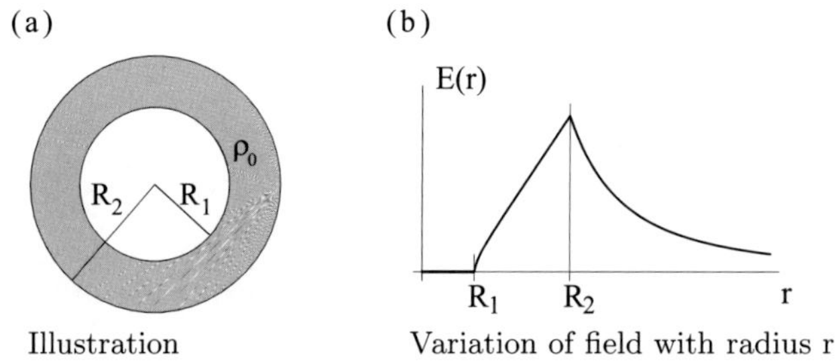

Fig. 1.26 Gauss' theorem: Hollow sphere

(the charge density varies with the distance from the centre of the sphere and not with the angles), then the calculation of the electric field is also simple. One still deals with a spherical situation and calculates the enclosed charge by

$$Q_{\text{in}} = \iiint_{\text{sphere(r)}} \rho(\boldsymbol{r}')\,\mathrm{d}V' = 4\pi \int_0^r \rho(r')\, r'^2 \mathrm{d}r' \, .$$

- A direct example is a hollow sphere (Fig. 1.26a) with the uniform charge distribution

$$\rho(\boldsymbol{r}) = \begin{cases} 0 & r < R_1 \\ \rho_0 & R_1 \le r \le R_2 \\ 0 & r > R_2 \end{cases} .$$

The charges enclosed in the three sections are

$$Q_{\text{in}} = \begin{cases} 0 & r < R_1 \\ \frac{4}{3}\pi\rho_0(r^3 - R_1^3) & R_1 \le r \le R_2 \\ \frac{4}{3}\pi\rho_0(R_2^3 - R_1^3) = Q & r > R_2 \end{cases} .$$

The magnitude of the field in radial direction is therefore

$$E(r) = \begin{cases} 0 & r < R_1 \\ \frac{4\pi}{3} k_e \rho_0 \left(r - \frac{R_1^3}{r^2} \right) & R_1 \le r \le R_2 \\ k_e \frac{Q}{r^2} & R_2 < r \end{cases} .$$

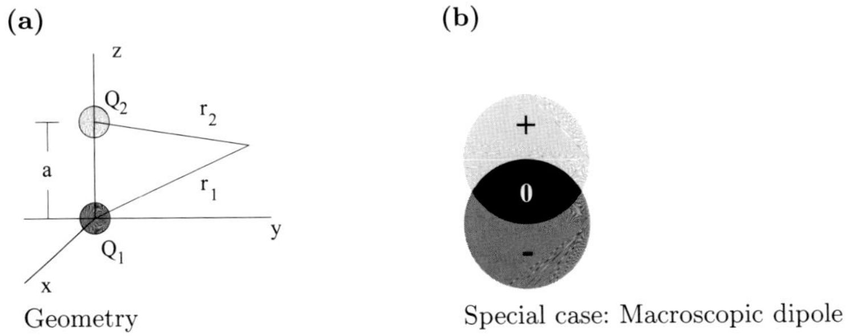

Fig. 1.27 Macroscopic dipole

The magnitude of the field as a function of r has the value zero in the interior section, it grows more slowly than a linear growth in the region with the charge density (up to the value $k_e Q/R_2^2$) and decreases according to r^{-2} in the exterior region (Fig. 1.26b). The total function $E(r)$ is continuous, but the derivative $\mathrm{d}E(r)/dr$ is not.

If the charge distributions do not show obvious symmetries, so that a suitable Gaussian surface can not be found, the following procedure can help: Apply the Gauss' theorem to part of the charges with symmetry and determine the total electric field by superposition of the fields of the different parts. An example of this kind of problem is:

- Calculate the electric field in the area outside of two uniformly charged spheres[8] with the radii R_1 and R_2 and the total charges Q_1 and Q_2, if the charge centres are separated by the distance a. Consider the special case $Q_1 = -Q_2 = Q$, which corresponds to a macroscopic dipole (Fig. 1.27a). For the solution one can use a coordinate system with the centres of the spheres at

$$M_1 = (0,\, 0,\, 0) \qquad \text{and} \qquad M_2 = (0,\, 0,\, a)\,.$$

It would not be a simple task to determine a symmetry surface in this case. If one begins, however, with two individual spheres, one finds for a point (x, y, z) outside the spheres the Cartesian decomposition of the

$$\boldsymbol{E}_1 = k_e \left(\frac{Q_1 x}{r_1^3},\, \frac{Q_1 y}{r_1^3},\, \frac{Q_1 z}{r_1^3} \right) \qquad \boldsymbol{E}_2 = k_e \left(\frac{Q_2 x}{r_2^3},\, \frac{Q_2 y}{r_2^3},\, \frac{Q_2 (z-a)}{r_2^3} \right)$$

with

$$r_1 = \sqrt{x^2 + y^2 + z^2} \qquad r_2 = \sqrt{x^2 + y^2 + (z-a)^2}\,.$$

[8] A calculation of the field in all sections of the space is possible, but requires the inclusion of all the different areas of the space.

The total field is then

$$\boldsymbol{E} = \boldsymbol{E}_1 + \boldsymbol{E}_2$$

$$= k_e \left(x \left(\frac{Q_1}{r_1^3} + \frac{Q_2}{r_2^3} \right),\ y \left(\frac{Q_1}{r_1^3} + \frac{Q_2}{r_2^3} \right),\ z \frac{Q_1}{r_1^3} + (z-a) \frac{Q_2}{r_2^3} \right).$$

This expression is also valid, if the two spheres overlap.

For the special case of a macroscopic dipole with equal spheres

$$R_1 = R_2 = R \qquad Q_1 = -Q_2 = Q$$

the following statements can be made: For $a > 2R$ one finds the macroscopic dipole. For $2R > a > 0$ one finds the field of a charge distribution, which consists of two partial spheres. The region of overlap is uncharged (as $\rho_1 = -\rho_2$ Fig. 1.27b).

For $a = 0$, the two oppositely charged spheres coincide ($r_1 = r_2$) completely, so that

$$\boldsymbol{E} = \boldsymbol{0}.$$

This example demonstrates that fields of rather complex charge distributions can be calculated by simple means but more effort.

- In the next example a surface charge distribution, which is basically a type of capacitor (also known as condenser), will be considered. It addresses a special phenomenon in the case of surface charges.

In this example the charges Q_1 and Q_2 are distributed uniformly over two concentric spherical surfaces with radii R_1 and R_2 (Fig. 1.28). The constant surface charge densities are σ_1 and σ_2. As the electric field has a constant value on any spherical surface and is oriented radially, the Gauss' theorem allows the following statements:

- For $0 \leq r < R_1$ one has $E(r) = 0$ as $Q_{\text{in}} = 0$.
- For $R_1 \leq r < R_2$ the enclosed charge is $Q_{\text{in}} = 4\pi\sigma_1 R_1^2 = Q_1$ and the radial field is

$$E(r) = k_e \frac{Q_1}{r^2}.$$

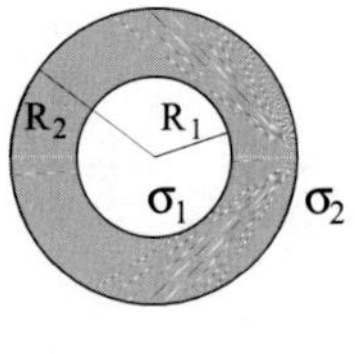

Fig. 1.28 Ideal spherical capacitor

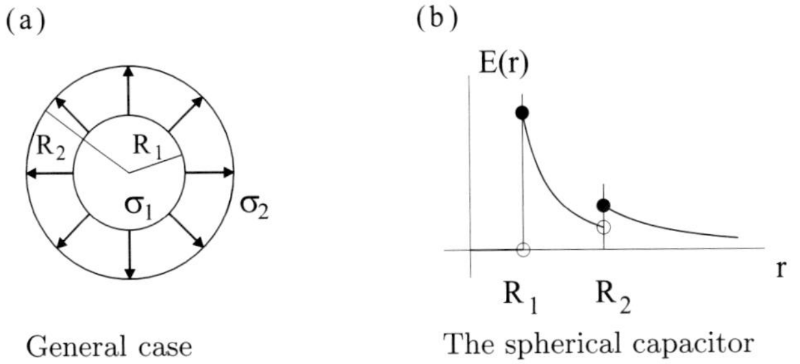

Fig. 1.29 The electric field of two uniformly charged spherical surfaces

- The enclosed charge is

$$Q_{\text{in}} = 4\pi(\sigma_1 R_1^2 + \sigma_2 R_2^2) = Q_1 + Q_2$$

for points with $r \geq R_2$ (the outside) and the corresponding field is

$$E(r) = k_e \frac{Q_1 + Q_2}{r^2} .$$

The function $E(r)$ has the value zero in the interior region. It changes suddenly to a value $k_e Q_1/R_1^2$ on passing the radius $r = R_1$ and decreases to the value $k_e Q_1/R_2^2$ in the intermediate region. At the radius $r = R_2$ it changes suddenly once more to the value $k_e(Q_1 + Q_2)/R_2^2$ (Fig. 1.29b).

The basic observation is: The electric field is not continuous when it passes through an infinitesimal layer of surface charges. The magnitude of the field changes suddenly. The magnitude of this jump is

from $R_1 - \varepsilon$ to $R_1 + \varepsilon$:

$$E(R_1 + \varepsilon) - E(R_1 - \varepsilon) = \frac{k_e Q_1}{R_1^2} = 4\pi\sigma_1$$

from $R_2 - \varepsilon$ to $R_2 + \varepsilon$:

$$E(R_2 + \varepsilon) - E(R_2 - \varepsilon) = \frac{k_e Q_2}{R_2^2} = 4\pi\sigma_2 .$$

For the special case with the charges $Q_1 = -Q_2 = Q$ one finds the result (ideal spherical capacitor)

$$E(r) = \begin{cases} 0 & 0 \le r < R_1 \\ k_e \dfrac{Q}{r^2} & R_1 \le r < R_2 \\ 0 & r \ge R_2 \end{cases} \cdot$$

The field exists only in the region between the two spherical shells (Fig. 1.29a). The function $E(r)$ changes abruptly from the value zero to the value $k_e Q/R_1^2$ at the radius R_1 and at the radius R_2 from $k_e Q/R_2^2$ back to zero. The first jump is larger than the second as $|\sigma_1| > |\sigma_2|$ and $R_2 > R_1$.

The sudden changes of the field, found here, is a general feature, which will be investigated more closely in Chap. 2.2. The first task is, however, to assemble the complete framework of electrostatics and to discuss the set of corresponding equations.

1.5 Details

1.5.1 Millikan's Experiment for the Determination of the Charge of the Electron

R.A. Millikan has undertaken a series of experiments in the years 1908 until 1913 in order to determine the (elementary) charge of the electron. The basic idea of these experiments was the observation and the analysis of the motion of charged droplets of fluids under the action of gravitation and an electric field in the opposite direction.

1.5.1.1 The Experiment

A liquid (oil, alcohol, water, etc.) has been atomised in the space between the horizontal plates of a plate condenser. Some of the droplets are ionised by the atomisation. The movement of the illuminated droplets under the influence of the competing forces was observed with a thread microscope. Four forces act on the droplets

gravity	F_g
frictional force	F_f
buoyancy force	F_b
electrical force	F_e

The charge of the droplets and the elementary charge can be determined by analysing the movement (see below).

1.5.1.2 Analysis of the Experiment

A droplet is characterized by the parameters (assuming spherical shape)

$$\text{mass } m_d\,, \text{ charge } q\,, \text{ radius } R\,, \text{ volume } V_d = 4\pi R^3/3\,.$$

The forces acting on the particle are:

- The simple force of gravity

$$\boldsymbol{F}_g = -m_d g\, \boldsymbol{e}_z\,,$$

  where the unit vector $\boldsymbol{e}_z$ points vertically upwards in z-direction. With the definition of density

$$\rho_d = \frac{m_d}{V_d}$$

$$\boldsymbol{F}_g = -\rho_d \frac{4\pi}{3} R^3 g\, \boldsymbol{e}_z$$

  can be written.
- The electric force (see Chap. 1.3) is given by the electric field strength and the charge of the droplet

$$\boldsymbol{F}_e = qE\, \boldsymbol{e}_z\,.$$

  In the homogeneous field area

$$E = \frac{U}{d}$$

  applies, if U is the voltage between the plates and d their distance. The product of charge and voltage is chosen to be positive.
- The particle moves in the 'tough' medium air. A frictional force then acts, which is applied according to Stokes' law

$$\boldsymbol{F}_f = 6\pi\, \eta_a R\, \boldsymbol{v}\,.$$

  The constant η_a denotes the viscosity coefficient of the air.
- Movement in a medium also causes a buoyancy force which, according to Archimedes, is given by

$$\boldsymbol{F}_b = m_a g\, \boldsymbol{e}_z\,.$$

The mass m_a is the mass of the air displaced by the droplet. With the value ρ_a for the density of the air follows

$$\boldsymbol{F}_b = \rho_a\, V_d\, g\, \boldsymbol{e}_z = \frac{4\pi}{3}\, R^3 \rho_a g\, \boldsymbol{e}_z\,.$$

If the electric field is switched off, an equilibrium situation is established after a short time under the influence of the remaining three forces. The force of gravity is compensated by the force of friction and buoyancy

$$\boldsymbol{F}_g = \boldsymbol{F}_f + \boldsymbol{F}_b\,.$$

(Remember the free fall experiment with friction.) The particle then sinks at a constant velocity v_0.

If the electric field is now switched on, there are four possible options for changing the droplet movement, depending on the direction and magnitude of the field:

The particle	
falls faster	$\boldsymbol{F}_e \propto \boldsymbol{F}_g$
slows down the falling motion or rises	$\boldsymbol{F}_e \propto -\boldsymbol{F}_g$
floats	compensation of all forces
does not react	droplet is uncharged, neutral.

After a short time again an equilibrium of the forces is restored, which is described by

$$\boldsymbol{F}_g = \boldsymbol{F}_f + \boldsymbol{F}_b + \boldsymbol{F}_e\,. \tag{1.12}$$

In the equilibrium of forces the velocity of fall is constant and has the value v_m. Solving (1.12) according to the charge of the droplet gives

$$\frac{4\pi}{3}\, \rho_d\, R^3 g = 6\pi\, \eta_a R\, v_m + \frac{4\pi}{3}\, R^3 \rho_a g + qE$$

$$\Longleftrightarrow$$

$$q = \frac{4\pi}{3}\, R^3\, \frac{g}{E}\, (\rho_d - \rho_a) - 6\pi\, \eta_a\, \frac{R\, v_m}{E}\,.$$

The variables g, E, ρ_d, ρ_a, η_a in this equation are known. The droplet radius R and the velocity v_m still have to be measured. The droplet radius can be obtained from the equilibrium condition without field:

$$F_g = F_f + F_b$$

$$\frac{4\pi}{3}\,\rho_d\,R^3 g = 6\pi\,\eta_a R\,v_0 + \frac{4\pi}{3}\,R^3\rho_a\,g$$

resp.

$$R = 3\left[\frac{v_0\,\eta_a}{2g\,(\rho_d - \rho_a)}\right]^{1/2} .$$

With the same droplet (you must ensure that you do not lose the droplet from the field of observation), the falling velocity must be determined. If this quantity has been measured, the charge can be calculated from the relation

$$\begin{aligned} q &= \left(\frac{4\pi}{3}\,R^2 g\,(\rho_d - \rho_a) - 6\pi\,\eta_a R\,v_m\right)\frac{R}{E} \\ &= \left(\frac{4\pi}{3}\,R^2 g\,(\rho_d - \rho_a) - 6\pi\,\eta_a R\,v_m\right)\frac{3}{E}\left[\frac{v_0\,\eta_a}{2g\,(\rho_d - \rho_a)}\right]^{1/2} \\ &= 9\pi\,\sqrt{2}\,\eta_a\frac{(v_0 - v_m)}{E}\left[\frac{v_0\,\eta_a}{g\,(\rho_d - \rho_a)}\right]^{1/2} \\ &= q(R)\,. \end{aligned}$$

If there is a smallest charge quantity, the measured charge must be an integer multiple of this elementary charge

$$q = N\,e_0 \qquad N = 1, 2, 3, \ldots .$$

Figure 1.30 shows the (first 25) results schematically. The equidistant lines each represent a multiple of the elementary charge and e_0 can be read off.

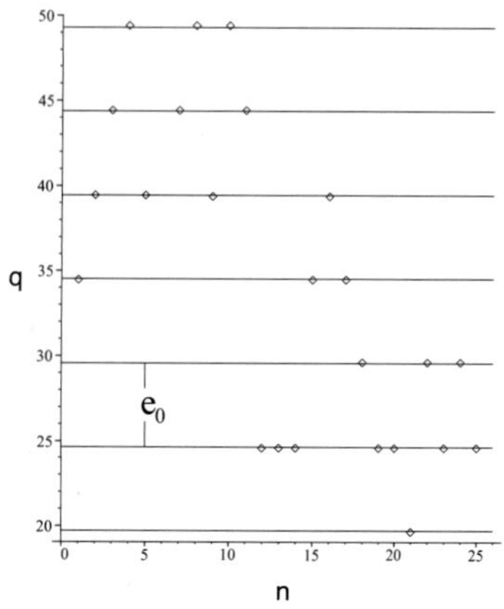

Fig. 1.30 Millikan's experimental data

1.5.1.3 Practical Aspects of Carrying Out the Experiment

For the results of a series of measurements with several droplets (or a series of measurements with a droplet that has been recharged by the effect of weak radiation)

$$q_1, q_2, q_3, \ldots, q_m$$

one must be able to find unique integer divisors N_i. If the elementary charge cannot be determined unambiguously despite numerous measurements, the search for possible sources of error is called for. Millikan was forced to investigate the following possibilities in the course of his series of experiments and eliminate them if necessary:

- The electric field is not homogeneous, as assumed, because the capacitor plates are not flat enough.
- Water and alcohol evaporate too quickly during the experiment; oil is more suitable.
- Natural recharging takes place during the measurement (e.g. due to the radiation background), the droplet became neutral.
- Turbulence occurs due to temperature fluctuations in the apparatus. To avoid this, the entire apparatus is enclosed in an oil bath and thus thermally insulated.
- Turbulence occurs due to collisions with other droplets. This could be remedied by using a sealable opening to severely limit the number of aerosol particles between the condenser plates.
- The observation microscope must have a precisely known resolution so that the length component of the velocity measurement can be determined accurately enough.
- The same accuracy requirements apply to time measurement.
- If the air is too contaminated, the constants η_a and ρ_a are not known accurately enough.
- The temperature dependence of the viscosity must be taken into account.
- The temperature dependence of the density of the oils used and the corresponding inaccuracies must be investigated.

- If the droplets are too large, the friction can no longer be calculated according to Stokes' law.
- The droplets can deform in the electric field. The assumption of a spherical shape is then incorrect.

Millikan published the results of his research (with step by step refined experimental methods) in the following journals

Philosophical Magazine, Vol. 19 (1910), p. 209,
The Physical Review, Vol. 32 (1911), p. 349,
The Physical Review, Vol. 2 (2nd series) (1913), p. 109,

which make very interesting reading.

The values, which Millikan found, are

$$e_0 = \{4.65\ (1910),\ 4.92\ (1911),\ 4.77\ (1913)\}10^{-10}\text{statcoul}\,.$$

The value accepted today is $e_0 = 4.803\ldots\ 10^{-10}$statcoul. The value of the elementary charge obtained by Millikan lead to more precise values for other basic constants of nature, as the Planck constant and the Avogadro number. Millikan received the Nobel prize for his work in 1923.

1.5.2 Calculation of the Asymptotic Electric Dipole Field

This is an often used formula for the magnitude of a dipole field at large distances from the dipole.

The electric dipole is characterised by the data: The charge $+q$ is at the position $(0, 0, a)$, the charge $-q$ at the position $(0, 0, -a)$, (see Fig. 1.31). The formula in question is

$$|\boldsymbol{E}| = k_e \frac{p}{r^3}\left[1 + 3\cos^2\theta\right]^{1/2}\,,$$

where r indicates a sufficiently large distance from the middle of the dipole and θ is the polar angle between the z-axis and the direction of the field point.

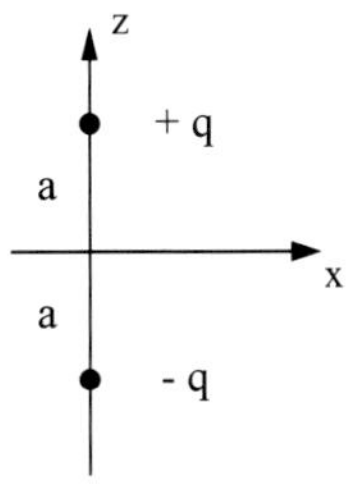

Fig. 1.31 Dipole geometry

The starting point is the general equation for the dipole field

$$
\boldsymbol{E}(\boldsymbol{r}) = \begin{pmatrix} E_x \\ E_y \\ E_z \end{pmatrix} = \begin{pmatrix} k_e\, q\, \rho \cos\varphi \left(\dfrac{1}{r_+^3} - \dfrac{1}{r_-^3} \right) \\[2ex] k_e\, q\, \rho \sin\varphi \left(\dfrac{1}{r_+^3} - \dfrac{1}{r_-^3} \right) \\[2ex] k_e\, q \left(\dfrac{(z-a)}{r_+^3} - \dfrac{(z+a)}{r_-^3} \right) \end{pmatrix} \tag{1.13}
$$

with the quantities

$$
\rho = \left[x^2 + y^2\right]^{1/2}
$$

$$
r_+ = \left[x^2 + y^2 + (z-a)^2\right]^{1/2} \qquad r_- = \left[x^2 + y^2 + (z+a)^2\right]^{1/2}
$$

and

$$
|\boldsymbol{E}| = \sqrt{E_x^2 + E_y^2 + E_z^2}\,.
$$

The aim is to find a formula for $|\boldsymbol{E}|$, valid for points with $x^2 + y^2 + z^2 \gg a^2$. It is sufficient to calculate the field in the x - z plane due to the rotational symmetry with respect to the z-axis ($\varphi = 0 \implies y = 0$)

$$
|\boldsymbol{E}| = \sqrt{E_x^2 + E_z^2}\,.
$$

In order to concentrate on points with $r^2 = x^2 + z^2 \gg a^2$, the separation of the field point from the position of the charges $1/r_+$ and $1/r_-$ is expanded in terms of powers of a/r. The first steps are

$$
\begin{aligned}
\frac{1}{r_-^3} = \frac{1}{\left[x^2 + (z+a)^2\right]^{3/2}} &\approx \frac{1}{\left[x^2 + z^2 + 2az\right]^{3/2}} = \frac{1}{\left[r^2 + 2az\right]^{3/2}} \\
&= \frac{1}{r^3} \frac{1}{\left[1 + 2az/r^2\right]^{3/2}}
\end{aligned}
$$

and also

$$
\frac{1}{r_+^3} = \frac{1}{\left[x^2 + (z-a)^2\right]^{3/2}} \approx \frac{1}{r^3} \frac{1}{\left[1 - 2az/r^2\right]^{3/2}}\,.
$$

The binomial formula for $[1 \pm t]^{-3/2}$ with $|t| < 1$

$$\frac{1}{[1 \pm t]^{3/2}} = 1 \mp \frac{3}{2}t + \frac{3 \cdot 5}{2 \cdot 4}t^2 \mp \frac{3 \cdot 5 \cdot 7}{2 \cdot 4 \cdot 6}t^3 + \ldots$$

allows the extraction of approximations

$$\frac{1}{r^3}\frac{1}{\left[1 \pm 2az/r^2\right]^{3/2}} = \frac{1}{r^3}\left(1 \mp \frac{3}{2}\frac{2az}{r^2} + \ldots\right) \approx \frac{1}{r^3}\left(1 \mp \frac{3az}{r^2}\right),$$

which yield the corresponding approximation for the x-component of the $\boldsymbol{E}$-field

$$E_x = k_e q x \left(\frac{1}{r_+^3} - \frac{1}{r_-^3}\right) = k_e \frac{qx}{r^3}\left(\frac{6az}{r^2}\right) = 3k_e p\,\frac{z\,x}{r^5},$$

as well as the z-component

$$E_z = k_e q \left(\frac{(z-a)}{r_+^3} - \frac{(z+a)}{r_-^3}\right) = 2k_e \frac{aq}{r^5}(3z^2 - r^2) = k_e \frac{p}{r^5}(2z^2 - x^2).$$

The square of the absolute value of the field in this approximation is

$$\begin{aligned} |\boldsymbol{E}|^2 = E_x^2 + E_z^2 &= \left(k_e \frac{p}{r^5}\right)^2 \left((3xz)^2 + (2z^2 - x^2)^2\right) \\ &= \left(k_e \frac{p}{r^5}\right)^2 \left(4z^4 + 5z^2x^2 + x^4\right). \end{aligned}$$

With

$$x = r\sin\theta \qquad y = r\cos\theta$$

one finds

$$\begin{aligned} |\boldsymbol{E}|^2 &= \left(k_e \frac{p}{r^5}\right)^2 r^4 \Big(4\cos^4\theta + 5\sin^2\theta\cos^2\theta + \sin^4\theta\Big) \\ &= \left(k_e \frac{p}{r^3}\right)^2 \Big(+4\cos^4\theta + 5\cos^2\theta(1-\cos^2\theta) + (1-\cos^2\theta)^2\Big) \\ &= \left(k_e \frac{p}{r^3}\right)^2 \left(1 + 3\cos^2\theta\right) \end{aligned}$$

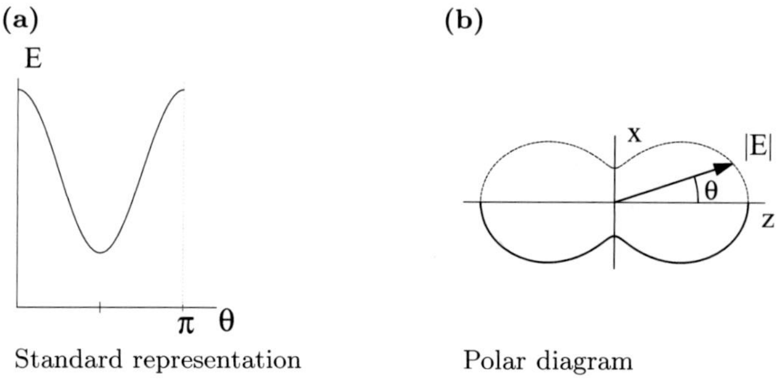

Fig. 1.32 Far dipole field

and thus the final result

$$|\boldsymbol{E}| = k_e \frac{p}{r^3} \left[1 + 3\cos^2\theta \right]^{1/2} .$$

It shows the radial variation of the far dipole field, which decreases more quickly than the field of a point charge ($1/r^2$). This result is also valid for points with $\boldsymbol{r} = (x, y, z)$, as equation (1.13) indicates, that $E_x^2 + E_y^2 + E_z^2$ with $\varphi \neq 0$ equals $E_x^2 + E_z^2$ with $\varphi = 0$. In addition one has

$$r^2 = (x^2 + y^2 + z^2)_{(\varphi \neq 0)} = (x^2 + z^2)_{(\varphi = 0)} ,$$

because of the rotational symmetry with respect to the z-axis.

The illustration 1.32 shows the function $|E(\theta)|$ (for a fixed value of r) in the standard representation (Fig. 1.32.a) and in the form of a polar diagram (Fig. 1.32.b). The polar diagram shows the absolute value $|E(\theta)|$ of the field as the length of a ray in the direction of θ, in order to demonstrate the cylinder symmetry for $\varphi = 0$ (grey) and $\varphi = \pi$ (black). Both representations indicate, that the magnitude of the far field is minimal in the x-direction (the x- y plane, $\theta = \pi/2$)) and maximal in the z-direction ($\theta = 0,\ \pi$).

1.5.3 The Electric Field of a Circular Disk and of a Sphere (Both Uniformly Charged)

The electric field of a uniformly charged sphere can be obtained in different ways with elementary means. One possibility is the decomposition of the sphere into infinitesimal disks. The field of such a disk on an axis perpendicular through the centre has to be calculated first. The contributions of the disks can be summed up in the next step. In view of the symmetry of the problem the position of the axis through the centre of the sphere need not be addressed in the second problem.

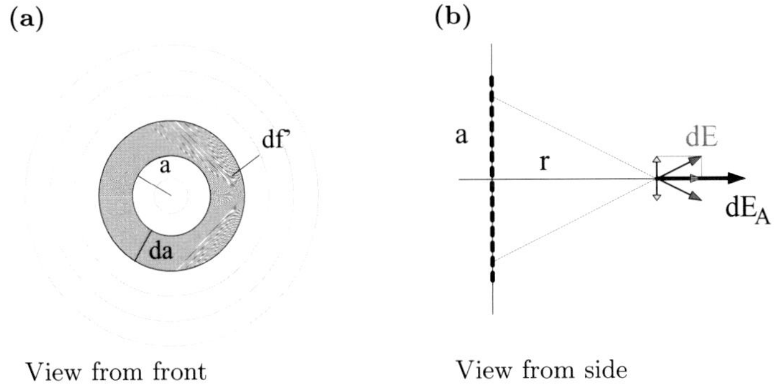

Fig. 1.33 Decomposition of a uniformly charged disk

1.5.3.1 Calculation for the Disk

The circular disk (radius R) is, as shown in Fig. 1.33a, decomposed into infinitesimal rings with the radius a and the thickness da. As the uniform surface charge density of a disk is

$$\sigma = \frac{Q}{\pi R^2}$$

and the surface of each ring is

$$\mathrm{d}f' = 2\pi\, a\, \mathrm{d}a$$

one finds, that each ring carries the charge

$$\mathrm{d}q' = \sigma\, \mathrm{d}f' = \frac{Q}{\pi R^2} 2\pi a\, \mathrm{d}a = \frac{2Q}{R^2} a\, \mathrm{d}a\,.$$

The contribution of each ring to the field on the axis is according to Fig. 1.33b (and Chap. 1.3)

$$\mathrm{d}E_A = \frac{k_e\, r\, \mathrm{d}q'}{(r^2 + a^2)^{3/2}} = \frac{2k_e Q r}{R^2} \frac{a}{\left[r^2 + a^2\right]^{3/2}} \mathrm{d}a\,.$$

The total field of each disk is obtained by integration from the centre of the disk to its outer border

$$E_A = \int \mathrm{d}E_A = \frac{2k_e Q r}{R^2} \int_0^R \frac{a}{\left[r^2 + a^2\right]^{3/2}} \mathrm{d}a$$

$$= \frac{2k_e Q}{R^2} \left[1 - \frac{r}{\left[r^2 + R^2\right]^{1/2}} \right].$$

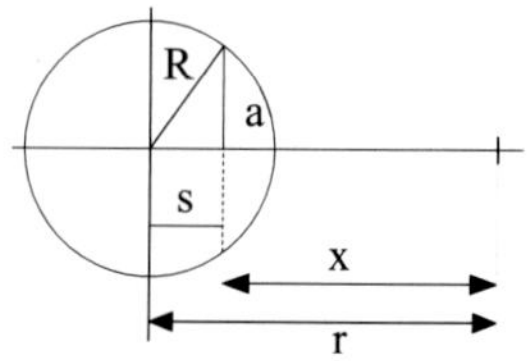

Fig. 1.34 Decomposition of the sphere into circular disks

1.5.3.2 Calculation for the Uniformly Charged Sphere

The electric field of a uniformly charged sphere with the radius R and the spatial charge density

$$\rho = \frac{Q}{\frac{4}{3}\pi R^3} = \text{const.}$$

can be calculated by the following steps: Decompose the sphere into infinitesimal circular disks of thickness ds and add up the contributions of all disks, using the result of the previous calculation. For this purpose, one divides the sphere into thin disks with the position s $(-R \leq s \leq R$, Fig. 1.34), so that the radius a of the disk is given by $a^2 = R^2 - s^2$ and its charge is

$$\mathrm{d}q' = \rho\pi a^2 \mathrm{d}s\,.$$

The field of each disk points in the direction the axis of the disk.

The field of a circular disk with radius a and a uniform surface density in a point on the axis with a distance x from the centre of the disk is

$$E_A = \frac{2k_e Q}{a^2}\left(1 - \frac{x}{\left[x^2 + a^2\right]^{1/2}}\right).$$

In order to use this result for the present problem, one has to express the quantity x by the distance of the point from the centre of the sphere R. With the relation $r = s + x$ and the replacement of Q by dq one can write for the contribution to the field of the sphere by a circular, infinitesimal disk on the axis with $R^2 = a^2 + s^2$

$$\mathrm{d}E_A = \frac{2k_e}{a^2}\rho\pi a^2 \mathrm{d}s\left(1 - \frac{(r-s)}{\left[(r-s)^2 + a^2\right]^{1/2}}\right)$$

$$= 2\pi\, k_e\rho\left(1 - \frac{(r-s)}{\left[r^2 - 2rs + R^2\right]^{1/2}}\right)\mathrm{d}s\,.$$

Further evaluation has to distinguish between points with $r < R$ and with $r > R$ (points in the interior and the exterior of the sphere).

The integration can be carried out directly for $r > R$

$$E_A = \int \mathrm{d}E_A = 2\pi\, k_e \rho \int_{-R}^{R} \left(1 - \frac{(r-s)}{\left[r^2 - 2rs + R^2\right]^{1/2}}\right) \mathrm{d}s$$

$$= 2\pi\, k_e \rho (I_1 + I_2) \quad \text{with}$$

$$I_1 = \int_{-R}^{R} \mathrm{d}s \qquad \text{and} \qquad I_2 = -\int_{-R}^{R} \frac{(r-s)}{\left[r^2 - 2rs + R^2\right]^{1/2}} \mathrm{d}s\,.$$

The first integral is trivial ($I_1 = 2R$), the second needs the substitution

$$y = \left[r^2 - 2rs + R^2\right]^{1/2}\,.$$

In order to apply the substitution one uses

$$\frac{\mathrm{d}y}{\mathrm{d}s} = -\frac{r}{y} \qquad \Longrightarrow \qquad \mathrm{d}s = -\frac{y}{r}\mathrm{d}y\,,$$

from

$$y^2 = r^2 - 2rs + R^2 \quad \text{follows} \quad s = \frac{r^2 + R^2 - y^2}{2r}$$

and

$$r - s = \frac{r^2 - R^2 + y^2}{2r}\,.$$

The limits of the integrals are

$$s = -R \quad \Longrightarrow \quad y = r + R$$
$$s = +R \quad \Longrightarrow \quad y = r - R\,.$$

The remaining integral

$$I_2 = \int_{r+R}^{r-R} \frac{r^2 - R^2 + y^2}{2r^2} \mathrm{d}y$$

gives

$$I_2 = \frac{1}{2r^2}\left((r^2 - R^2)y + \frac{1}{3}y^3\right)\Bigg|_{r+R}^{r-R} = \frac{2}{3}\frac{R^3}{r^2} - 2R\,.$$

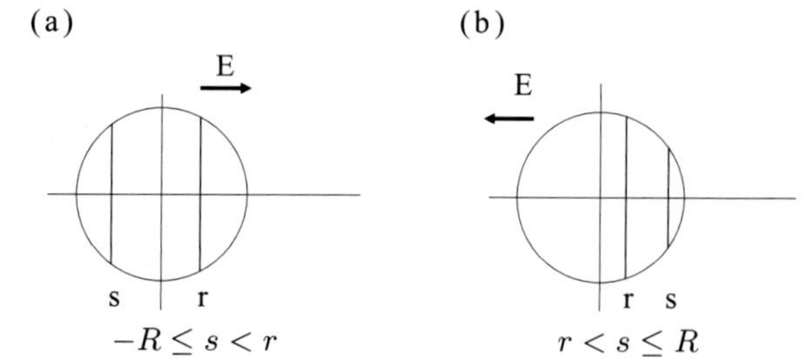

Fig. 1.35 Integration over the variable s

The electric field on an axis through the centre of the sphere for $r > R$ is therefore

$$E_A = \frac{4\pi}{3} k_e \rho \frac{R^3}{r^2} = k_e \frac{Q}{r^2} \, .$$

For $r < R$ one has to distinguish points r on the right side of the point s from points on the left side. The field vector $\boldsymbol{E}$ points in the direction of the 'positive' axis, if the point r is on the right side of the disk, that is for $-R \le s < r$ (Fig. 1.35a). The vector $\boldsymbol{E}$ points in the opposite direction for $r < s \le R$ (Fig. 1.35b). The two cases are sorted correctly by the sign of the different contributions towards E_A.

$$E_A = 2\pi \, k_e \rho \left\{ \int_{-R}^{r} \left(1 - \frac{(r-s)}{\left[r^2 - 2rs + R^2\right]^{1/2}} \right) \mathrm{d}s \right.$$

$$\left. - \int_{r}^{R} \left(1 - \frac{(s-r)}{\left[r^2 - 2rs + R^2\right]^{1/2}} \right\} \mathrm{d}s \right)$$

$$= 2\pi \, k_e \rho (I_1 + I_2) \quad \text{with}$$

$$I_1 = \int_{-R}^{r} \mathrm{d}s - \int_{r}^{R} \mathrm{d}s \qquad \text{and}$$

$$I_2 = \int_{-R}^{r} \frac{(s-r)}{\left[r^2 - 2rs + R^2\right]^{1/2}} \mathrm{d}s - \int_{r}^{R} \frac{(r-s)}{\left[r^2 - 2rs + R^2\right]^{1/2}} \mathrm{d}s$$

$$= \int_{-R}^{R} \frac{(s-r)}{\left[r^2 - 2rs + R^2\right]^{1/2}} \mathrm{d}s \, .$$

The integral I_1 is again trivial

$$I_1 = 2r \, ,$$

for the second integral one can also use the substitution

$$y = \left[r^2 - 2rs + R^2\right]^{1/2}$$

$$ds = -\frac{y}{r}dy\,, \qquad s - r = \frac{R^2 - r^2 - y^2}{2r}$$

with

$$\begin{aligned} s = -R \quad &\Longrightarrow \quad y = r + R \\ s = +R \quad &\Longrightarrow \quad y = R - r\,. \end{aligned}$$

The integral can be evaluated in the following manner

$$\begin{aligned} I_2 &= \frac{1}{2r^2}\int_{-R}^{R}(y^2 - R^2 + r^2)\,dy \\ &= \frac{1}{2r^2}\left((r^2 - R^2)y + \frac{1}{3}y^3\right)\Big|_{-R}^{R} = -\frac{4}{3}r\,. \end{aligned}$$

The electric field on an axis through the centre of the sphere is in the case $r < R$

$$E_A = \frac{4\pi}{3}k_e\rho r = k_e Q\frac{r}{R^3}\,.$$

The symmetry guarantees that the calculation yields the same result for *any* axis through the centre of the sphere

$$E(r) = \begin{cases} k_e Q\dfrac{r}{R^3} & r \leq R \quad \text{(interior)} \\ k_e\dfrac{Q}{r^2} & r > R \quad \text{(exterior)}\,. \end{cases} \tag{1.14}$$

The field is radial and points in the outward direction for a positive charge Q. It vanishes at the centre of the sphere, increases in a linear fashion up to the point $r = R$, is continuous at the surface and decreases in the exterior as $1/r^2$.

2 Basic Equations of Electrostatics

Besides the Gauss' law, through which the sources and sinks of the stationary electric fields of electrostatics are introduced, a second set of equations characterises the absence of vortex structures for such fields. Both sets of equations can be written in integral or in differential form. The integral form allows a general discussion of the behaviour of fields passing through layers of charge and the distribution of charges in conducting materials. The differential form leads, after the introduction of the concept of the electric potential and the proper representation of charge densities with the aid of distributions (especially the δ-function), to a compact version of electrostatics in terms of the Poisson/Laplace equations. The introduction of the electric potential will be investigated in detail for simple charge arrangements. The final part of the second chapter addresses the ability of the electric field to store energy using the practical concept of the energy density.

2.1 Basic Equations of Electrostatics

One of the basic equations of electrostatics (in vacuum) is Gauss' law (1.10)

$$\oint\!\!\!\oint \boldsymbol{E} \cdot \mathbf{d}\boldsymbol{f} = 4\pi\, k_e\, Q_{\text{in}}\,.$$

The introduction of a second basic equation starts also with the consideration of point charges. The field of one point charge (at the origin) has the form

$$\boldsymbol{E}_{\text{p}}(\boldsymbol{r}) = k_e \frac{q}{r^2}\, \boldsymbol{e}_{\text{r}} = k_e q \left(\frac{x}{r^3}, \frac{y}{r^3}, \frac{z}{r^3}\right)\,.$$

© The Author(s), under exclusive license to Springer-Verlag GmbH, DE, part of Springer Nature 2024

R. M. Dreizler, C. S. Lüdde, *Electrostatics and Magnetostatics*,
https://doi.org/10.1007/978-3-662-69933-1_2

One can check directly, that this vector function has the property[1]

$$\operatorname{rot} \boldsymbol{E}_{\mathrm{p}}(\boldsymbol{r}) = \nabla \times \boldsymbol{E}_{\mathrm{p}}(\boldsymbol{r}) = \mathbf{0}\,.$$

The electric field of a point charge at the origin is vortex free. For a field $\boldsymbol{E}_{\mathrm{p}}(\boldsymbol{r}, \boldsymbol{r}_i)$ of a point charge at the point $\boldsymbol{r}_i$ one finds also

$$\nabla_{\boldsymbol{r}} \times \boldsymbol{E}_{\mathrm{p}}(\boldsymbol{r}, \boldsymbol{r}_i) = \mathbf{0}\,.$$

The field of an arbitrary (stationary) charge distribution is generated by a finite set of point charges and by charge distributions

$$\boldsymbol{E}(\boldsymbol{r}) = \sum_i \boldsymbol{E}_i(\boldsymbol{r}) + \int \mathbf{d}\boldsymbol{E}(\boldsymbol{r})\,.$$

As the second term on the right hand side represents the contribution of specific assemblies of point charges, this equation is (under suitable conditions concerning differentiability) the general form for a stationary electric field. On the basis of this atomistic view, one can state, that the general stationary electric field is also characterised by the property

$$\operatorname{rot} \boldsymbol{E}(\boldsymbol{r}) = \nabla \times \boldsymbol{E}(\boldsymbol{r}) = \mathbf{0}\,. \tag{2.1}$$

This constitutes a second set of basic equations of electrostatics. As one of the basic equations is in integral form, the other set in differential form, it is useful to aim for sets of the two equations in either of the two forms.

2.1.1 Integral Form

The condition, that the electric field is vortex free, can be expressed by the theorem of Stokes. This theorem of differential geometry states

$$\iint_F (\nabla \times \boldsymbol{E}(\boldsymbol{r})) \cdot \mathbf{d}\boldsymbol{f} = \oint_{R(F)} \boldsymbol{E}(\boldsymbol{r}) \cdot \mathbf{d}\boldsymbol{s}\,.$$

The surface integral of the rotation of a vector field $\boldsymbol{E}$ over an open surface F equals the line integral of the field along an oriented boundary of the surface (Fig. 2.1). In the present situation one has $\operatorname{rot} \boldsymbol{E} = \mathbf{0}$ for each point in space. It follows,

[1] Consult the theorems of vector analysis given in Appendix C.1.

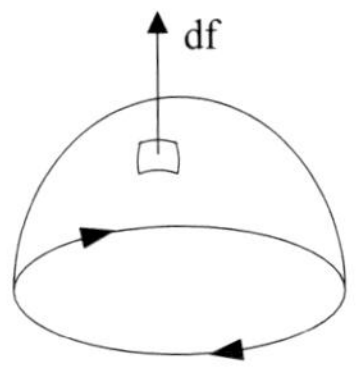

Fig. 2.1 Decomposition of the surface integral

that the line integral vanishes for any arbitrary closed path. The basic equations of electrostatics in integral form are therefore

$$\begin{aligned} &\oint\!\!\!\oint_F \boldsymbol{E}(\boldsymbol{r}) \cdot \mathbf{d}\boldsymbol{f} = 4\pi\, k_e\, Q_{\text{in}} && \text{for any closed surface } F \\ &\oint_K \boldsymbol{E}(\boldsymbol{r}) \cdot \mathbf{d}\boldsymbol{s} = 0 && \text{for any closed curve } K\,. \end{aligned} \tag{2.2}$$

In words they state: Sources and sinks of the (stationary) electric field are the charges. The field is vortex free.

Practical aspects concerning the electric field can be deduced from the integral form. However, the differential form proves to be more flexible for many applications.

2.1.2 Differential Form

In order to find a differential form of the basic equations, one has to reformulate the Gauss' theorem. This is easy if space charges are present. For a discussion of point- and surface charges the concept of a mathematical function has to be extended to include distributions. Distributions are actually tailored for the discussion of point charges. This concept can also be used for situations with surface charges, even if they are generally treated in a different fashion (see Chap. 2.4). Consider the case that space charges are found in a volume V_0, which is embedded in a volume $V(F)$, which in turn is characterised by a closed surface F. In this case, that is for $V_0 \in V(F)$, one can state

$$\oint\!\!\!\oint_F \boldsymbol{E}(\boldsymbol{r}) \cdot \mathbf{d}\boldsymbol{f} = 4\pi\, k_e\, Q_{\text{in}} = 4\pi\, k_e \iiint_{V_0} \rho(\boldsymbol{r})\, \mathrm{d}V\,.$$

The volume integral on the right hand side can be taken over the total volume enclosed by the surface F, as $\rho(\boldsymbol{r})$ is supposed to vanish in the difference of the two volumina $V(F) - V_0$

$$= 4\pi\, k_e \iiint_{V(F)} \rho(\boldsymbol{r})\, \mathrm{d}V\,.$$

The surface integral is recast as a volume integral with the aid of the divergence theorem

$$\oiint_F \boldsymbol{E}(\boldsymbol{r}) \cdot \mathbf{d}\boldsymbol{f} = \iiint_{\mathrm{V(F)}} (\boldsymbol{\nabla} \cdot \boldsymbol{E}(\boldsymbol{r}))\, \mathrm{d}V\,,$$

which states: The flux of the electric field through a closed surface F can be calculated by integration of the scalar function $\boldsymbol{\nabla} \cdot \boldsymbol{E}$ over the enclosed volume. A condition for the validity of the divergence theorem is

$$\left| \iiint_{\mathrm{V(F)}} \boldsymbol{\nabla} \cdot \boldsymbol{E}(\boldsymbol{r})\, \mathrm{d}V \right| < \infty\,.$$

The combination of the Gauss' and the divergence theorem leads to

$$\iiint_{\mathrm{V(F)}} \{\boldsymbol{\nabla} \cdot \boldsymbol{E}(\boldsymbol{r}) - 4\pi\, k_e\, \rho(\boldsymbol{r})\}\, \mathrm{d}V = 0\,.$$

As the surface F and the volume $V(F)$ can be arbitrarily chosen, this integral can only vanish if the integrand has the value zero.

The basic equations of electrostatics can therefore also be written in the form

$$\begin{aligned} \operatorname{div} \boldsymbol{E}(\boldsymbol{r}) &= \boldsymbol{\nabla} \cdot \boldsymbol{E}(\boldsymbol{r}) = 4\pi\, k_e\, \rho(\boldsymbol{r}) \\ \operatorname{rot} \boldsymbol{E}(\boldsymbol{r}) &= \boldsymbol{\nabla} \times \boldsymbol{E}(\boldsymbol{r}) = \mathbf{0}\,. \end{aligned} \tag{2.3}$$

The interpretation does not change: The first equation still states, that charges are sources or sinks of stationary electric fields. The second statement says: Stationary electric fields are vortex free.

Additional remarks are:

(1) The basic equations represent a set of coupled differential equations of first order. With the explicit definition of the operators div and rot in Cartesian coordinates they are

$$\boldsymbol{\nabla} \cdot \boldsymbol{E}(x,\, y,\, z) = \frac{\partial E_{\mathrm{x}}}{\partial x} + \frac{\partial E_{\mathrm{y}}}{\partial y} + \frac{\partial E_{\mathrm{z}}}{\partial z} = 4\pi\, k_e\, \rho(x,\, y,\, z)$$

$$\boldsymbol{\nabla} \times \boldsymbol{E}(x,\, y,\, z) = \mathbf{0} \quad \Longrightarrow$$

$$\frac{\partial E_{\mathrm{z}}}{\partial y} - \frac{\partial E_{\mathrm{y}}}{\partial z} = 0, \qquad \frac{\partial E_{\mathrm{x}}}{\partial z} - \frac{\partial E_{\mathrm{z}}}{\partial x} = 0, \qquad \frac{\partial E_{\mathrm{y}}}{\partial x} - \frac{\partial E_{\mathrm{x}}}{\partial y} = 0\,.$$

A representation of the operators in the form of curvilinear coordinates is possible and necessary for applications. For the moment the Cartesian decomposition is sufficient.

(2) The discussion above is restricted to spatial charge distributions. The problem in the case of point charges is the following: In order to calculate the integral defining Q_i for a point charge at the position $\boldsymbol{r}_i$ requires some extraordinary properties of the "point charge density" ρ_p

$$4\pi\, k_e\, Q_{\rm i} \;=\; 4\pi\, k_e \iiint_{\rm V(F)} \rho(\boldsymbol{r}, \boldsymbol{r}_i)\, {\rm d}V$$

$$= \begin{cases} 4\pi\, k_e\, q & \text{if } q \text{ in V(F)} \\ 0 & \text{if } q \text{ not in V(F)} \end{cases} .$$

(a) In order to represent the structure of the source correctly one needs

$$\rho_{\rm p}(\boldsymbol{r},\, \boldsymbol{r}_{\rm i}) = 0 \quad \text{for } \boldsymbol{r} \neq \boldsymbol{r}_{\rm i}\,.$$

(b) As the volume integral over the charge density should yield the total charge, one requires (for one point charge q in the complete space)

$$\iiint_{\text{vicinity of } r_i} \rho_{\rm p}(\boldsymbol{r},\, \boldsymbol{r}_{\rm i})\, {\rm d}V = \iiint_{\text{full space}} \rho_{\rm p}(\boldsymbol{r},\, \boldsymbol{r}_{\rm i})\, {\rm d}V = q\,.$$

Such properties can not be expected to be described by a function. One needs, as stated before, a new mathematical concept, distributions, which are introduced and discussed in Chaps. 2.3 and 2.4.

These differential equations do, however, not define the electric field completely. In order to achieve this, one has to note that the field $\boldsymbol{E}$ can be defined as the gradient of a scalar function $V(\boldsymbol{r})$

$$\boldsymbol{E}(\boldsymbol{r}) = -\,\text{grad}\, V(\boldsymbol{r}) = -\nabla V(\boldsymbol{r})\,, \tag{2.4}$$

which is a solution of the differential equation (the sign is a question of convention)

$$\nabla \times \boldsymbol{E}(\boldsymbol{r}) = -\nabla \times \nabla V(\boldsymbol{r}) = -\,\text{rot}\,\text{grad}\, V(\boldsymbol{r}) = \boldsymbol{0}\,.$$

The quantity $V(\boldsymbol{r})$ is called the **electric** (actually in the present context electrostatic) **potential**.

If one inserts the representation of the electric field by a potential into the first of the basic differential equations, one finds

$$\nabla \cdot \nabla V(\boldsymbol{r}) = \operatorname{div}(\operatorname{grad} V(\boldsymbol{r})) = -4\pi\, k_e\, \rho(\boldsymbol{r})\,.$$

The operator div grad is the **Laplace operator**

$$\operatorname{div}\operatorname{grad} = \nabla \cdot \nabla = \Delta\,.$$

Its form in Cartesian coordinates is

$$\Delta = \frac{\partial^2}{\partial x^2} + \frac{\partial^2}{\partial y^2} + \frac{\partial^2}{\partial z^2}\,.$$

With the aid of the potential it is possible to replace the two partial differential equations of first order for a vector function by a partial differential equation of second order. The explicit form of this equation in Cartesian coordinates is

$$\frac{\partial^2 V(x,\, y,\, z)}{\partial x^2} + \frac{\partial^2 V(x,\, y,\, z)}{\partial y^2} + \frac{\partial^2 V(x,\, y,\, z)}{\partial z^2} = -4\pi\, k_e\, \rho(x,\, y,\, z)\,.$$

The usual short form in the two systems of units (see below) is

$$\begin{aligned} \Delta V(\boldsymbol{r}) &= -4\pi\, \rho(\boldsymbol{r}) && \text{in the CGS-system} \\ \Delta V(\boldsymbol{r}) &= -\frac{\rho(\boldsymbol{r})}{\varepsilon_0} && \text{in the SI-system}\,. \end{aligned} \tag{2.5}$$

This equation is the **Poisson equation**. In contrast to the formulation of electrostatics in integral form, the formulation with the aid of a differential equation is a local statement:

For a charge distribution in an arbitrary volume V_0 with

$$\rho(\boldsymbol{r}) \begin{cases} \neq 0 & \text{for} \quad \boldsymbol{r} \in V_0 \\ = 0 & \text{for} \quad \boldsymbol{r} \notin V_0 \end{cases}$$

one has

$$\Delta V(\boldsymbol{r}) = \begin{cases} -4\pi\, k_e\, \rho(\boldsymbol{r}) & \text{for} \quad \boldsymbol{r} \in V_0 \\ 0 & \text{for} \quad \boldsymbol{r} \notin V_0 \end{cases}.$$

The differential equation for the charge free section of space is called **Laplace equation**. The general solution is naturally not $V(\boldsymbol{r}) = 0$. If the differential equation for the potential function $V(\boldsymbol{r})$ has been solved, the electrical field can be obtained by differentiation

$$\boldsymbol{E}(\boldsymbol{r}) = -\nabla V(\boldsymbol{r}) .$$

The mathematical problems, which have to be faced in using the formulation of the theory in the form of differential equations, are:

(a) How can one obtain the general solution of the Poisson- respectively the Laplace equation?
(b) How can one find the special solution for a special situation?

These questions will be addressed in Chaps. 3 and 4. Before one deals with these practical aspects, one should, however, look at some of the points, which have not been treated properly during the assembly of the basic equations of electrostatics. These are:

(1) Which additional information can be extracted from the integral form of the basic equations of electrostatics?
(2) Discuss the theory of distributions and their employment in electrostatics in order to complete the assembly of the basic equations.
(3) Discuss the potential concept in a more general context than in association with the Poisson/Laplace equation.

2.2 Additional Discussion of Electrostatics in Terms of Integrals

This section has a specific goal, a closer look at statements about the distribution of the electric field in conducting materials. For this purpose one may look at the following situation: Consider an arbitrary surface F, which is covered by surface charges $\sigma(\boldsymbol{r})$ and concentrate your attention on an infinitesimal section of this surface. Place, as indicated in Fig. 2.2a, an infinitesimal 'Gaussian box' around

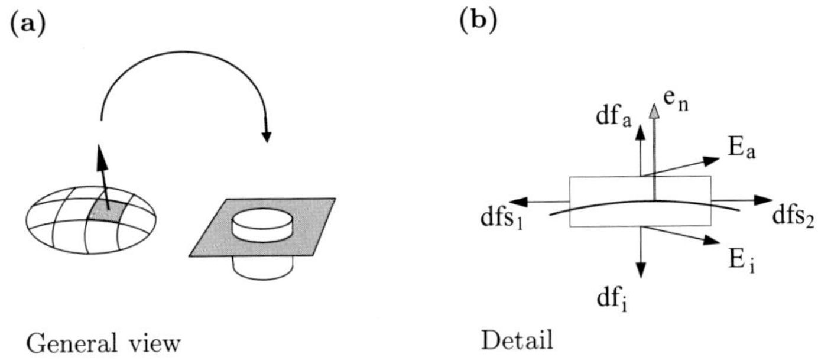

Fig. 2.2 Normal components: Gaussian box

this section. If this box (GB) serves as a boundary of a flux integral the following specification is possible

$$\oiint_{\text{GB}} \boldsymbol{E} \cdot \mathbf{d}\boldsymbol{f} \approx \boldsymbol{E}(\boldsymbol{r}_a) \cdot \mathbf{d}\boldsymbol{f}_a + \boldsymbol{E}(\boldsymbol{r}_i) \cdot \mathbf{d}\boldsymbol{f}_i + \iint_{\text{S}} \boldsymbol{E} \cdot \mathbf{d}\boldsymbol{f} \,.$$

The contribution of the surface due to the sides can be made arbitrarily small by choosing arbitrarily flat boxes (Fig. 2.2b). For an infinitesimal, infinitesimally flat Gaussian box one has

$$\oiint_{\text{GB}} \boldsymbol{E} \cdot \mathbf{d}\boldsymbol{f} \approx \boldsymbol{E}(\boldsymbol{r}_a) \cdot \mathbf{d}\boldsymbol{f}_a + \boldsymbol{E}(\boldsymbol{r}_i) \cdot \mathbf{d}\boldsymbol{f}_i \,,$$

where the top and the bottom surfaces are characterised by the unit vectors of the normal of the surface $\boldsymbol{e}_{\text{n}}$

$$\mathbf{d}\boldsymbol{f}_a = \mathrm{d}f\,\boldsymbol{e}_{\text{n}} \qquad \mathbf{d}\boldsymbol{f}_i = -\mathrm{d}f\,\boldsymbol{e}_{\text{n}} \,.$$

If the surface F is part of a closed volume V, the normal direction points to the exterior. If this is not the case, the direction of the normal of the surface has to be defined explicitly, e.g. by the sense of curve around the edge of the box or with the right hand rule. The integral over the Gaussian box can then be written in the form

$$\oiint_{\text{GB}} \boldsymbol{E} \cdot \mathbf{d}\boldsymbol{f} = (\boldsymbol{E}(\boldsymbol{r}_a) - \boldsymbol{E}(\boldsymbol{r}_i)) \cdot \boldsymbol{e}_{\text{n}} \mathrm{d}f \,.$$

The charge enclosed in the box is, with the same approximation as for the flux integral, equal to

$$Q_{\text{in}} = \iint \sigma(\boldsymbol{r})\,\mathrm{d}f \approx \sigma(\boldsymbol{r})\,\mathrm{d}f \,.$$

The theorem of Gauss (2.1) thus leads to the relation

$$(\boldsymbol{E}(\boldsymbol{r}_a) - \boldsymbol{E}(\boldsymbol{r}_i)) \cdot \boldsymbol{e}_{\text{n}} = 4\pi\, k_e\, \sigma$$

or more precisely to

$$\lim_{\boldsymbol{\varepsilon} \to 0} (\boldsymbol{E}_a(\boldsymbol{r} + \boldsymbol{\varepsilon}) - \boldsymbol{E}_i(\boldsymbol{r} - \boldsymbol{\varepsilon})) \cdot \boldsymbol{e}_{\text{n}} = 4\pi\, k_e\, \sigma(\boldsymbol{r}) \,, \tag{2.6}$$

provided one approaches a point of surface from both sides. The field $\boldsymbol{E}_a$ is the field on the exterior part of the surface, which is marked by the direction of the surface

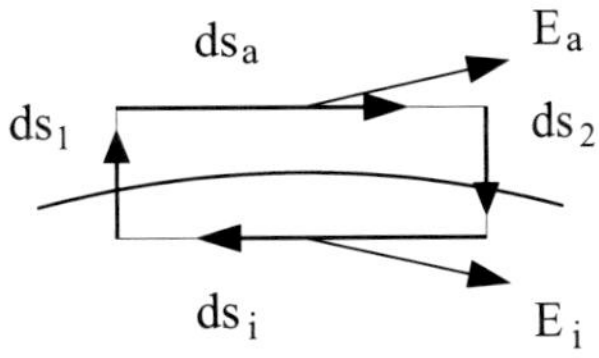

Fig. 2.3 Tangential component: Stokes line

normal. The field $\boldsymbol{E}_i$ is the field on the inner side. The scalar product $\pm\boldsymbol{E}_s \cdot \boldsymbol{e}_{\mathrm{s}}$ defines the normal component of the electric field on each side. For this reason one can also write

$$\left(\boldsymbol{E}_{\mathrm{a,n}}(\boldsymbol{r}) - \boldsymbol{E}_{\mathrm{i,n}}(\boldsymbol{r})\right) = 4\pi\, k_e\, \sigma(\boldsymbol{r})\,. \tag{2.7}$$

The **normal component of an electric field** is not continuous, if one passes through a charged surface. It experiences a jump of the magnitude $4\pi\, k_e\, \sigma(\boldsymbol{r})$ at the point of passage.

In order to obtain a statement on the behaviour of the tangential component, one can use the second basic equation of (2.2). One places, as indicated in Fig. 2.3, an infinitesimal 'Stokes curve' (SC) and finds

$$\oint_{\mathrm{SC}} \boldsymbol{E}\cdot \mathbf{d}\boldsymbol{s} = \boldsymbol{E}(\boldsymbol{r}_a)\cdot \mathbf{d}\boldsymbol{s}_a + \boldsymbol{E}(\boldsymbol{r}_i)\cdot \mathbf{d}\boldsymbol{s}_i + \boldsymbol{E}_{\mathrm{S}1}\cdot \mathbf{d}\boldsymbol{s}_1 + \boldsymbol{E}_{\mathrm{S}2}\cdot \mathbf{d}\boldsymbol{s}_2\,.$$

The contributions of the sides can again be made arbitrarily small by using an arbitrarily flat curve. The contributions in the tangential direction are

$$\boldsymbol{E}(\boldsymbol{r}_a)\cdot \mathbf{d}\boldsymbol{s}_a + \boldsymbol{E}(\boldsymbol{r}_i)\cdot \mathbf{d}\boldsymbol{s}_i \approx (\boldsymbol{E}(\boldsymbol{r}_a) - \boldsymbol{E}(\boldsymbol{r}_i))\cdot \boldsymbol{e}_{\mathrm{t}}\, \mathrm{d}s\,.$$

As the line integral vanishes, one finds

$$(\boldsymbol{E}(\boldsymbol{r}_a) - \boldsymbol{E}(\boldsymbol{r}_i))\cdot \boldsymbol{e}_{\mathrm{t}} = 0$$

or more precisely

$$\lim_{\boldsymbol{\varepsilon}\to 0}\left(\boldsymbol{E}_{a,\mathrm{t}}(\boldsymbol{r}+\boldsymbol{\varepsilon}) - \boldsymbol{E}_{i,\mathrm{t}}(\boldsymbol{r}-\boldsymbol{\varepsilon})\right) = \boldsymbol{E}_{a,\mathrm{t}}(\boldsymbol{r}) - \boldsymbol{E}_{i,\mathrm{t}}(\boldsymbol{r}) = 0\,. \tag{2.8}$$

The **tangential components of the electric field** are continuous at a passage through a charged surface.

These statements correspond exactly to the result that was found in the discussion of the spherical capacitor (Chap. 1.4): The tangential components on both sides of a surface charge have the value zero

$$\boldsymbol{E}_{a,\mathrm{t}}(\boldsymbol{r}) = \boldsymbol{E}_{i,\mathrm{t}}(\boldsymbol{r}) = 0\,.$$

The normal (radial) components change by an amount

$$\boldsymbol{E}_{a,\mathrm{n}}(\boldsymbol{r}) - \boldsymbol{E}_{i,\mathrm{n}}(\boldsymbol{r}) = 4\pi\, k_e\, \sigma(\boldsymbol{r})\,.$$

With Eqs. (2.6) and (2.8) an explicit idea of the distribution of the electric field in conductors can be obtained.

2.2.1 Electric Fields in Conductors

A certain percentage of the electrons in conductors can move freely. The following brief argument shows that the charges in a piece of metal with total charge zero must be distributed in a fashion, so that on the average the enclosed total charge in every interior volume element must be zero (Fig. 2.4a). If this were not the case, there would exist, according to Gauss' theorem, an electric field. This field would exert a force $\boldsymbol{F} = -e_0\boldsymbol{E}$ on the free charges and displace them. The displacement would continue until the field and the force has vanished. Therefore one finds in the interior of a conducting material always

$$\boldsymbol{E}_{\text{interior}}(\boldsymbol{r}) = \boldsymbol{0} \qquad \text{(on the average)}\,.$$

If the amount of charge on the conductor is increased (Fig. 2.4b), one still finds for any closed Gaussian box in the interior: There will be a displacement of the free charge carriers, if the enclosed charge is not equal to zero. The equilibrium situation $\boldsymbol{E}_{\text{interior}}(\boldsymbol{r}) = \boldsymbol{0}$ must also be present in a charged conductor. The additional charges can only be found at the surface of the metallic object, which is covered by a charge layer with the thickness of atomic dimensions. If the total charge of the metallic object is negative, additional electrons constitute this layer. If the total charge is positive, the layer is due to missing electrons in the surface layer.

The gist of the argument is: If the carriers of charge in a metallic object can move freely, it is possible to have a field free equilibrium situation in the interior. If this statement is combined with the fact, that electric fields, which traverse a layer of

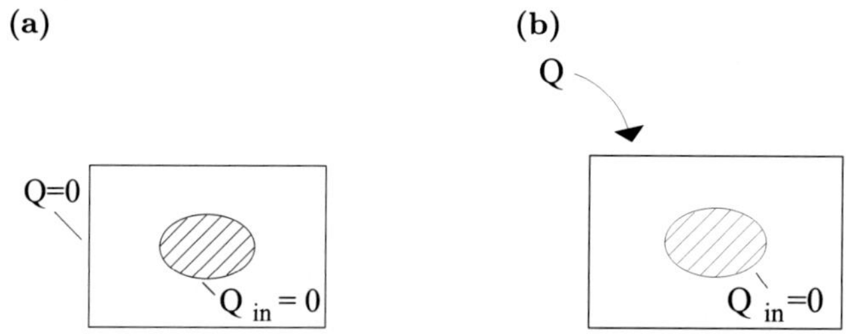

Fig. 2.4 Charge distribution in conductors

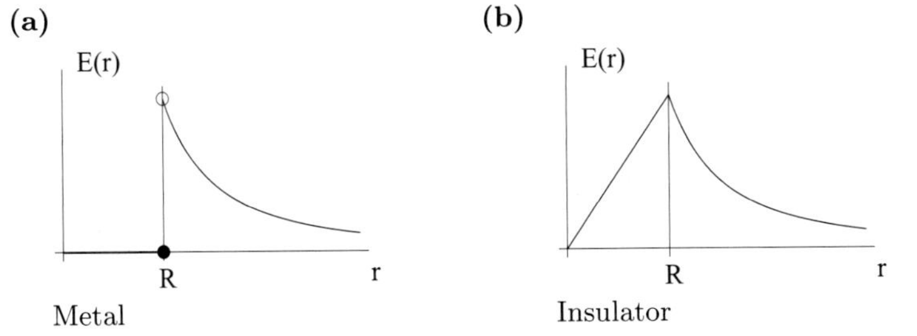

Fig. 2.5 Electric field of a sphere

charge, have only a normal component, one finds for conductors in general[2]

in the interior	at the surface
$E_\text{n} = 0$	$E_\text{n} = 4\pi\, k_e\, \sigma$
$E_\text{t} = 0$	$E_\text{t} = 0$

The vector of the electric field (the field lines) of a charged metallic object is (are) always perpendicular to the surface. The total charge of this object can be calculated according to

$$Q = \oiint_{\text{F(conductor)}} \sigma(\boldsymbol{r})\, \mathrm{d}f\ .$$

In order to illustrate this point one may look at the electric fields of a metallic sphere and a dielectric sphere (insulator) with equal total charge and equal radius (Fig. 2.5). The metallic sphere carries a surface charge, the dielectric sphere, without freely movable charges, a distribution of space charges. On the outside both types of material exhibit, due to the symmetry, the same uniform field. The fields in the interior differ.

A second phenomenon, which can occur in the presence of freely moving charges is **electrostatic induction**. This can also be understood in the case of simple situations with the aid of the theorem of Gauss, e.g. for the example of a spherical capacity. A point charge $q > 0$ is placed in the centre of an uncharged, metallic hollow sphere (Fig. 2.6a) (ignore the question, how this is done). In order to discuss the electric field one considers three spherical Gaussian spheres: (1) in the hollow interior (2) in the metallic part, (3) in the exterior (Fig. 2.6b).

For the surface (1) one has:

$$\oiint \boldsymbol{E}_1 \cdot \mathbf{d}\boldsymbol{f} = 4\pi k_e q$$

[2] The normal of a closed surface always points to the outside.

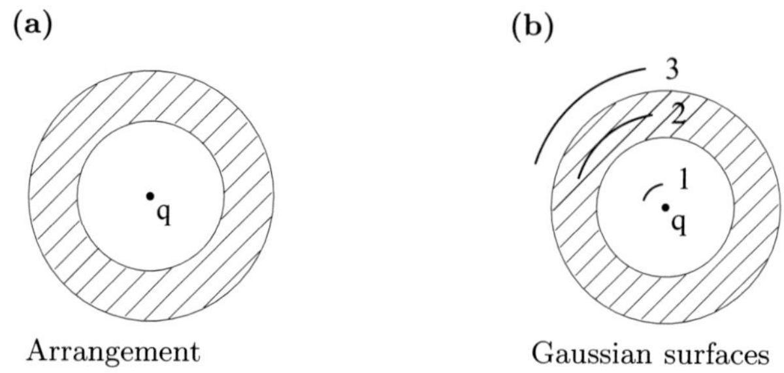

Fig. 2.6 A point charge in a hollow metallic sphere

and therefore

$$\boldsymbol{E}_1(\boldsymbol{r}) = k_e \frac{q}{r^2} \boldsymbol{e}_\mathrm{r} \, .$$

The electric field in the metallic shell around the point charge has the value zero. This implies, that the charge enclosed by surface (2) has to be zero. This is only possible, if the inner surface of the metal carries a charge $-q$. Due to the symmetry (there is no tangential component) one has a uniform distribution of this surface charge. As no charge was added to or subtracted from the metallic section, there must be a surface charge $+q$ distributed over its exterior surface (Fig. 2.7a). The presence of a point charge leads to a separation of the charge in the metallic shell. This is **electric induction**. The uniform surface charge density on the exterior metallic surface is (with respect to the absolute value) smaller than on the interior surface

$$|\sigma_i| = \frac{|q|}{4\pi\, R_i^2} > |\sigma_a| = \frac{|q|}{4\pi\, R_a^2} \, ,$$

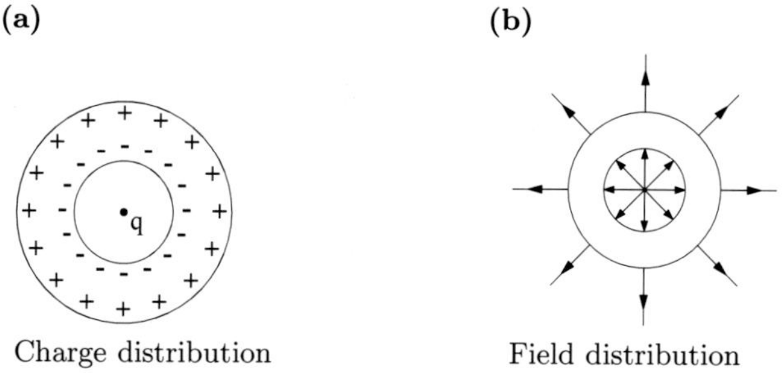

Fig. 2.7 Point charge in a hollow metallic sphere

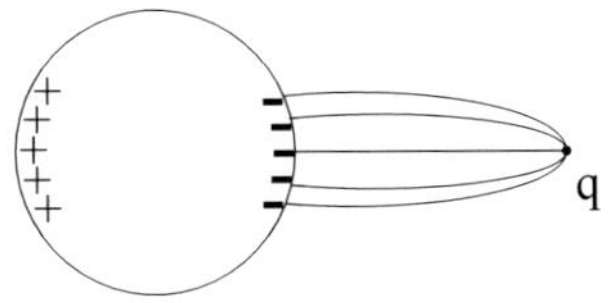

Fig. 2.8 Induction: General situation

as the outer radius R_a is larger than the inner radius R_i. For the Gaussian surface (3) one has again

$$\oiint \boldsymbol{E}_3 \cdot \mathbf{d}\boldsymbol{f} = 4\pi k_e q \,,$$

so that the field in the exterior region is

$$\boldsymbol{E}_3(\boldsymbol{r}) = k_e \frac{q}{r^2} \boldsymbol{e}_{\mathrm{r}} \,.$$

The situation can be described in terms of field lines as: They begin at the point charge $q > 0$ and end at the inner induced charges. There is no field in the interior of the metallic shell. The exterior field begins at the outer induced charges (Fig. 2.7b). Viewed from the outside it is not possible, to decide whether the charge q is distributed over a full metallic sphere or sits in the centre of an uncharged hollow sphere.

A similar separation of the electric charge can be observed, if a charge is placed in the vicinity of any metallic object (e.g. a charge outside a metallic sphere or plane). The distribution of the induced charges is, naturally, not uniform (Fig. 2.8).

For this reason it is not possible to argue with the Gauss' theorem alone, one needs the full mathematical machinery of electrostatics (see Chaps. 3 and 4).

Induction can also be observed to a certain degree in dielectrics. But in this case it is the **polarisation** of the atoms or molecules of the material and not a displacement of freely movable charges. The effect is for this reason less drastic.

The induction in metals leads to a number of practical applications. Two examples are:

(1) The **Faraday cage**. If one introduces a piece of metal into an area with an electric field, one observes the induced layer of surface charges. The inner part of the metallic object is field free. This statement is also correct for a metal cage (Fig. 2.9a). There is no electric field in the interior of the cage, so that one is able to protect electronic devices, which are disturbed by electric fields. Such Faraday cages can also be used as a protection against lightning storms.
(2) The **Plimpton-Lawton Experiment** was performed to check the validity of the $1/r^2$ Coulomb law (compare Chap. 1.2). The apparatus consisted of two concentric, metallic spherical shells, which are connected by a highly sensitive galvanometer (Fig. 2.9b). A change of the charge of the outer spherical shell should only take place on this shell. There should be no current through the

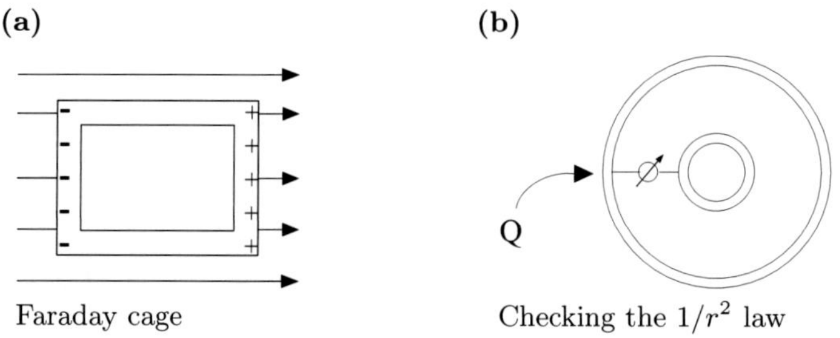

Fig. 2.9 Practical aspects of charge separation

connection of the two shells. As these arguments are based on the validity of Gauss' law and thus indirectly on the validity of the Coulomb law for point charges, the fact that no current is observed, is an indirect confirmation of the Coulomb law. The sensitivity of the galvanometer leads to the errors quoted before for the exponent α of the $1/r^{\alpha}$ law.

The discussion of electrostatics should actually be based on Poisson's differential equation. In order to do this properly, it is necessary to introduce the mathematical concept of **distributions**. Only the use of the delta-function, a special distribution, allows to handle the source term of this equation for all situations of interest. Distributions are introduced in Chap. 2.3, so that their application can start in Chap. 2.4.

2.3 Distributions

The mathematical concept of functions can be extended with the introduction of distributions. A main motivation for addressing this topic in the discussion of electrodynamics is the wish to represent an abstract object like a point charge by a charge distribution in proper mathematical terms. The situation, that a point charge q is located at the origin of the coordinate system $\boldsymbol{r} = 0$, can be handled with the Gauss' theorem by the statement, that the surface integral of the scalar product of the electric field $\boldsymbol{E}$ and the infinitesimal surface element $\mathbf{d}\boldsymbol{f}$ yields for any closed surface S about the origin the result

$$\oint\!\!\!\oint_S \boldsymbol{E} \cdot \boldsymbol{D}\boldsymbol{f} = 4\pi\, k_e q\,.$$

The divergence theorem and the representation of the point charge by a charge density $\rho_{\mathrm{p}}(\boldsymbol{r})$ leads to the statement

$$\iiint_{\mathrm{V(S)}} (\nabla \cdot \boldsymbol{E})\, \mathrm{d}V = 4\pi\, k_e q \int_{\mathrm{V(S)}} \rho_{\mathrm{p}}(\boldsymbol{r})\, \mathrm{d}V\,,$$

which should be valid for any volume $V(F)$, which contains the surface with the point charge. In order to represent the point charge, the charge distribution has to satisfy the condition

$$\rho_{\mathrm{p}}(\boldsymbol{r}) = 0 \qquad \text{for} \qquad \boldsymbol{r} \neq \boldsymbol{0}\,.$$

On the other side the statement

$$\iiint_{\text{space}} \rho_{\mathrm{p}}(\boldsymbol{r})\,\mathrm{d}V = 1$$

must be valid, so that the two variants of the Gauss' theorem agree. It is indicated below, that these conditions overstrain the concept of functions. The mathematical concept, which is addressed here, is a special case of a distribution, the δ-distribution in three space dimensions, which is used in many areas of Theoretical Physics. This distribution is none the less quite generally called the δ-function. It was introduced about 1930 by P.A.M. Dirac in the early phase of quantum mechanics, a mathematically correct foundation of the theory of distributions was only given by L. Schwartz in the years 1950/1951. One can try to introduce the δ-function in a pragmatic manner, e.g. as a limit of ordinary functions. It turns out, however, that this road has to be traversed with caution. One easily ends in mathematically dubious situations, even though some of the properties of the δ-function can be derived in this fashion.

2.3.1 Pragmatic Approach

As the δ-function in three space dimensions is a simple extension of this function in one dimension, it is sufficient to look at the one-dimensional case. The two requirements (with a point charge at the position $x = 0$) are in this case

$$\rho_{\mathrm{p}}(x) = 0 \qquad \text{for} \qquad x \neq 0$$

$$\int_{-\infty}^{\infty} \rho_{\mathrm{p}}(x)\,\mathrm{d}x = 1$$

or for a point charge at the position $x = x_0$

$$\rho_{\mathrm{p}}(x - x_0) = 0 \qquad \text{for} \qquad x \neq x_0$$

$$\int_{-\infty}^{\infty} \rho_{\mathrm{p}}(x - x_0)\,\mathrm{d}x = 1\,.$$

It is possible to show, that the limiting function (and others)

$$\rho_p(x) = \frac{1}{\pi} \lim_{\varepsilon \to 0} \frac{\varepsilon}{(x^2 + \varepsilon^2)} \tag{2.9}$$

is a candidate to satisfy the properties required. Other possible candidates are

$$\rho_p(x) = \frac{1}{\pi} \lim_{\varepsilon \to 0} \left[\frac{1}{\sqrt{\varepsilon}} e^{-x^2/\varepsilon} \right]$$

$$\rho_p(x) = \frac{1}{\pi} \lim_{k \to \infty} \left[\frac{\sin kx}{x} \right].$$

The proof for (2.9) goes as follows. For the function of two variables

$$f(\varepsilon, x) = \frac{\varepsilon}{x^2 + \varepsilon^2}$$

the limit for all values of $x \neq 0$ is

$$\lim_{\substack{\varepsilon \to 0 \\ x \neq 0}} f(\varepsilon, x) = 0 .$$

On the other side on finds

$$f(\varepsilon, 0) = \frac{1}{\varepsilon} \qquad \text{so that} \quad \lim_{\varepsilon \to 0} f(\varepsilon, 0) \longrightarrow \infty .$$

The actual calculation of the improper integral with the function $\varepsilon/(x^2 + \varepsilon^2)$ requires the steps

$$\int_{-\infty}^{\infty} \frac{\varepsilon}{(x^2 + \varepsilon^2)} \mathrm{d}x = \lim_{a \to \infty} \int_{-a}^{a} \frac{\varepsilon}{(x^2 + \varepsilon^2)} \mathrm{d}x = \lim_{a \to \infty} \int_{-a/\varepsilon}^{a/\varepsilon} \frac{1}{(1 + z^2)} \mathrm{d}z$$

$$= \lim_{a \to \infty} 2 \arctan \left(\frac{a}{\varepsilon} \right) = \pi ,$$

so that

$$\lim_{\varepsilon \to 0} \frac{1}{\pi} \int_{-\infty}^{\infty} f(\varepsilon, x) \, \mathrm{d}x = 1 .$$

The following statement

$$\lim_{\varepsilon \to 0} \lim_{a \to \infty} \left[\int_{-a}^{a} \frac{\varepsilon}{(x^2 + \varepsilon^2)} \mathrm{d}x \right] = \lim_{a \to \infty} \lim_{\varepsilon \to 0} \left[\int_{-a}^{a} \frac{\varepsilon}{(x^2 + \varepsilon^2)} \mathrm{d}x \right],$$

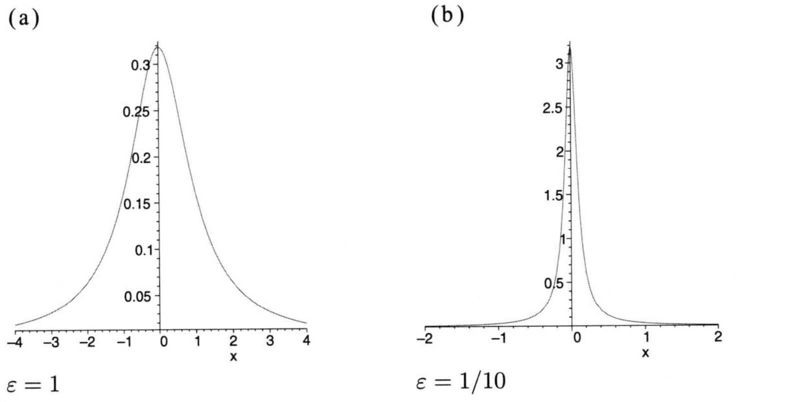

Fig. 2.10 Pragmatic definition of the δ-function

is also possible, provided one evaluates the proper integral before executing the limits. In this sense one can say, that the two limiting processes $\varepsilon \to 0$ and $a \to \infty$ can be interchanged.

It is possible, to follow the behaviour of the function $f(\varepsilon, x)$ up to the limit $\varepsilon \to 0$. The function shrinks with decreasing ε to an ever smaller surrounding of the point $x = 0$ (Fig. 2.10). In the same fashion, the value of the function at $t = 0$ grows, so that the area under the curve remains constant (Fig. 2.11). It is not possible to represent the limit properly.

On the basis of this argumentation one can state: The 'function' $\delta(x)$ can be considered to be the limiting function of $f(\varepsilon, x) = \varepsilon/(x^2 + \varepsilon^2)$ in the limit $\varepsilon \to 0$. The limit does not exist in the strict sense. The limit of the improper integral involved is, however, defined.

For a point charge ($q = 1$) at the position x_0 one uses the definition

$$\rho_{\mathrm{p}} \equiv \delta(x - x_0) = \lim_{\varepsilon \to 0} \frac{1}{\pi} \left[\frac{\varepsilon}{(x - x_0)^2 + \varepsilon^2} \right] .$$

Fig. 2.11 Pragmatic definition of the δ-function, $\varepsilon = 1/100$

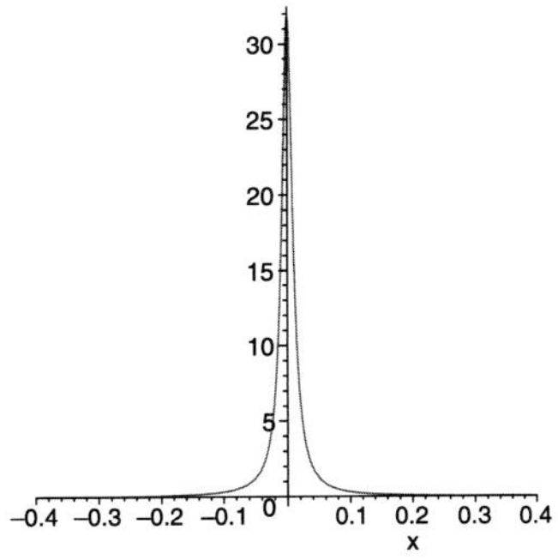

2.3.2 Properties of the δ-Function

Some of the properties of the function $\delta(x - x_0)$ can be demonstrated on the basis of the pragmatic definition. Examples are:

(1) The relation

$$\int_{-\infty}^{\infty} \delta(x - x_0)\, \mathrm{d}x = \int_{-\infty}^{\infty} \delta(x_0 - x)\, \mathrm{d}x \quad (= 1)$$

or the short form

$$\delta(x - x_0) = \delta(x_0 - x) \qquad \text{resp.} \quad \delta(x) = \delta(-x)$$

is valid. The δ-function is symmetric. This reflects the fact, that the function $f(\varepsilon, x)$ is symmetric in x and that this property is not lost in the limiting process.

(2) The relation

$$\int_{-\infty}^{\infty} \delta(x - x_0) f(x)\, \mathrm{d}x = f(x_0)$$

is valid, provided the function $f(x)$ can be expanded in a power series around the position x_0. This property expresses the fact, that the δ-function 'sees' only the value of the function f at the position x_0. The proof of this property is more lengthy, it is therefore restricted to the case $x_0 = 0$, writing

$$I = \int_{-\infty}^{\infty} \delta(x) f(x)\, \mathrm{d}x = \lim_{a\to\infty} \int_{-a}^{a} \delta(x) f(x)\, \mathrm{d}x = \lim_{a\to\infty} I(a)\,,$$

with the definition

$$I(a) = \frac{1}{\pi} \lim_{\varepsilon\to 0} \int_{-a}^{a} \frac{\varepsilon}{(x^2 + \varepsilon^2)} f(x)\, \mathrm{d}x\,.$$

If the function $f(x)$ can be expanded in a Taylor series about the position $x = 0$, one has

$$I(a) = \frac{1}{\pi} \lim_{\varepsilon\to 0} \sum_{n0}^{\infty} \frac{1}{n!} f^{(n)}(0) \int_{-a}^{a} \frac{\varepsilon x^n}{(x^2 + \varepsilon^2)}\, \mathrm{d}x\,.$$

The next step is the calculation of the integrals

$$I_n(\varepsilon) = \frac{1}{\pi} \int_{-a}^{a} \frac{\varepsilon x^n}{(x^2 + \varepsilon^2)}\, \mathrm{d}x\,,$$

or with the substitution $x = \varepsilon t$

$$I_n(\varepsilon) = \frac{\varepsilon^n}{\pi} \int_{-a/\varepsilon}^{a/\varepsilon} \frac{t^n}{(1+t^2)}\,\mathrm{d}t\,.$$

The integrand is an odd function for odd n and therefore is

$$I_n(\varepsilon) = 0 \qquad \text{for } n \text{ odd}\,.$$

For even n one extents the integrand

$$\begin{aligned}
I_n(\varepsilon) &= \frac{\varepsilon^n}{\pi} \int_{-a/\varepsilon}^{a/\varepsilon} \frac{\left(t^{n-2}(1+t^2) - t^{n-2}\right)}{(1+t^2)}\,\mathrm{d}t \\
&= \frac{\varepsilon^n}{\pi} \int_{-a/\varepsilon}^{a/\varepsilon} t^{n-2}\,\mathrm{d}t - \varepsilon^2 I_{n-2}(\varepsilon)\,.
\end{aligned}$$

and obtains the recursion formula

$$I_n(\varepsilon) = \left(\frac{2a^{n-1}}{(n-1)\pi}\right)\varepsilon - \varepsilon^2 I_{n-2}(\varepsilon)\,.$$

The recursion can be started with (see above)

$$I_0(\varepsilon) = \frac{2}{\pi}\arctan\left(\frac{a}{\varepsilon}\right)\,.$$

If one assumes, that the limiting processes

$$\lim_{\varepsilon\to 0} \qquad \text{and} \qquad \sum_{n=0}^{\infty}$$

can be interchanged, there follows

$$\begin{aligned}
I_0(\varepsilon) &\xrightarrow{\varepsilon\to 0} 1 \\
I_2(\varepsilon) &= \left(\frac{2a}{\pi}\right)\varepsilon - \frac{2}{\pi}\varepsilon^2 \arctan\left(\frac{a}{\varepsilon}\right) \xrightarrow{\varepsilon\to 0} 0 \\
&\vdots \\
I_n(\varepsilon) &= \left(\frac{2a^{n-1}}{(n-1)\pi}\right)\varepsilon - \varepsilon^2 I_{n-2}(\varepsilon) \xrightarrow{\varepsilon\to 0} 0\,.
\end{aligned}$$

The final result is then

$$
\begin{aligned}
I &= \lim_{a\to\infty} \lim_{\varepsilon\to 0} \sum_{n=0}^{\infty} \frac{1}{n!} f^{(n)}(0)\, I_n(\varepsilon) \\
&\overset{!}{=} \lim_{a\to\infty} \sum_{n=0}^{\infty} \frac{1}{n!} f^{(n)}(0) \lim_{\varepsilon\to 0} I_n(\varepsilon) \\
&= \lim_{a\to\infty} f(0) = f(0)\,.
\end{aligned}
$$

The proof, that the limiting processes can be interchanged is non-trivial, it will not be discussed.

(3) The relation ($a \neq 0$)

$$\int_{-\infty}^{\infty} \delta(ax) f(x)\, \mathrm{d}x = \frac{1}{|a|} \int_{-\infty}^{\infty} \delta(x) f(x)\, \mathrm{d}x$$

or in short form

$$\delta(ax) = \frac{1}{|a|}\, \delta(x)$$

is valid. For a proof, the substitution, e.g. for $a > 0$, gives

$$
\begin{aligned}
\int_{-\infty}^{\infty} f(x)\, \delta(ax)\, \mathrm{d}x &= \frac{1}{a} \int_{-\infty}^{\infty} f\left(\frac{t}{a}\right) \delta(t)\, \mathrm{d}t \\
&= \frac{1}{a} f(0) = \frac{1}{a} \int_{-\infty}^{\infty} f(x)\, \delta(x)\, \mathrm{d}x\,.
\end{aligned}
$$

The case $a < 0$ follows from the symmetry of the δ-function

$$
\begin{aligned}
\int_{-\infty}^{\infty} f(x)\, \delta(ax)\, \mathrm{d}x &= \int_{-\infty}^{\infty} f(x)\, \delta(-|a|x)\, \mathrm{d}x \\
&= \int_{-\infty}^{\infty} f(x)\, \delta(|a|x)\, \mathrm{d}x\,.
\end{aligned}
$$

2.3.3 The Derivatives of the δ-Function

An alternative definition of the δ-function is based on the derivative of the step function $\Theta(x)$. This function can be given by

$$\Theta(x) = \begin{cases} 0 & \text{for} \quad x < 0 \\ 1 & \text{for} \quad x > 0 \end{cases} .$$

Variants with a step to the right or the left are possible. The integral

$$I = \int_{-a_1}^{a_2} \frac{\mathrm{d}\Theta(x)}{\mathrm{d}x} f(x)\,\mathrm{d}x ,$$

with the positive numbers a_1 and a_2, can be evaluated via partial integration with the result

$$\begin{aligned} I &= \Theta(x)\, f(x)\Big|_{-a_1}^{a_2} - \int_{-a_1}^{a_2} \Theta(x)\, f'(x)\,\mathrm{d}x \\ &= f(a_2) - \int_0^{a_2} f'(x)\,\mathrm{d}x = f(0) . \end{aligned}$$

The result can also be written as

$$\int_{-a_1}^{a_2} \frac{\mathrm{d}\Theta(x)}{\mathrm{d}x} f(x)\,\mathrm{d}x = f(0) = \int_{-\infty}^{\infty} \delta(x)\, f(x)\,\mathrm{d}x .$$

The derivative of the step function represents an alternative definition of the δ-function.

In applications higher order derivatives of the δ-function play a role. The argumentation becomes problematic, if one attempts to prove the formula

$$\int_{-\infty}^{\infty} \delta^{(n)}(x - x_0) f(x)\,\mathrm{d}x = (-1)^n f^n(x_0)$$

for the derivatives of the δ-function. This formula can be suggested with some reasonable arguments, but a full proof was not possible. One finds e.g. for the first derivative by partial integration (for piece-wise continuous functions)

$$\int_{-\infty}^{\infty} \delta'(x) f(x)\,\mathrm{d}x = \delta(x) f(x)\Big|_{-\infty}^{\infty} - \int_{-\infty}^{\infty} \delta(x) f'(x)\,\mathrm{d}x$$

and argues: The first term vanishes, as $\delta(x) = 0$ for $x \neq 0$. Therefore follows

$$\int_{-\infty}^{\infty} \delta'(x) f(x)\,\mathrm{d}x = -f'(0)\,.$$

If this argument is repeated for derivatives of higher order, e.g. for

$$\int_{-\infty}^{\infty} \delta''(x) f(x)\,\mathrm{d}x = \delta'(x) f(x)\Big|_{-\infty}^{\infty} - \int_{-\infty}^{\infty} \delta'(x) f'(x)\,\mathrm{d}x\,,$$

there is no ready argument to deal with either $\delta'(x)$ or $\delta'(x)f(x)$, not to speak of higher derivatives.

If one attempts to approach the relation with the derivatives of the δ-function in terms of its definition by limiting processes, one experiences larger difficulties. The reason is: Only integrals with derivatives of the δ-function are well defined. The derivatives themselves can not be defined in a consistent fashion. The problems with the definition in terms of limiting processes can be overcome with the program of Schwartz, which is based on the concept of generalised functions.

2.3.4 Generalised Functions

The starting point of the program is the concept of a functional in the form

$$T[\rho, f] = \int_a^b \rho(x) f(x)\,\mathrm{d}x\,.$$

Each pair of functions is associated by a set of rules with a number T

$$\{\rho(x), f(x)\} \longrightarrow T\,[\rho, f]\,.$$

The question, which leads finally to a generalisation of the concept of functions, is: Can this association be inverted? Can one define a function $\rho(x)$ with the specification of a set of functions $f_n(x)$ and a set of numbers T_n with e.g. $n = 1, 2, \ldots$?

A hint for an answer to these questions can be gleaned from the theory of representation of functions by series, as e.g. Fourier series. The usual statement concerning Fourier series is: The Fourier representation of a function $\rho(x)$ in the interval $[0, 2\pi]$ has the form

$$\rho(x) = \sum_{n=-\infty}^{\infty} T_n\,\mathrm{e}^{\mathrm{i}nx}\,,$$

where the coefficients T_n are calculated by

$$T_n = \frac{1}{2\pi}\int_0^{2\pi} \rho(x)\,\mathrm{e}^{-inx}\,\mathrm{d}x\,.$$

The series represents the function $\rho(x)$ uniquely, if it converges uniformly.

One can express the same fact by shifting the accents slightly and introduce the language, which is used in the discussion of generalised functions. By a set of test functions

$$f_n(x) = \frac{1}{2\pi}\,\mathrm{e}^{-inx} \qquad (0 \le x \le 2\pi,\ n = 0, \pm 1, \dots)$$

and the specification of a set of numbers

$$T_n = T_n[\rho, f_n] = \int_0^{2\pi} \rho(x)\, f_n(x)\,\mathrm{d}x$$

is it possible to define a function $\rho(x)$ in the interval $[0, 2\pi]$ uniquely. All properties of the function $\rho(x)$ can be discussed with the set of numbers $\{T_n\}$, and the series, in which these quantities feature as expansion coefficients.

This is the fact, that L. Schwartz used as the back ground for the generalisation of the concept of functions. The actual implementation of this program requires the following steps:

(1) Define a suitable set of test functions $\{f_n(x)\}$. Enumeration is not necessary, but is more convenient to argue on the basis of such a set instead of a general set $\{f(x), \dots\}$. A suitable test function is characterised by the following statements:

(1a) All derivatives of f_n exist and are continuous. This property is not needed for the definition of a generalised function. It is, however, needed for the discussion of the derivatives of the generalised functions.

(1b) There exists a number $a > 0$, so that for all $f_n(x)$

$$f_n(x) = 0$$

for $x^2 \ge a$ holds. This property guaranties, that the integral (the functional)

$$\int_{-\infty}^{\infty} \rho(x)\, f_n(x)\,\mathrm{d}x$$

exists for every continuous function ρ.

(1b') The test functions f_n have to decrease fast enough in the asymptotic region, e.g. they and their derivatives have to approach zero faster than any power of $1/|x|$ for $|x| \to \infty$.

One can then prove,

(2) that for every set of test functions (1) a sufficient number of test functions exist and that the mapping

$$\{\rho, f_n\} \longrightarrow T_n = T[\rho, f_n]$$

can be inverted in a unique fashion

$$\{T_n, f_n\} \longrightarrow \rho(x) .$$

The set of functionals $\{T_n\}$ for a well defined space of test functions is called a generalised function.

Instead of defining a function in the usual manner by a mapping of the form $x \longrightarrow f(x)$, one can characterise them by a set of functionals. In the case of functions there is no difference, which definition is employed. The second option covers a larger set of mathematical objects. Two examples will illustrate this statement.

In the first example one chooses the test functions

$$f_n(x) = \mathrm{e}^{-n^2x^2} \qquad n = 1, 2, \dots ,$$

which satisfy the conditions (1a) and (1b'). If the set of functionals

$$T_n = \frac{\sqrt{\pi}}{4}\frac{1}{n^3} \Longrightarrow \int_0^\infty \rho(x) f_n(x)\,\mathrm{d}x ,$$

is added, a generalised function is defined. One can readily verify, that

$$T_n = \int_0^\infty x^2 f_n(x)\,\mathrm{d}x$$

is valid. The two statements define (in an unusual way) the continuous function $\rho(x) = x^2$. If $\rho(x)$ is continuous, one calls this function a regular generalised (that is a normal) function.

In the second example one demands

$$T[\rho, f] = f(0)$$

for *every* admissible test function $f(x)$. Every test function is mapped on its value at the position $x = 0$. In this example, one has not defined a normal function, but the

δ-function and this completely and uniquely. All additional properties of generalised functions can be obtained from the set of functionals, as e.g.

$$T[\rho', f] = -f'(0)$$

for the δ-function as well as other properties.

2.3.5 Some Additional Properties of the δ-Function

A correct proof of the properties of the *delta*-function and its derivatives can only be given, if an appropriate sequences of test functions is used. All the properties, listed in Chap. 2.3.2 can be confirmed in this way

(1) For every interval I about the position x_0 and for every function $f(x)$, which is differentiable in this interval, one has

$$\int_I f(x)\,\delta(x - x_0)\,\mathrm{d}x = f(x_0)\,,$$

in particular

$$\int_I \delta(x - x_0)\,\mathrm{d}x = 1\,.$$

(2) If a function $g(x)$ possesses simple zeros at the positions $x_1, \ldots, x_n$, one finds for an interval, which encompasses all zeros

$$\int_I f(x)\,\delta(g(x))\,\mathrm{d}x = \sum_{i=1}^{n} \frac{1}{|g'(x_i)|} \int_I f(x)\,\delta(x - x_i)\,\mathrm{d}x$$
$$= \sum_{i=1}^{n} \frac{1}{|g'(x_i)|} f(x_i)\,.$$

The function $g'(x_i)$ is the derivative of the function $g(x)$ at the position x_i. In particular one finds.

$$\int_I f(x)\delta(x^2 - x_0^2)\,\mathrm{d}x = \frac{1}{2|x_0|} \int_I f(x)(\delta(x - x_0) + \delta(x + x_0))\,\mathrm{d}x$$
$$= \frac{1}{2|x_0|}(f(x_0) + f(-x_0))\,.$$

(3) The relation for the n-th derivative of the δ-function is

$$\int_I f(x)\,\delta^{(n)}(x - x_0)\,\mathrm{d}x = (-1)^n f^{(n)}(x_0)\,.$$

(4) A much used representation of the δ-function is the integral

$$\delta(x - x') = \frac{1}{2\pi}\int_{-\infty}^{\infty} \mathrm{d}k\,\mathrm{e}^{\mathrm{i}\,k(x-x')}\,.$$

The reason is the fact, that this representation can be extended to higher space dimensions as a consequence of the factorisation of the exponential function, as e.g.

$$\begin{aligned}
\delta^{(3)}(\boldsymbol{r} - \boldsymbol{r}') &= \delta(x - x')\,\delta(y - y')\,\delta(z - z') \\
&= \left[\frac{1}{2\pi}\int_{-\infty}^{\infty} \mathrm{d}k_x\,\mathrm{e}^{\mathrm{i}\,k_x(x-x')}\right]\left[\frac{1}{2\pi}\int_{-\infty}^{\infty} \mathrm{d}k_y\,\mathrm{e}^{\mathrm{i}\,k_y(y-y')}\right] \\
&\quad\times\left[\frac{1}{2\pi}\int_{-\infty}^{\infty} \mathrm{d}k_z\,\mathrm{e}^{\mathrm{i}\,k_z(z-z')}\right] \\
&= \left[\frac{1}{(2\pi)^3}\iiint \mathrm{d}k^3\,\mathrm{e}^{\mathrm{i}\,\boldsymbol{k}\cdot(\boldsymbol{r}-\boldsymbol{r}')}\right].
\end{aligned}$$

The triple integral involves the complete $\boldsymbol{k}$-space.

2.4 Representation of Charge Densities by Distributions

A point charge with the value q at the position $\boldsymbol{r}_i$ in three-dimensional space can be represented by a product of three delta-functions

$$\rho_{\mathrm{p}}(\boldsymbol{r}, \boldsymbol{r}_i) = q\,\delta(x - x_i)\,\delta(y - y_i)\,\delta(z - z_i)\,. \tag{2.10}$$

This representation has the properties required. It describes a strictly localised charge

$$\rho_{\mathrm{p}}(\boldsymbol{r}, \boldsymbol{r}_i) = 0 \qquad \text{for} \quad \boldsymbol{r} \neq \boldsymbol{r}_i\,.$$

The volume integral over the charge distribution is equal to the value of the point charge, if the charge is enclosed in the volume

$$\iiint_V \rho_p(\boldsymbol{r},\, \boldsymbol{r}_i)\,\mathrm{d}V = q\int \delta(x - x_i)\,\mathrm{d}x \int \delta(y - y_i)\,\mathrm{d}y \int \delta(z - z_i)\,\mathrm{d}z$$

$$= \begin{cases} q & \text{if } q \text{ in } V \\ 0 & \text{if } q \text{ not in } V\,. \end{cases}$$

The product of a three-dimensional δ-function can be factorised in the form

$$\frac{\rho_p(\boldsymbol{r},\, \boldsymbol{r}_i)}{q} = \delta(\boldsymbol{r} - \boldsymbol{r}_i) = \delta^{(3)}(\boldsymbol{r} - \boldsymbol{r}_i)\,. \tag{2.11}$$

The factorisation of the three-dimensional δ-function can also be expressed in terms of other sets of curvilinear coordinates. The most important are cylinder and spherical coordinates. In the case of cylinder coordinates ϱ, φ, z one has

$$\delta(\boldsymbol{r} - \boldsymbol{r}_i) = \frac{1}{\varrho}\,\delta(\varrho - \varrho_i)\,\delta(\varphi - \varphi_i)\,\delta(z - z_i)\,. \tag{2.12}$$

The justification of this decomposition is simple. Only with this form does one obtain

$$\begin{aligned} \iiint & \delta(\boldsymbol{r} - \boldsymbol{r}_i)\,\mathrm{d}V \\ &= \int_0^\infty \frac{1}{\varrho}\,\delta(\varrho - \varrho_i)\varrho\,\mathrm{d}\varrho \int_0^{2\pi} \delta(\varphi - \varphi_i)\mathrm{d}\varphi \int_{-\infty}^{\infty} \delta(z - z_i)\mathrm{d}z \\ &= 1\,. \end{aligned}$$

The infinitesimal volume element in spherical coordinates r, θ, φ is

$$\mathrm{d}V = r^2\,\mathrm{d}r\,\sin\theta\,\mathrm{d}\theta\,\mathrm{d}\varphi$$

or equivalently

$$\mathrm{d}V = r^2\,\mathrm{d}r\,\mathrm{d}\cos\theta\,\mathrm{d}\varphi\,.$$

This leads to the possible decomposition

$$\delta(\boldsymbol{r} - \boldsymbol{r}_i) = \frac{1}{r^2\sin\theta}\delta(r - r_i)\,\delta(\theta - \theta_i)\,\delta(\varphi - \varphi_i) \tag{2.13}$$

or

$$\delta(\boldsymbol{r}-\boldsymbol{r}_i)=\frac{1}{r^2}\delta(r-r_i)\,\delta(\cos\theta-\cos\theta_i)\,\delta(\varphi-\varphi_i)\,. \tag{2.14}$$

Surface or linear charge distributions can also, if desired, be expressed in terms of δ-functions. For example the expression for an arbitrary surface charge distribution on a sphere about the origin of the coordinate system with radius R is

$$\rho_{\text{sphere}}(\boldsymbol{r})=\sigma(r,\,\theta,\,\varphi)\,\delta(r-R)\,.$$

The (one-dimensional) δ-function restricts the integration to the integration over the surface of the sphere, so that one obtains for the total charge of the sphere

$$Q_{\text{spher.surf.}}=\iiint\rho_{\text{spher.surf.}}(\boldsymbol{r})\,\mathrm{d}V=R^2\iint\mathrm{d}\Omega\,\sigma(R,\,\theta,\,\varphi)\,.$$

For a uniform distribution $\sigma=\sigma_0$ the result is

$$Q_{\text{spher.surf.}}=4\pi\,R^2\,\sigma_0\,.$$

A uniformly charged plane through the x-axis would be represented by (Fig. 2.12a)

$$\rho_{\text{E}}(\boldsymbol{r})=\sigma_0\,\delta(z-y\tan\alpha)\,.$$

Such representations of surface charge distributions are in general not required. In dealing with the Poisson equation, the surface charges (which are not really specified but arise through the action of a distribution of point and/or space charges on metallic surfaces by induction) are treated in terms of boundary conditions.

Linear charge distributions can also be represented, if desired, by δ-functions. One example is a charged ring (radius R) in the x-y plane (Fig. 2.12b). If one uses

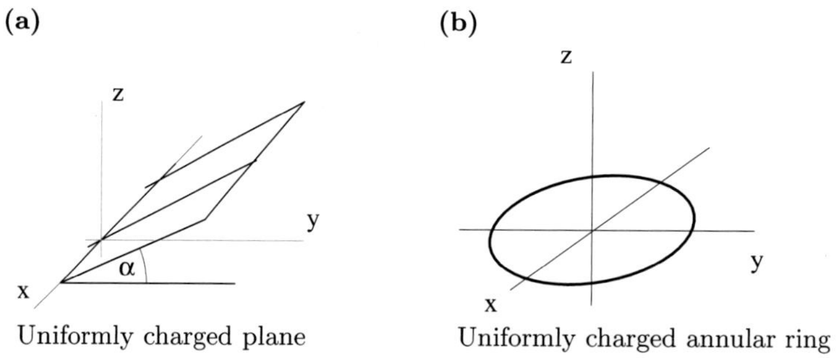

Fig. 2.12 Definition of charge distributions with the aid of δ-functions

cylindrical coordinates, which are well adapted to this symmetry, one would write

$$\rho_{\text{ring}}(\boldsymbol{r}) = \lambda(\varphi)\,\delta(r-R)\,\delta(z)\,.$$

The function $\lambda(\varphi)$ represents an arbitrary linear charge distribution along the ring, the distribution $\delta(r-R)$ represents the mantle of the cylinder about the z-axis, which is intersected by the x- y-plane $\delta(z)$. The total charge of the ring is

$$Q_{\text{ring}} = \iiint \rho_{\text{ring}}(\boldsymbol{r})\,\mathrm{d}V = \int_0^{2\pi} \lambda(\varphi)\,\mathrm{d}\varphi \int_0^{\infty} \delta(r-R)\,r\,\mathrm{d}r \int_{-\infty}^{\infty} \delta(z)\,\mathrm{d}z$$

$$= R\int_0^{2\pi} \lambda(\varphi)\,\mathrm{d}\varphi$$

or for a uniform distribution $\lambda(\varphi) = \lambda_0$

$$Q_{\text{ring}} = 2\pi R\lambda_0\,.$$

A general, explicitly specified charge distribution contains

$$\begin{aligned}
&\text{Point charges} &&= \rho_{\text{P}}(\boldsymbol{r}) = \textstyle\sum_{i=1}^{N} q_i\,\delta(\boldsymbol{r}-\boldsymbol{r}_i)\\
&\text{Space charges} &&= \rho(\boldsymbol{r}) = f(\boldsymbol{r})\\
&\text{Surface charges} &&= \rho_{\text{S}}(\boldsymbol{r}) = \sigma(\boldsymbol{r})\,\delta(\text{Geometry})\\
&\text{Linear charges} &&= \rho_{\text{lin}}(\boldsymbol{r}) = \lambda(\boldsymbol{r})\,\delta(\text{Geometry 1})\,\delta(\text{Geometry 2})\,.
\end{aligned}$$

If the starting point is the integral form of the Gauss's theorem, one would use the differential equation of first order

$$\nabla\cdot\boldsymbol{E}(\boldsymbol{r}) = 4\pi\,k_e\left(\rho(\boldsymbol{r}) + \rho_{\text{P}}(\boldsymbol{r}) + \rho_{\text{S}}(\boldsymbol{r}) + \rho_{\text{lin}}(\boldsymbol{r})\right)\,.$$

If the electric potential is the object of interest rather than the field

$$\Delta V(\boldsymbol{r}) = -4\pi\,k_e\left(\rho(\boldsymbol{r}) + \rho_{\text{P}}(\boldsymbol{r}) + \rho_{\text{S}}(\boldsymbol{r}) + \rho_{\text{lin}}(\boldsymbol{r})\right)\,,$$

one has to solve a partial differential equation of second order with boundary conditions, mostly without linear charges and surface charges. Linear charges are a fiction rather than reality and surface charges, which are the result of induction can be handled by the specification of appropriate boundary conditions.

As the second of these options offers some advantages, one can state: The basic problem of electrostatics is the solution of the partial differential equation

$$\Delta V(\boldsymbol{r}) = -4\pi\, k_e \left(\rho(\boldsymbol{r}) + \rho_\mathrm{p}(\boldsymbol{r})\right) \tag{2.15}$$

with boundary conditions (still to be discussed)

Before the discussion of methods for the actual solution of the Laplace and Poisson problems in different situations (in vacuum, in the presence of conductors and/or dielectric materials) a better, separate understanding of the electric potential is required. This is the goal of the next section.

2.5 The Electric Potential

The relation (see (2.4))

$$\boldsymbol{E}(\boldsymbol{r}) = -\boldsymbol{\nabla}\, V(\boldsymbol{r}) = -\,\mathrm{grad}\; V(\boldsymbol{r})$$

allows the calculation of an electric field function for a given potential function by differentiation. The inverse operation, the calculation of a potential function from a given vector field requires the evaluation of a line integral, as illustrated in Fig. 2.13

$$V(\boldsymbol{r}) = -\int^{\boldsymbol{r}} \boldsymbol{E}(\boldsymbol{r}') \cdot \mathbf{d}\boldsymbol{s}'\,.$$

The integral does not depend on a particular path of integration. This means, that the calculation can be simplified by the choice of a suitable path. The lower limit of the integration is not specified, so that the potential is only determined up to an arbitrary constant.

There exist two possibilities to characterise the modification of the space by the presence of a point (or other) charge. Each point of space can be provided with a vector (3 components, that is 3 functions of the 3 space coordinates) or by a scalar quantity (1 function of the 3 space coordinates). The second option is generally the simpler one. However, the vector field, which can be obtained by differentiation of the scalar function, is needed, if one wishes to discuss the action of electric forces

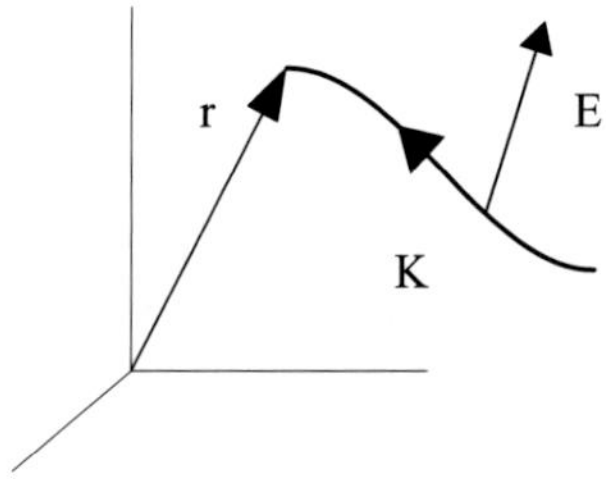

Fig. 2.13 Definition of the electric potential

on charges

$$\boldsymbol{F}_{\text{on}\,q'}(\boldsymbol{r}) = q'\boldsymbol{E}(\boldsymbol{r})\,.$$

On the other hand, the potential energy W of a charge q', located in an electric field $\boldsymbol{E}(\boldsymbol{r})$, is directly related to the potential

$$W(\boldsymbol{r}) = q'V(\boldsymbol{r}) \qquad \leftrightarrow \qquad \boldsymbol{F}(\boldsymbol{r}) = -\operatorname{grad} W(\boldsymbol{r})\,.$$

The definition of the (static) electric potential as a line integral of the electric field determines the units of the potential

$$[V] = \left[\text{Field} \cdot \text{Length}\right] = \left[\frac{\text{Force} \cdot \text{Length}}{\text{Charge}}\right] = \left[\frac{\text{Energy}}{\text{Charge}}\right]$$

or explicitly in the two systems of units

$$\begin{aligned}
&\text{SI-system} && : [V] = \frac{\text{Joule}}{\text{Coulomb}} = \text{Volt} \\
&\text{CGS-system} && : [V] = \frac{\text{erg}}{\text{statcoul}} = \text{statvolt}\,.
\end{aligned}$$

The conversion of the units requires

$$1\,\text{Joule} = 10^7\,\text{erg} \quad \text{and} \quad 1\,\text{C} \approx 3 \cdot 10^9\,\text{statcoul}\,,$$

so that one finds

$$1\ \text{Volt} = \frac{1}{300}\ \text{statvolt}$$

or more exactly

$$1\ \text{Volt} = 3.335635 \cdot 10^{-3}\ \text{statvolt}\,.$$

The general voltage in European electricity systems is about 220 Volt. The corresponding statement in the CGS-system reads: The general voltage is 0.735 statvolt.

The calculation of the potential of a given distribution of charges is the problem discussed in the next two chapters. For the case of a point charge one can write down the result directly. Using the relation

$$\nabla \frac{1}{|\boldsymbol{r} - \boldsymbol{r}_i|} = -\frac{(\boldsymbol{r} - \boldsymbol{r}_i)}{|\boldsymbol{r} - \boldsymbol{r}_i|^3}$$

and the obvious ansatz

$$V(\boldsymbol{r}) = \frac{k_e\, q}{|\boldsymbol{r} - \boldsymbol{r}_i|} + \text{const.}$$

one can regain the result

$$-\nabla V(\boldsymbol{r}) = \frac{k_e\, q(\boldsymbol{r} - \boldsymbol{r}_i)}{|\boldsymbol{r} - \boldsymbol{r}_i|^3} = \boldsymbol{E}(\boldsymbol{r})$$

for a point charge. Naturally, this result can also be obtained by explicit line integration.

Usually one chooses the constant, so, that the potential vanishes in the asymptotic region

$$V(\boldsymbol{r}) \xrightarrow{r \to \infty} 0\,,$$

that is, one sets const. $= 0$.

The fact, that one may choose the constant in this fashion, is corroborated by experiment. Only potential differences (voltages) play a role in nature.

The potential of a given distribution of point charges and space charges can be calculated directly with the superposition principle

$$V(\boldsymbol{r}) = k_e \sum_{i=1}^{N} \frac{q_i}{|\boldsymbol{r} - \boldsymbol{r}_i|} + k_e \iiint \rho(\boldsymbol{r}') \frac{1}{|\boldsymbol{r} - \boldsymbol{r}'|}\, \mathrm{d}V'\,, \tag{2.16}$$

if no inductive or polarisable materials are present.

The associated electric field can be regained by evaluating the gradient of this potential. The gradient acts on the unprimed coordinates in (2.16) and can therefore be exchanged (in most cases) with the integration. If there are surface or line charges specified (and not part of the solution of the problem), one can extend (2.16) to

$$\ldots + k_e \iint \sigma(\boldsymbol{r}') \frac{\mathrm{d}f'}{|\boldsymbol{r} - \boldsymbol{r}'|} + k_e \int \lambda(\boldsymbol{r}') \frac{\mathrm{d}s'}{|\boldsymbol{r} - \boldsymbol{r}'|}\,.$$

The calculation of the potential function is in general more simple than the direct calculation of the field function, as the corresponding integrals are more accessible. In addition, one has to calculate only one instead of three integrals. For this reason the standard approach to the calculation of (stationary) electric fields is

(i) Calculate the potential function (by integration in simpler situations or by solution of the Poisson equation, a differential equation).
(ii) Calculate the corresponding electric field by differentiation.

Some examples are used to illustrate this procedure. The **first** example is another brief look at the point dipole (Fig. 2.14a) with the charges $\pm q$ $(q > 0)$ at the position $\boldsymbol{r} = (0, 0, \pm a)$. With the formulae for the separation of the charges and a point in space $(x\ y\ z)$

$$r_+ = \left[x^2 + y^2 + (z-a)^2\right]^{1/2}$$

$$r_- = \left[x^2 + y^2 + (z+a)^2\right]^{1/2}$$

one finds for the potential

$$V(x,\ y,\ z) = \frac{k_e q}{r_+} - \frac{k_e q}{r_-} = k_e q \left(\frac{r_- - r_+}{r_+ r_-}\right) .$$

This function of x, y and z is not particularly simple, but can provide sufficient insight into the properties of the dipole.

(i) The equipotential surfaces, the surfaces for which the potential has the same value, are characterised by

$$\frac{r_-(x,\ y,\ z) - r_+(x,\ y,\ z)}{r_+(x,\ y,\ z) r_-(x,\ y,\ z)} = \text{const.}$$

The relation $\boldsymbol{E} = -\nabla V$ indicates that the field vectors (and thus the field lines) are perpendicular to the equipotential surfaces. These can be characterised as follows: Their projection on the x-z plane contains the x-axis and closed, circle-like curves around the position of the charges. They are better circles the closer one gets to the charges. One can imagine the surfaces themselves, if the projection is rotated about the z-axis (Fig. 2.14b). The field lines can also be gleaned from this picture.

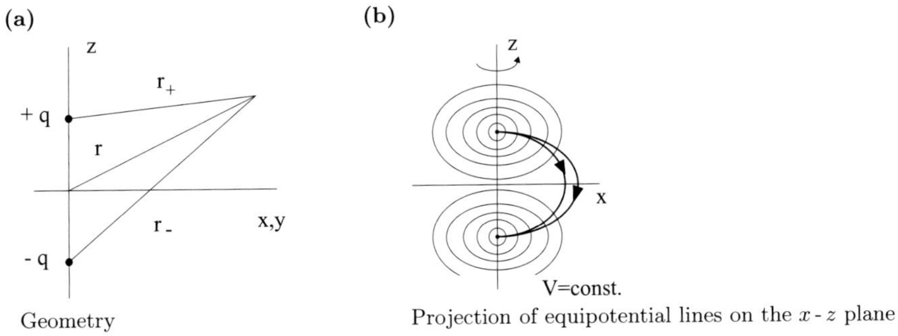

Fig. 2.14 A second look at the electric dipole

(ii) For the discussion of the limit $r^2 = x^2 + y^2 + z^2 \gg a^2$ one expands the expressions for the radial coordinates

$$r_+ \approx \left[x^2 + y^2 + z^2 - 2az\right]^{1/2} = r\left[1 - \frac{2az}{r^2}\right]^{1/2} \approx r - \frac{az}{r} + \dots$$

$$r_- \approx r\left[1 + \frac{2az}{r^2}\right]^{1/2} \approx r + \frac{az}{r} + \dots$$

and obtains for the differences and the products

$$r_- - r_+ \approx \frac{2az}{r} + O\left(\frac{a^2}{r^2}\right) \approx 2a\cos\theta$$

$$r_+ r_- \approx r^2 + O\left(\frac{a^2}{r^2}\right) \approx r^2\,,$$

as well as for the potential

$$V(\boldsymbol{r}) \xrightarrow{r \gg a} \frac{2k_e aq\cos\theta}{r^2} = k_e\frac{p\,z}{r^3}\,.$$

Using the vectorial dipole moment

$$\boldsymbol{p} = (0,\ 0,\ 2aq)$$

for the selected coordinate system (the vector $\boldsymbol{r}$ points from the charge $-q$ to the charge $+q$) allows to write the result as

$$V(\boldsymbol{r}) \xrightarrow{r \gg a} k_e\frac{\boldsymbol{p}\cdot\boldsymbol{r}}{r^3}\,. \tag{2.17}$$

This form with a scalar product is independent of the chosen coordinate system. It is valid for any orientation of the dipole.

(iii) The calculation of the electric field requires the calculation of the gradient. This will only be considered for the far field $r \gg a$.
With the decomposition of the gradient in Cartesian coordinates

$$\boldsymbol{\nabla} = \boldsymbol{e}_{\mathrm{x}}\partial_{\mathrm{x}} + \boldsymbol{e}_{\mathrm{y}}\partial_{\mathrm{y}} + \boldsymbol{e}_{\mathrm{z}}\partial_{\mathrm{z}}\,,$$

one finds

$$E_{\mathrm{x}} = -\frac{\partial V}{\partial x} = \frac{3k_e pxz}{r^5}\,, \qquad E_{\mathrm{y}} = -\frac{\partial V}{\partial y} = \frac{3k_e pyz}{r^5}\,,$$

$$E_{\mathrm{z}} = -\frac{\partial V}{\partial z} = \frac{k_e p(2z^2 - x^2 - y^2)}{r^5}\,.$$

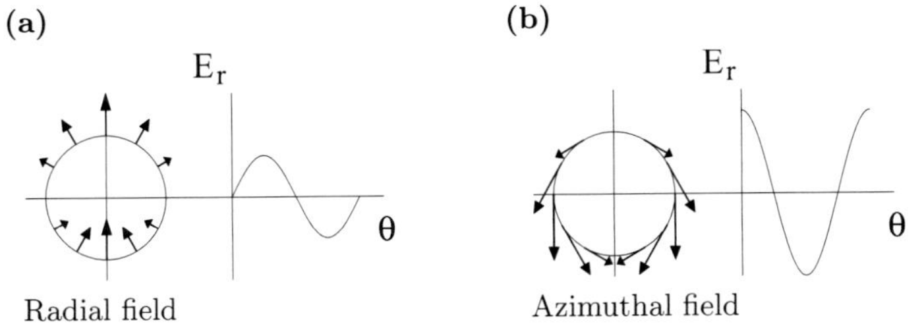

Fig. 2.15 The far field of the electric dipole in spherical coordinates

A glance at this result shows that further discussion will be more transparent in spherical coordinates.[3] The decomposition of the gradient operator in spherical coordinates is

$$\nabla = \boldsymbol{e}_r \frac{\partial}{\partial r} + \boldsymbol{e}_\theta \frac{1}{r} \frac{\partial}{\partial \theta} + \boldsymbol{e}_\varphi \frac{1}{r \sin\theta} \frac{\partial}{\partial \varphi},$$

so that one finds for the components of the field in the limit $r \gg a$

$$E_{\mathrm{r}} = -\frac{\partial}{\partial r} V = -\frac{\partial}{\partial r} \frac{k_e p \cos\theta}{r^2} = +2k_e p \frac{\cos\theta}{r^3}$$

$$E_\theta = -\frac{1}{r} \frac{\partial V}{\partial \theta} = +k_e p \frac{\sin\theta}{r^3}$$

$$E_\varphi = \frac{1}{r \sin\theta} \frac{\partial V}{\partial \varphi} = 0\,.$$

One notices, that both the components in radial and azimuthal direction decrease as $1/r^3$. In addition there is a variation with the angle θ. The radial field has the value zero on the x-axis and has its maximal value on the z-axis (Fig. 2.15a). The azimuthal field is zero for points on the z-axis and maximal for points in the x- y plane (Fig. 2.15b). The far dipole field can also be given in vector form by beginning with the corresponding form of the potential (2.17) and using the product rule for the differentiation

$$\boldsymbol{E}(\boldsymbol{r}) = -\nabla\left(k_e \frac{\boldsymbol{p} \cdot \boldsymbol{r}}{r^3}\right) = 3k_e \left(\frac{(\boldsymbol{p} \cdot \boldsymbol{r})\,\boldsymbol{r}}{r^5}\right) - k_e \frac{\boldsymbol{p}}{r^3} \qquad (r \gg a)\,.$$

[3] The differential operators of vector analysis decomposed in different coordinates are found in Appendix C.5.

This result, which is valid for any orientation of the dipole, is used in many areas of physics.

The **second** problem is the calculation of the potential of a uniformly charged sphere about the origin. The radius is R, the charge density ρ_0. In order to obtain this potential, one has to evaluate the integral

$$V(\boldsymbol{r}) = k_e \rho_0 \iiint_{\text{sphere}} \frac{\mathrm{d}\,V'}{|\boldsymbol{r} - \boldsymbol{r}'|} \,.$$

The symmetry allows the restriction of the calculation of the potential to a point on the z-axis

$$\boldsymbol{r} = (0,\ 0,\ r)\,.$$

As the distance between the two vectors $\boldsymbol{r}$ and $\boldsymbol{r}'$ in spherical coordinates is

$$|\boldsymbol{r} - \boldsymbol{r}'| = \left[r^2 + r'^2 - 2rr' \cos\theta'\right]^{1/2}$$

and the volume element

$$\mathrm{d}\,V' = r'^2 \mathrm{d}r' \mathrm{d}\varphi' \mathrm{d}\cos\theta'\,,$$

one is faced with the evaluation of the triple integral

$$V(\boldsymbol{r}) = k_e \rho_0 \int_0^{2\pi} \mathrm{d}\varphi' \int_0^R r'^2 \mathrm{d}r' \int_{-1}^{1} \frac{\mathrm{d}\cos\theta'}{\left[r^2 + r'^2 - 2rr' \cos\theta'\right]^{1/2}} \,.$$

The integral over the angle φ' is trivial, the elementary integration over $\cos\theta'$ and r' can be performed with little effort (see Detail 2.7.1.1). The result is

$$V(\boldsymbol{r}) = \begin{cases} k_e \dfrac{Q}{r} & r \geq R \\[2ex] k_e \dfrac{Q}{2}\left(\dfrac{3}{R} - \dfrac{r^2}{R^3}\right) & r \leq R \end{cases} \,.$$

There is a parabolic decrease from the value at the origin to the value $k_e Q/R$ at the surface. The potential decreases as $k_e Q/r$ in the outside (Fig. 2.16).

The **last** problem is the calculation of the potential of a uniformly charged thin ring of radius R for all points of space. The electric field for this problem has been calculated already in Chap. 1.3, but only for points on the axis of the ring. The present solution will demonstrate, why the first attempt was restricted to these special points.

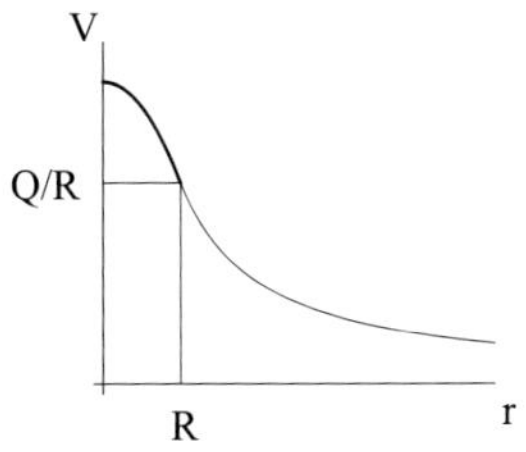

Fig. 2.16 Electric potential of a uniformly charged sphere

The ring can be placed into the x- y plane. For any point of space (Fig. 2.17) one has to deal with the line integral

$$V(\boldsymbol{r}) = k_e \lambda_0 \oint_{\text{ring}} \frac{\mathrm{d}s'}{|\boldsymbol{r} - \boldsymbol{r}'|} \,.$$

The geometry demands cylinder or spherical coordinates. The relation between Cartesian and spherical coordinates is

$$\boldsymbol{r} = \ (x,\ y,\ z) \ = (r\cos\varphi\sin\theta,\ r\sin\varphi\cos\theta,\ r\cos\theta)$$

$$\boldsymbol{r}' = (x',\ y',\ z') = (R\cos\varphi',\ R\sin\varphi',\ 0)\,.$$

The distance between two points in space is therefore

$$|\boldsymbol{r} - \boldsymbol{r}'| = \left[(x - x')^2 + (y - y')^2 + z^2\right]^{1/2}$$

$$= \left[r^2 + R^2 - 2rR\sin\theta\cos(\varphi' - \varphi)\right]^{1/2}\,.$$

In addition the infinitesimal line element is $\mathrm{d}s' = R\mathrm{d}\varphi'$, so that the angular integral

$$V(\boldsymbol{r}) = k_e \lambda_0 R \int_0^{2\pi} \frac{\mathrm{d}\varphi'}{\left[r^2 + R^2 - 2rR\sin\theta\cos(\varphi' - \varphi)\right]^{1/2}}$$

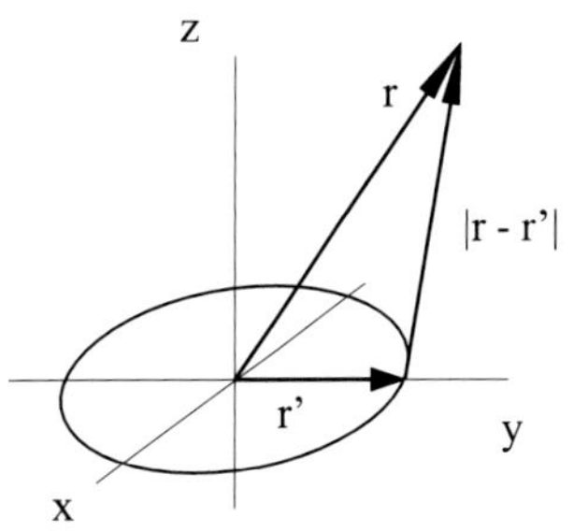

Fig. 2.17 Uniformly charged ring

has to be calculated. The symmetry allows a simplification. The final result cannot depend on the angle φ, so that one can set $\varphi = 0$. The integral can be transformed with the substitution

$$\varphi' = 2\eta \qquad d\varphi' = 2d\eta \qquad 0 \leq \eta \leq \pi$$

and the relation

$$\cos 2\eta = 1 - 2\sin^2\eta\,.$$

The intermediate statement is then

$$V(r,\,\theta) = 2k_e\lambda_0 R \int_0^{\pi} \frac{d\eta}{\left[r^2 + R^2 - 2rR\sin\theta + 4rR\sin\theta\sin^2\eta\right]^{1/2}}\,.$$

The root can be simplified by extracting the terms independent of η. Define the quantity

$$\kappa^2 = \kappa^2(r,\,\theta) = \frac{-4rR\sin\theta}{r^2 + R^2 - 2rR\sin\theta}$$

and obtain for the final integral

$$V(r,\,\theta) = \frac{2k_e\lambda_0 R}{\left[r^2 + R^2 - 2rR\sin\theta\right]^{1/2}} \int_0^{\pi} \frac{d\eta}{\left[1 - \kappa^2\sin^2\eta\right]^{1/2}}\,.$$

Here one recognises the difficulty. The potential can not be represented by an elementary function. One finds a complete elliptic integral of the first kind. The standard form, which follows from the symmetry of the integrand, is given in Detail 2.7.1.2

$$K(\kappa) = \int_0^{\pi/2} \frac{d\eta}{\left[1 - \kappa^2\sin^2\eta\right]^{1/2}} = \frac{1}{2}\int_0^{\pi} \frac{d\eta}{\left[1 - \kappa^2\sin^2\eta\right]^{1/2}}\,.$$

The evaluation is simple for points on the z-axis ($\theta = 0,\,\pi$). For other points there exist two possibilities. One can use standard Tables of the elliptic integrals[4] or one can calculate the integral directly by expansion of the integrand in a power series and integrate term by term with the usual warning concerning the validity of this step. This means, that one has to answer the question, for which values of κ^2 is this expansion a reasonable procedure. For the expansion one can use the binomial

[4] Consult the detailed references cited in Appendix A.5.

formula

$$\left[1-\kappa^2\sin^2\eta\right]^{-1/2} = \sum_{n=0}^{\infty}\binom{-1/2}{n}(-\kappa^2)^n\,(\sin\eta)^{2n}$$

$$= 1+\frac{1}{2}\kappa^2\sin^2\eta+\frac{3}{8}\kappa^4\sin^4\eta-\ldots .$$

The explicit expression for the binomial coefficients is

$$\binom{-1/2}{n} = (-)^n\frac{1\cdot3\cdot5\ldots(2n-1)}{2\cdot4\cdot6\ldots2n} .$$

For the term by term integration one needs the integrals

$$I_{2n} = \int_0^{\pi}(\sin\eta)^{2n}\,\mathrm{d}\eta\,,$$

for which one can find the recursion relation with the help of partial integration

$$I_{2n} = \frac{(2n-1)}{2n}\,I_{2n-2}\,.$$

Evaluation of the recursion with

$$I_0 = \pi$$

results in

$$I_2 = \frac{1}{2}\pi \qquad I_4 = \frac{1}{2}\frac{3}{4}\pi \quad \ldots \quad I_{2n} = (-)^n\binom{-1/2}{n}\pi\,.$$

The representation of the elliptic integrals by a power series is obtained, if these statements are combined[5]

$$K(\kappa) = \frac{\pi}{2}\sum_{n=0}^{\infty}\left[\binom{-1/2}{n}\right]^2\kappa^{2n}$$

$$= \frac{\pi}{2}\left\{1+\left(\frac{1}{2}\right)^2\kappa^2+\ldots\right.$$

$$\left.+\left(\frac{1\cdot3\cdot5\ldots(2n-1)}{2\cdot4\cdot6\ldots(2n)}\right)^2\kappa^{2n}+\ldots\right\}.$$

[5] All properties of the elliptic integrals can be obtained on the basis of the power series.

The elliptic integral is a special case of a more general function, which is one of the functions from the set of *Special Functions of Mathematical Physics*. The function, under which the elliptic integrals are subsumed, is the hypergeometric function $F(a, b, c; x)$. The quantities a, b and c are specified constants, the variable is x.

The hypergeometric function can be defined by the power series

$$F(a, b, c; x) = 1 + \frac{ab}{c}\frac{x}{1!} + \frac{a(a+1)\,b\,(b+1)}{c(c+1)}\frac{x^2}{2!} + \cdots .$$

The series becomes a polynomial, if the constants are negative whole numbers. The function is not defined, if c is a whole negative number, except if $c = -n$ and a or b are equal to $-m$ with $m < n$. If a, b, c are not equal to zero and not a whole negative number, then the series converges for all values of x with $|x| < 1$.

The elliptic integral encountered above can also be written in terms of the hypergeometric function as

$$K(\kappa) = \frac{\pi}{2} F\left(\frac{1}{2}, \frac{1}{2}, 1; \kappa^2\right) .$$

If one uses the fact, that the total charge of the ring is

$$Q = 2\pi R\lambda_0 ,$$

the potential can be written in the form

$$V(r, \theta) = \frac{k_e Q}{\left[r^2 + R^2 - 2rR\sin\theta\right]^{1/2}} \; F\left(\frac{1}{2}, \frac{1}{2}, 1; \kappa^2\right) .$$

Special points are:

For points on the z-axis one has $\sin\theta = 0$. The previous result (Fig. 2.18a) is reproduced with $F(1/2, 1/2, 1; 0) = 1$

$$V(r, \theta = (0, \pi)) = \frac{k_e Q}{[r^2 + R^2]^{1/2}} .$$

For points in the x-y plane with $\sin\theta = 1$ the potential is

$$V(r, \theta = \frac{\pi}{2}) = \frac{k_e Q}{|r - R|} \; F\left(\frac{1}{2}, \frac{1}{2}, 1; \frac{-4rR}{(r-R)^2}\right) .$$

The function $V(r, \theta = \pi/2)$ has the value $k_e Q/R$ for $r = 0$. It increases strongly with r and approaches for $r \longrightarrow R$ the value ∞ (Fig. 2.18b). The singularity indicates that locally one meets a point charge if one traverses the ring in the x-y plane.

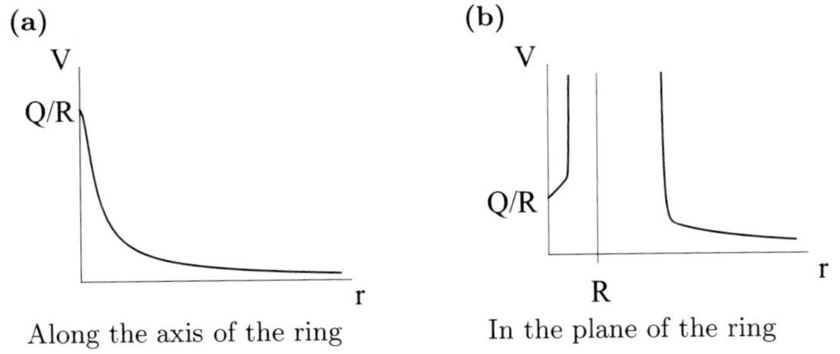

Fig. 2.18 The potential of the circular ring

For $r \gg R$, one finds for the argument of the hypergeometric function

$$\frac{4\,r\,R}{(r-R)^2} \xrightarrow{r\to\infty} 4\frac{R}{r} - 8\frac{R^2}{r^2} + \ldots ,$$

so that the function behaves as

$$F \xrightarrow{r\to\infty} 1 - \frac{R}{r} + \ldots .$$

This leads to the behaviour of the potential

$$V \xrightarrow{r\to\infty} k_e \frac{Q}{r} + \ldots$$

in this limit. The ring looks as a point charge if viewed from a large distance.

The electric field of the ring can be obtained by calculating the gradient. The derivative of the hypergeometric function is

$$\frac{\mathrm{d}}{\mathrm{d}x} F\,(a,\,b,\,c;\;x) = \frac{ab}{c} F\,(a+1,\,b+1,\,c+1;\;x)$$

and the radial component turns out to be

$$\begin{aligned} E_r(r,\theta) &= -\frac{\partial V(r,\theta)}{\partial r} \\ &\quad + k_e Q \frac{(r - R\sin\theta)}{[r^2 + R^2 - 2rR\sin\theta]^{3/2}}\, F\left(\frac{1}{2},\frac{1}{2},1;\kappa^2\right) \\ &\quad + k_e Q \frac{R(R^2 - r^2)\sin\theta}{[r^2 + R^2 - 2rR\sin\theta]^{5/2}}\, F\left(\frac{3}{2},\frac{3}{2},2;\kappa^2\right) . \end{aligned}$$

One sees explicitly, why a direct elementary calculation of the field would have run into difficulties. The field on the axis of the ring with $\sin\theta = 0$ is, as obtained in Chap. 1.3,

$$E_r(r, \theta = (0, \pi)) = \frac{k_e Q r}{[r^2 + R^2]^{3/2}} .$$

Corresponding to the discussion of the gravitational potential in mechanics one finds that the electric potential is directly associated with the potential energy, which is stored in the electric field. This point is the topic of the next section.

2.6 The Electric Field and the Storage of Energy

A point charge q_1 at the position $\boldsymbol{r}_1$ generates an electric potential

$$V(\boldsymbol{r}) = \frac{k_e q_1}{|\boldsymbol{r} - \boldsymbol{r}_1|}$$

in the surrounding space, if the space is free of materials. If a second point charge q_2 is placed at the position $\boldsymbol{r}_2$ from an infinite distance, work, which is equal to the potential energy W_{12}, has to be expended

$$W_{12} = \frac{k_e q_1 q_2}{|\boldsymbol{r}_1 - \boldsymbol{r}_2|} .$$

If a third point charge is brought (again from ∞) to the position $\boldsymbol{r}_3$ (Fig. 2.19), additional work is required

$$W_{13} + W_{23} = k_e \left(\frac{q_1 q_3}{|\boldsymbol{r}_1 - \boldsymbol{r}_3|} + \frac{q_2 q_3}{|\boldsymbol{r}_2 - \boldsymbol{r}_3|} \right) .$$

At this stage it is possible to state, how much work is necessary, in order to assemble a set of point charges, which are initially very far apart into a chosen final

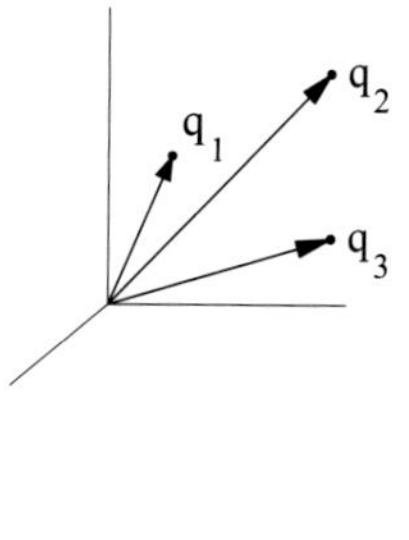

Fig. 2.19 The energy content of a distribution of point charges

configuration

$$W = k_e \sum_{j=1}^{N} \sum_{i<j} \frac{q_i q_j}{|\boldsymbol{r}_i - \boldsymbol{r}_j|} \, .$$

Due to the symmetry of the individual terms one can also write

$$W = \frac{k_e}{2} \sum_{\substack{i,j \\ i \neq j}} \frac{q_i q_j}{|\boldsymbol{r}_i - \boldsymbol{r}_j|} \, .$$

The contributions with $i = j$ have to be omitted.

The work necessary to set up a continuous charge distribution is obtained by replacing the charge q by $\rho(\boldsymbol{r})\mathrm{d}V$ and each of the sums by a triple integral. The result is

$$W = \frac{k_e}{2} \iiint \mathrm{d}V \iiint \mathrm{d}V' \, \frac{\rho(\boldsymbol{r})\rho(\boldsymbol{r}')}{|\boldsymbol{r} - \boldsymbol{r}'|} \, . \tag{2.18}$$

This argument has to be regarded with caution, as points with $\boldsymbol{r} = \boldsymbol{r}'$ are not excluded any more. As will be seen below, this point does not lead to any problem.

One of the triple integrals equals a potential, e.g.

$$V(\boldsymbol{r}) = k_e \iiint \frac{\rho(\boldsymbol{r}')}{|\boldsymbol{r} - \boldsymbol{r}'|} \, \mathrm{d}V' \, .$$

For this reason one finds for the work

$$W = \frac{1}{2} \iiint \rho(\boldsymbol{r}) V(\boldsymbol{r}) \, \mathrm{d}V \, . \tag{2.19}$$

This equation states, that the work, which is necessary to assemble a charge distribution, can be calculated, if one multiplies half of the charge distribution with the potential and integrates over the complete space.

An additional equation is found, if the density is replaced with the Poisson equation

$$\rho(\boldsymbol{r}) = -\frac{1}{4\pi k_e} \Delta V(\boldsymbol{r}) \qquad \longrightarrow \qquad W = -\frac{1}{8\pi k_e} \iiint V(\boldsymbol{r}) \Delta V(\boldsymbol{r}) \, \mathrm{d}V \, .$$

Subsequent application of the first Green's theorem gives

$$\iiint_{\mathrm{B}} [V(\boldsymbol{r}) \Delta V(\boldsymbol{r}) + \nabla V(\boldsymbol{r}) \cdot \nabla V(\boldsymbol{r})] \, \mathrm{d}V = \oiint_{\mathrm{O(B)}} V(\boldsymbol{r}) \frac{\partial V(\boldsymbol{r})}{\partial n} \, \mathrm{d}f \, .$$

The surface term does not contribute for a very large sphere with $R \to \infty$ and the standard boundary conditions. Therefore the final result is

$$W = \frac{1}{8\pi k_e} \iiint (\nabla V(\boldsymbol{r}) \cdot \nabla V(\boldsymbol{r}))\, \mathrm{d}V$$

or

$$W = \frac{1}{8\pi k_e} \iiint (\boldsymbol{E}(\boldsymbol{r}) \cdot \boldsymbol{E}(\boldsymbol{r}))\, \mathrm{d}V\,. \tag{2.20}$$

The integrand in this expression can be interpreted as an energy density

$$W = \iiint w(\boldsymbol{r})\, \mathrm{d}V \qquad \text{with} \quad w(\boldsymbol{r}) = \frac{1}{8\pi k_e} (\boldsymbol{E}(\boldsymbol{r}) \cdot \boldsymbol{E}(\boldsymbol{r}))\,.$$

This result says: Energy is stored in every point $\boldsymbol{r}$ of space by the field $\boldsymbol{E}(\boldsymbol{r})$. The energy in a volume element $\mathrm{d}V$ at the position $\boldsymbol{r}$ is

$$\mathrm{d}W = w(\boldsymbol{r})\, \mathrm{d}V\,.$$

The mechanical equivalent is the straining of a spring. In order to do this, work A has to be expended. The energy A is stored in the strained spring.

An application of these considerations is illustrated by the problem: Calculate the work, which is necessary to assemble a uniformly charged sphere in a space, which is free of matter. The field of such a sphere is

$$\boldsymbol{E}_{\mathrm{i}} = k_e Q \frac{r}{R^3}\, \boldsymbol{e}_{\mathrm{r}} \qquad r \leq R$$

$$\boldsymbol{E}_{\mathrm{a}} = \frac{k_e Q}{r^2}\, \boldsymbol{e}_{\mathrm{r}} \qquad r \geq R\,.$$

The energy content of the sphere is therefore

$$W = \frac{k_e}{8\pi} \left[\iiint_{\text{in the inside}} \frac{Q^2 r^2}{R^6}\, \mathrm{d}V + \iiint_{\text{on the outside}} \frac{Q^2}{r^4}\, \mathrm{d}V \right].$$

Integration over the solid angle leads to a factor of 4π for each of the integrals. The remaining radial integrals

$$= \frac{k_e Q^2}{2} \left(\int_0^R \frac{r^4}{R^6} \mathrm{d}r + \int_R^\infty \frac{\mathrm{d}r}{r^2} \right)$$

are elementary, so that the result is

$$W = \frac{k_e Q^2}{2}\left(\frac{1}{5R} + \frac{1}{R}\right) = \frac{3k_e}{5}\frac{Q^2}{R}.$$

The energy content of the exterior of the sphere is larger by a factor of five than the energy content of the interior. This example shows, that use of the relation (2.20) instead of Eq. (2.18) does not lead to problems. The correct evaluation of (2.18) would not lead to any difficulties as well. If one stays on the other hand at the level of point charges, one has to exclude the contributions with $i = j$. Such contributions correspond to the *self-energy* of a point charge. A point charge at the position $\boldsymbol{r}_i$ can not be placed at this spot again.

2.7 Details

2.7.1 Calculation of the Electric Potential of a Sphere and a Spherical Ring, Both Uniformly Charged

The electric field of a uniformly charged sphere can also be obtained by first calculating the potential and then its gradient. For the calculation of the potential, the sphere is decomposed into infinitesimal volume elements using spherical coordinates, as suggested by the symmetry. This is complemented by a short remark on the calculation of the potential of a spherical ring.

2.7.1.1 The Potential of a Uniformly Charged Sphere

The potential of a uniformly charged sphere with radius R can be obtained by evaluating the integral

$$V(\boldsymbol{r}) = k_e \rho_0 \iiint_{\text{sphere}} \frac{\mathrm{d}V'}{|\boldsymbol{r} - \boldsymbol{r}'|}.$$

The constant charge density within the sphere is

$$\rho_0 = \frac{3Q}{4\pi R^3}.$$

The symmetry allows the use of a point on the z-axis, provided the origin of the coordinate system is placed into the centre of the sphere. The two vectors, involved in the equation above, can then be represented in spherical coordinates as

$$\boldsymbol{r} = (0,\, 0,\, r) \qquad \text{and} \quad \boldsymbol{r}' = (r'\cos\varphi'\sin\theta',\, r'\sin\varphi'\sin\theta',\, r'\cos\theta'),$$

so that the distance of the end points of the two vectors is

$$|\boldsymbol{r} - \boldsymbol{r}'| = \left[r^2 + r'^2 - 2rr'\cos\theta'\right]^{1/2} .$$

With the volume element

$$\mathrm{d}V' = r'^2 \mathrm{d}r' \mathrm{d}\varphi' \mathrm{d}\cos\theta'$$

the triple integral to be evaluated is

$$V(\boldsymbol{r}) = k_e \rho_0 \int_0^{2\pi} \mathrm{d}\varphi' \int_0^R r'^2 \mathrm{d}r' \int_{-1}^{1} \frac{\mathrm{d}\cos\theta'}{\left[r^2 + r'^2 - 2rr'\cos\theta'\right]^{1/2}} .$$

The integration over φ' can be done directly. The integral over the angle θ' can be simplified with the substitution $x = \cos\theta'$, so that

$$\int \frac{\mathrm{d}x}{[a + bx]^{1/2}} = \frac{2}{b}\sqrt{a + bx} .$$

The result after this step is

$$\int_{-1}^{1} \frac{\mathrm{d}\cos\theta'}{\left[r^2 + r'^2 - 2rr'\cos\theta'\right]^{1/2}} = -\frac{1}{rr'}\left[\sqrt{r^2 + r'^2 - 2rr'\cos\theta'}\,\right]_{-1}^{1}$$

$$= -\frac{1}{rr'}\left[(r - r')^2\right]^{1/2} - \left[(r + r')^2\right]^{1/2} .$$

It is necessary to handle the resolution of the square root in the correct way at this step

$$\int_{-1}^{1} \frac{\mathrm{d}\cos\theta'}{\left[r^2 + r'^2 - 2rr'\cos\theta'\right]^{1/2}} = \begin{cases} -\dfrac{1}{rr'}(-2r') = \dfrac{2}{r} & \text{for} \quad r \geq r' \\[2ex] -\dfrac{1}{rr'}(-2r) = \dfrac{2}{r'} & \text{for} \quad r < r' \end{cases} .$$

In the exterior ($r \geq R$), there is always $r \geq r'$. This implies, that the r' integration can be executed directly

$$V(r) = k_e \rho_0 \, 2\pi \, \frac{2}{r} \int_0^R r'^2 \mathrm{d}r' = k_e \frac{4\pi}{3} \frac{\rho_0}{r} R^3 = k_e \frac{Q}{r} .$$

For $r < R$ one has to split the domain of integration into two domains with the limits $r' \leq r$ and $r' \geq r$

$$V(r) = k_e \rho_0\, 2\pi\, 2 \left\{ \frac{1}{r} \int_0^r r'^2 \mathrm{d}r' + \int_r^R r' \mathrm{d}r' \right\}$$

$$= 4\pi\, k_e \rho_0 \left\{ \frac{1}{3} r^2 + \frac{1}{2} R^2 - \frac{1}{2} r^2 \right\}$$

$$= \frac{4\pi\, k_e \rho_0}{6} \left\{ 3R^2 - r^2 \right\} = k_e \frac{Q}{2} \left\{ \frac{3}{R} - \frac{r^2}{R^3} \right\} .$$

The electric field is obtained by calculating the gradient $\boldsymbol{E}(\boldsymbol{r}) = -\nabla V(r)$.

2.7.1.2 Remark on the Potential of the Spherical Ring

For this problem one has to deal with the integral

$$\int_0^\pi \frac{\mathrm{d}\eta}{\left[1 - \kappa^2 \sin^2 \eta\right]^{1/2}} . \tag{2.21}$$

The integral is nearly a complete elliptic integral, but the upper limit is π instead of $\pi/2$. The following argument shows, however, that the integral can be reduced to a complete elliptic integral. The integrand $f(\eta)$ in (2.21) is a function of $\sin^2 \eta$. Therefore one finds $f(\eta) = f(\pi - \eta)$ and the integral can be split into

$$\int_0^\pi f(\sin^2 \eta)\, \mathrm{d}\eta = \int_0^{\pi/2} f(\sin^2 \eta)\, \mathrm{d}\eta + \int_{\pi/2}^\pi f(\sin^2 \eta)\, \mathrm{d}\eta .$$

The substitution in the second term leads to

$$= \int_0^{\pi/2} f(\sin^2 \eta)\, \mathrm{d}\eta + \int_{\pi/2}^0 f(\sin^2 \eta')\, (-\mathrm{d}\eta')$$

$$= 2 \int_0^{\pi/2} f(\sin^2 \eta)\, \mathrm{d}\eta = 2K(\kappa)$$

and therefore to

$$\int_0^\pi \frac{\mathrm{d}\eta}{\left[1 - \kappa^2 \sin^2 \eta\right]^{1/2}} = 2 \int_0^{\pi/2} \frac{\mathrm{d}\eta}{\left[1 - \kappa^2 \sin^2 \eta\right]^{1/2}} .$$

3 Solution of the Poisson Equation: Simple Boundary Conditions

It is useful to distinguish the following cases for the discussion of the solution of the Poisson or the Laplace equation:

(i) The general situation, in which one has to deal with a given number of space and point charges *as well* as charged or uncharged conducting and polarisable materials distributed in space. The solution is more demanding in this situation. It will be treated in Chap. 4.

(ii) The situation is simpler, if there are no other objects present except space and point charges. The solution of the Poisson and Laplace equations requires an appropriate formulation and application of boundary conditions in this case. For spherically symmetric problems the partial differential equations are reduced to ordinary differential equations. For problems with more complicated geometries one has to deal with boundary values in two or three space dimensions. Problems, which can be treated analytically, rely in most cases on the machinery of special functions of mathematical physics. A selection of these functions and a listing of the relevant literature are presented in a separate, mathematically oriented section of the Appendix. In this chapter, the main topic is the solution of the Poisson and Laplace equations in terms of spherical coordinates. A very useful tool is the multipole expansion, which is flexible enough to be used for quite a number of problems. The solution of the Poisson and the Laplace equations in cylindrical coordinates will, by contrast, only be touched briefly.

3.1 Problems with Spherical Symmetry

The task is the solution of the differential equation for a given distribution of space charges $\rho(\boldsymbol{r})$ and point charges $\rho_{\rm p}(\boldsymbol{r})$

$$\Delta V(\boldsymbol{r}) = -4\pi k_e \left(\rho(\boldsymbol{r}) + \rho_{\rm p}(\boldsymbol{r})\right). \tag{3.1}$$

© The Author(s), under exclusive license to Springer-Verlag GmbH, DE, part of Springer Nature 2024

R. M. Dreizler, C. S. Lüdde, *Electrostatics and Magnetostatics*,
https://doi.org/10.1007/978-3-662-69933-1_3

If the charges are contained in a finite volume about the origin of the coordinate system, one can demand, that the solution of (3.1) satisfies the (asymptotic) condition

$$V(\boldsymbol{r}) \xrightarrow{r\to\infty} 0\,.$$

This requirement is a **simple boundary condition**, for which an infinite sphere constitutes the boundary surface. A solution which satisfies the differential equation (3.1) and this condition, has already been found in Chap. 2.4.

$$V(\boldsymbol{r}) = k_e \sum_i \frac{q_i}{|\boldsymbol{r}-\boldsymbol{r}_i|} + k_e \int \frac{\rho(\boldsymbol{r}')}{|\boldsymbol{r}-\boldsymbol{r}'|}\,\mathrm{d}V'\,.$$

If one is able to evaluate the triple integral for a given space charge, the problem is solved. The question, whether the solution found in this fashion for the boundary specified, is the most general or whether it is unique, will be addressed in Chap. 4.2. It is, however, useful to search for more direct methods for the solution of the differential equation (3.1), as the evaluation of the integral might be cumbersome. A standard example for the application of direct integration is the uniformly charged sphere, in which one has to use the Poisson equation in the interior region and the Laplace equation in the exterior

$$\begin{aligned} r \le R: &\quad \Delta V_i(\boldsymbol{r}) = -4\pi k_e\,\rho_0 \\ r > R: &\quad \Delta V_a(\boldsymbol{r}) = 0\,. \end{aligned}$$

Here one recognises clearly the local character of the differential equations. The simplest way to the solution of the differential equation is the appeal to the symmetry of the problem, which calls in this case for the use of spherical coordinates. As one expects, that the potential depends only on the radial coordinate

$$V(r,\theta,\varphi) \longrightarrow V(r)\,,$$

the terms with partial derivatives with respect to the angular coordinates do not contribute. The partial differential equation is reduced to an ordinary differential equation for the function $V(r)$ with the Laplace operator

$$\Delta V(r) = \frac{1}{r^2}\frac{\mathrm{d}}{\mathrm{d}r}\left(r^2\frac{\mathrm{d}V(r)}{\mathrm{d}r}\right)\,. \tag{3.2}$$

The solution of the Poisson equation as well as the Laplace equation require only elementary integration. The differential equation for $r \le R$ is

$$\frac{\mathrm{d}}{\mathrm{d}r}\left(r^2\frac{\mathrm{d}V_i(r)}{\mathrm{d}r}\right) = -4\pi k_e\,\rho_0 r^2\,.$$

A first integration leads to

$$r^2 \frac{\mathrm{d}V_i(r)}{\mathrm{d}r} = -4\pi k_e \, \rho_0 \left(\frac{1}{3} r^3 + a_1 \right)$$

or

$$\frac{\mathrm{d}V_i(r)}{\mathrm{d}r} = -4\pi k_e \, \rho_0 \left(\frac{1}{3} r + \frac{a_1}{r^2} \right) .$$

A second integration gives then the result

$$V_i(r) = -4\pi k_e \, \rho_0 \left(\frac{1}{6} r^2 - \frac{a_1}{r} + a_2 \right) .$$

The general solution of an ordinary differential equation of second order contains (as expected) two constants of integration. For $r > R$ one starts with the differential equation

$$\frac{\mathrm{d}}{\mathrm{d}r} \left(r^2 \frac{\mathrm{d}V_a(r)}{\mathrm{d}r} \right) = 0 .$$

After the first integration resulting in

$$r^2 \frac{\mathrm{d}V_a(r)}{\mathrm{d}r} = b_1 ,$$

one obtains with the second integration

$$V_a(r) = -\frac{b_1}{r} + b_2 .$$

The solutions in the two regions contain together four constants of integration a_1, a_2, b_1, b_2. Their values are obtained with the conditions

(1) The general boundary condition on a very large sphere $V_a(r) \to 0$ demands

$$b_2 = 0 \qquad \text{for } r \to \infty .$$

(2) A second condition is the requirement, that the integral over the field in the exterior region taken over an arbitrary, closed surface around the charge, must satisfy

$$\oiint \boldsymbol{E}_a \cdot \mathbf{d}\boldsymbol{f} = 4\pi k_e \, Q \qquad \left(Q = \frac{4}{3} \pi \, \rho_0 R^3 \right) .$$

The field in the exterior region is

$$\boldsymbol{E}_a(r) = -\frac{\partial V_a}{\partial r}\boldsymbol{e}_{\rm r} = -\frac{b_1}{r^2}\boldsymbol{e}_{\rm r}\,,$$

so that one finds for a spherical shell around the charge

$$b_1 = -k_e Q\,.$$

The solution in the exterior region is therefore (as before)

$$V_a(r) = +k_e\frac{Q}{r}\,.$$

In order to find the solution V_i in the interior region one needs two additional conditions, which allow the connection of the potential in the interior with the one in the exterior. As conditions of this kind have to be used in many instances, it is useful to address this point in a general manner.

(3) The potential for any point $\boldsymbol{R}$ on this interface (not necessarily a spherical shell) can be calculated with the line integral

$$V(\boldsymbol{R}) = -\int^{\boldsymbol{R}} \boldsymbol{E}(\boldsymbol{r})\cdot \mathbf{d}\boldsymbol{s}\,.$$

As the field is irrotational, the line integral is independent of the path. It is possible for this reason to approach the point $\boldsymbol{R}$ from the exterior region as well as from the interior region (Fig.3.1). As the potential on the interface must be continuous, even if the charge densities in the two regions are different, one has

$$V_i(\boldsymbol{R}) = V_a(\boldsymbol{R})\,. \tag{3.3}$$

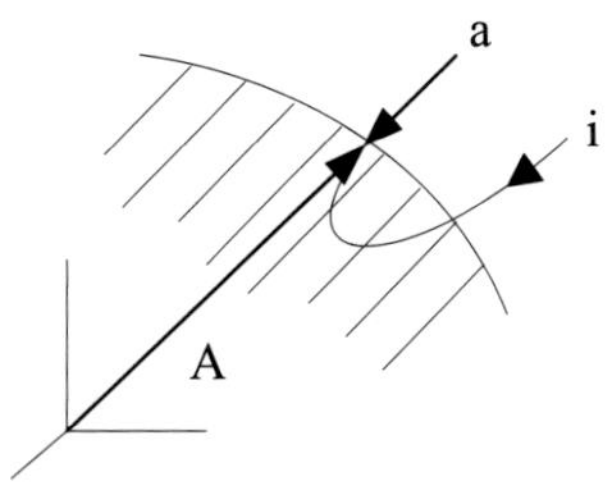

Fig. 3.1 Illustration of the condition for the behaviour of the potential on interfaces

(4) On the other hand the behaviour of the normal component of the electric field in the presence of an interface with surface charges is governed by (2.7)

$$E_{a,n}(\boldsymbol{R}) - E_{i,n}(\boldsymbol{R}) = 4\pi k_e\, \sigma(\boldsymbol{R})\,.$$

If there are no surface charges ($\sigma = 0$), one can argue as follows: The normal component of the field is calculated, according to the definition of the gradient, by the derivative of the potential in the direction of the normal to the surface

$$E_n = \boldsymbol{e}_n \cdot \boldsymbol{E} = -\boldsymbol{e}_n \cdot \boldsymbol{\nabla} V = -\frac{\partial V}{\partial n}\,.$$

For this reason follows

$$\left.\frac{\partial V_i}{\partial n}\right|_{\boldsymbol{R}} = \left.\frac{\partial V_a}{\partial n}\right|_{\boldsymbol{R}}\,. \tag{3.4}$$

The normal derivatives of the potential have to be continuous on an interface without surface charges.

In the example of a spherical charge distribution the surface is characterised by the radius $\boldsymbol{R} \to R$ alone, so that the evaluation of these conditions leads to

$$(3.3) \qquad -4\pi k_e\, \rho_0 \left\{\frac{1}{6}R^2 - \frac{a_1}{R} + a_2\right\} = k_e \frac{Q}{R} = \frac{4}{3}\pi k_e\, \rho_0 R^2$$

$$(3.4) \qquad -4\pi k_e\, \rho_0 \left\{\frac{1}{3}R + \frac{a_1}{R^2}\right\} = -k_e \frac{Q}{R^2} = -\frac{4}{3}\pi k_e\, \rho_0 R\,.$$

The second equation yields $a_1 = 0$, the first one to $a_2 = -R^2/2$. The solution in the interior ($r < R$) is therefore

$$V_i(r) = 2\pi k_e\, \rho_0 \left\{R^2 - \frac{1}{3}r^2\right\} \quad \text{or} \quad V_i(r) = \frac{k_e\, Q}{2}\left\{\frac{3}{R} - \frac{r^2}{R^3}\right\}\,. \tag{3.5}$$

This result can also be obtained, as shown in Chap. 2.4, by a direct evaluation of the integral

$$V(r) = k_e \rho_0 \iiint_{\text{sphere}} \frac{\mathrm{d}V'}{|\boldsymbol{r} - \boldsymbol{r}'|}\,.$$

The following remarks can be added to the discussion of problems with spherical symmetry:

(i) The steps, necessary for solving the Poisson equation with radial symmetry, can also be used for any specifications of the charge density $\rho(r)$, e.g. for an exponential charge distribution

$$\rho(r) = \rho_0\, \mathrm{e}^{-\mu r} \ .$$

The total charge of this distribution is

$$Q = 4\pi\, \rho_0 \int_0^{\infty} r^2\, \mathrm{e}^{-\mu r}\, \mathrm{d}r = \frac{8\pi \rho_0}{\mu^3} \ .$$

The Poisson equation is valid in the entire space. A first direct integration of the differential equation

$$\frac{\mathrm{d}}{\mathrm{d}r}\left(r^2 \frac{\mathrm{d}}{\mathrm{d}r} V(r)\right) = -4\pi k_e \rho_0 r^2\, \mathrm{e}^{-\mu r}$$

yields (use a Table of Integrals or partial integration)

$$r^2 \frac{\mathrm{d}V(r)}{\mathrm{d}r} = -4\pi k_e\, \rho_0 \left\{ \left(-\frac{r^2}{\mu} - \frac{2r}{\mu^2} - \frac{2}{\mu^3}\right) \mathrm{e}^{-\mu r} + a_1 \right\} ,$$

the second

$$V(r) = -4\pi k_e\, \rho_0 \left\{ \left(\frac{1}{\mu^2} + \frac{2}{\mu^3 r}\right) \mathrm{e}^{-\mu r} - \frac{a_1}{r} + a_2 \right\} .$$

The boundary condition and the condition concerning the charge enclosed lead to

$$(1) \longrightarrow \quad a_2 = 0 \qquad (2) \longrightarrow \quad a_1 = \frac{Q}{4\pi\, \rho_0} = \frac{2}{\mu^3}$$

for an infinite sphere. The potential for an exponential distribution of charges has the form

$$V(r) = -k_e Q \left\{ \frac{\mu}{2}\, \mathrm{e}^{-\mu r} - \frac{\left(1 - \mathrm{e}^{-\mu r}\right)}{r} \right\} ,$$

so that one finds

$$V(r) \xrightarrow{r \to \infty} k_e \frac{Q}{r}$$

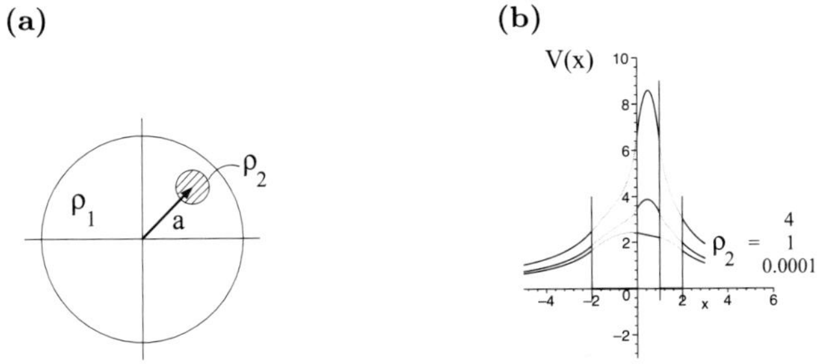

Geometry

Potential along a straight line through the centres of the two spheres

Fig. 3.2 The potential of a sphere embedded in a sphere

in the asymptotic region $r \to \infty$. The exponential function decreases faster than $1/r$.

(ii) For charge distributions, which are composed of different charged spheres, one can calculate the potential of each sphere separately and obtain the total potential with the superposition principle, e.g. for the following problem. A smaller uniformly charged sphere $(R_2,\ \rho_2)$ is embedded in a large, uniformly charged sphere $(R_1,\ \rho_1)$. The separation of the charge centres is $\boldsymbol{a}$, so that $a + R_2 \le R_1$ (Fig. 3.2a). The task is: Calculate the potential of this charge distribution.

For the solution one chooses a coordinate system with the origin at the centre of the larger sphere. The potential of this system can be obtained by addition of the potentials of a sphere with the uniform charge density ρ_1 and radius R_1 and a sphere with the uniform charge density $\Delta\rho = \rho_2 - \rho_1$ and radius R_2. The centre of this sphere is at the position $\boldsymbol{a}$. The contribution of the charge distribution ρ_1 to the potential is

$$
\begin{aligned}
V_1^{(i)}(\boldsymbol{r}) &= 2\pi k_e\,\rho_1 \left(R_1^2 - \frac{1}{3} r^2\right) && r \le R_1 \\
V_1^{(a)}(\boldsymbol{r}) &= \frac{4}{3}\pi k_e \rho_1\, R_1^3\, \frac{1}{r} && r > R_1\,.
\end{aligned}
$$

The potential, which is due to the difference of the charge distributions, is

$$
\begin{aligned}
V_2^{(i)}(\boldsymbol{r}) &= 2\pi k_e\,\Delta\rho \left(R_2^2 - \frac{1}{3}|\boldsymbol{r} - \boldsymbol{a}|^2\right) && |\boldsymbol{r} - \boldsymbol{a}| \le R_2 \\
V_2^{(a)}(\boldsymbol{r}) &= \frac{4}{3}\pi k_e \Delta\rho\, R_2^3\, \frac{1}{|\boldsymbol{r} - \boldsymbol{a}|} && |\boldsymbol{r} - \boldsymbol{a}| > R_2\,.
\end{aligned}
$$

The potential of the total charge distribution in the different regions is therefore

within the small sphere:

$$|\boldsymbol{r} - \boldsymbol{a}| \leq R_2 \qquad V(\boldsymbol{r}) = V_1^{(i)} + V_2^{(i)},$$

within the large, but outside the small sphere:

$$r \leq R_1, \ |\boldsymbol{r} - \boldsymbol{a}| \leq R_2 \qquad V(\boldsymbol{r}) = V_1^{(i)} + V_2^{(a)},$$

outside the large sphere:

$$r \geq R_1 \qquad V(\boldsymbol{r}) = V_1^{(a)} + V_2^{(a)}.$$

The total potential along a straight line through the centres of the spheres is illustrated in Fig. 3.2b [for the parameters: $R_1 = 2$, $R_2 = 0.5$ (units of length), $\rho_1 = 0.1$ (units of charge of density) and $\rho_2 = 4\,,1$ and 0.0001 (units of charge density)]. Additional illustrations can be found in Detail 3.5.1.

3.2 Examples with Azimuthal Symmetry

The calculation of the potential of a charge distribution, which is a function of two variables, e.g. a function of the radius and the azimuthal angle

$$\rho(\boldsymbol{r}) = \rho(r, \theta)$$

can be handled in the same fashion, but is more involved. A typical example is the potential of a uniformly charged, rotational ellipsoid, which is concentrated about the origin (Fig. 3.3a). The surface is described by

$$\frac{x^2}{a^2} + \frac{y^2}{a^2} + \frac{z^2}{c^2} = 1$$

($c > a$ can be assumed without essential restriction). Cuts of the figure with the planes z = const. are circles, cuts with the planes x =const. or y =const. are ellipses. The equation of the surface in spherical coordinates is

$$R^2 \left(\frac{\sin^2\theta}{a^2} + \frac{\cos^2\theta}{c^2} \right) = 1 .$$

Resolution with respect to $R(\theta)$ yields a description of the surface in the form

$$R(\theta) = c \left[1 + \frac{(c^2 - a^2)}{a^2} \sin^2\theta \right]^{-1/2} .$$

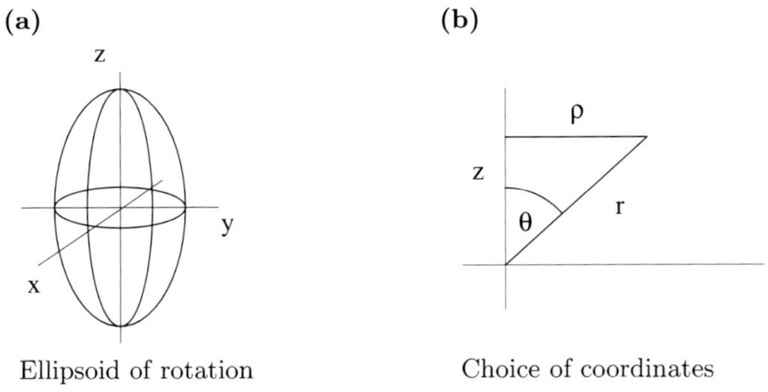

Fig. 3.3 Example for a problem with azimuthal symmetry

The uniformly charged ellipsoid can thus be characterised by the charge distribution

$$\rho(\boldsymbol{r}) = \begin{cases} \rho_0 & r \leq R(\theta) \\ 0 & r > R(\theta) \end{cases} .$$

The dependence on the variable θ is expressed by the limits. An ellipsoid with an arbitrary charge distribution can be specified by the statement

$$\rho(\boldsymbol{r}) = \begin{cases} \rho(r,\ \theta) & r \leq R(\theta) \\ 0 & r \geq R(\theta) \end{cases} ,$$

so that the Poisson equation for the ellipsoid takes the form

$$\Delta V_i(\boldsymbol{r}) = -4\pi k_e\, \rho(r,\ \theta)\,. \tag{3.6}$$

The Laplace equation

$$\Delta V_a(\boldsymbol{r}) = 0 \qquad\qquad r > R(\theta) \tag{3.7}$$

is responsible for the characterisation of the potential on the outside of the ellipsoid.

It is advantageous to start the discussion of the solution with the Laplace equation. The symmetry allows the assumption that the potential does not need to depend on the polar angle φ. Further discussion can be based on spherical (r, θ) or on cylindrical $(\rho,\ z)$ coordinates (Fig. 3.3b). Concerning the choice of coordinates, one might expect that spherical coordinates are a better option, provided the ellipsoid is not very long. In the case that $c \gg a$ or even $c \to \infty$ (a cylinder of infinite length) cylindrical coordinates are the preferable choice, but the discussion can be expected to be more involved in the case of azimuthal symmetry.

With the choice of spherical coordinates one is faced with the Laplace equation

$$\Delta V_a(\boldsymbol{r}) = \frac{1}{r^2}\frac{\partial}{\partial r}\left(r^2\frac{\partial V_a(\boldsymbol{r})}{\partial r}\right) + \frac{1}{r^2\sin\theta}\frac{\partial}{\partial\theta}\left(\sin\theta\frac{\partial V_a(\boldsymbol{r})}{\partial\theta}\right) = 0\,, \tag{3.8}$$

or after multiplication with r^2

$$\frac{\partial}{\partial r}\left(r^2\frac{\partial V_a(\boldsymbol{r})}{\partial r}\right) + \frac{1}{\sin\theta}\frac{\partial}{\partial\theta}\left(\sin\theta\frac{\partial V_a(\boldsymbol{r})}{\partial\theta}\right) = 0\,.$$

Introduction of the operators

$$\Delta_{\mathrm{r}} = \frac{\partial}{\partial r}\left(r^2\frac{\partial}{\partial r}\right) = r^2\frac{\partial^2}{\partial r^2} + 2r\frac{\partial}{\partial r} \tag{3.9}$$

$$\Delta_\theta = \frac{1}{\sin\theta}\frac{\partial}{\partial\theta}\left(\sin\theta\frac{\partial}{\partial\theta}\right) = \frac{\partial^2}{\partial\theta^2} + \cot\theta\frac{\partial}{\partial\theta}\,, \tag{3.10}$$

allows the abbreviated notation

$$\Delta_{\mathrm{r}} V_a(r,\,\theta) + \Delta_\theta V_a(r,\,\theta) = 0\,.$$

The task, which is expressed with this equation, is: Find a general solution of the partial differential equation (3.8) as well as any special solutions of interest. A summary of the relevant points is: The first step is the application of the method of the **separation of variables**. In order to find the solution of a Laplace equation in two variables, e.g. the radius and the azimuthal angle

$$(\Delta_{\mathrm{r}} + \Delta_\theta)\, V_a(r,\,\theta) = 0$$

requires an ansatz of the form (note that R stands here for a function of the radial coordinate r)

$$V_a(r,\,\theta) = R(r)P(\theta)\,.$$

The partial differential equation can now be sorted in the form

$$\frac{\Delta_{\mathrm{r}} R(r)}{R(r)} + \frac{\Delta_\theta P(\theta)}{P(\theta)} = 0\,.$$

As each term is only a function of one of the variables, they must be constants (referred to as **separation constants**), which add up to the value zero. With this argument one is able to partition the partial differential equation into two ordinary differential equations

$$\frac{\mathrm{d}}{\mathrm{d}r}\left(r^2\,\frac{\mathrm{d}R(r)}{\mathrm{d}r}\right) = \kappa R(r) \qquad \text{(radial equation)} \tag{3.11}$$

$$\frac{1}{\sin\theta}\frac{\mathrm{d}}{\mathrm{d}\theta}\left(\sin\theta\frac{\mathrm{d}P(\theta)}{\mathrm{d}\theta}\right) = -\kappa P(\theta) \qquad \text{(angular equation)}\,. \tag{3.12}$$

In the next step one tries to find the general solution of these ordinary differential equations of second order. This offers the possibility to obtain information on possible values (or ranges of values) for the separation constants.

The ordinary differential equations are homogeneous and linear, but in general with variable coefficients (also called coefficient functions), as e.g. the radial equation

$$r^2\frac{\mathrm{d}^2R(r)}{\mathrm{d}r^2} + 2r\frac{\mathrm{d}R(r)}{\mathrm{d}r} - \kappa R(r) = 0\,.$$

A general solution is a combination of two linearly independent solutions, that is

$$R(r) = c_1 R_1(r) + c_2 R_2(r)\,,$$

if the Wronski determinant of the two particular solutions is not equal to zero (a general criterion of linear independence)

$$W(R_1,\, R_2) = \begin{vmatrix} R_1(r) & R_2(r) \\ \dfrac{\mathrm{d}R_1(r)}{\mathrm{d}r} & \dfrac{\mathrm{d}R_2(r)}{\mathrm{d}r} \end{vmatrix} \neq 0\,.$$

It might be possible to guess two useful particular solutions, if the functions, which occur in the differential equation, are simple. If this is not the case, it is necessary to proceed in a more methodical fashion. Often it is possible to use an ansatz in the form of a power series

$$R(r) = r^{\beta}\sum_{n=0}^{\infty} a_n r^n\,.$$

The factor r^{β} is introduced in order to control the behaviour of the solution at the origin (the solution could e.g. behave as $r^{1/2}$ in the vicinity of the origin). The expansion coefficients are determined by insertion of the ansatz into the differential equation and comparing the coefficients. This leads in general to recursion relations for the coefficients. The structure of the radial differential equation of the present

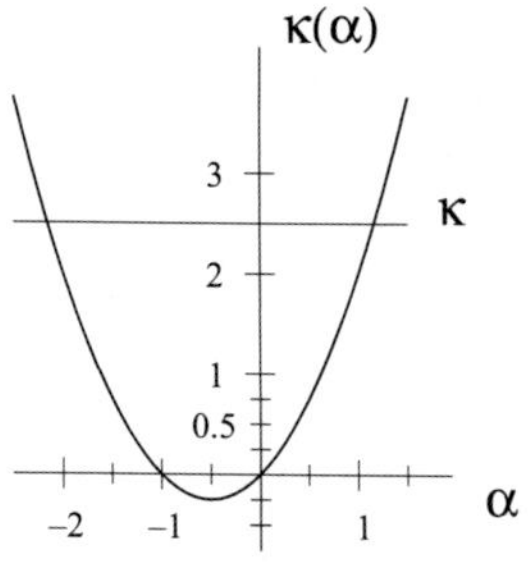

Fig. 3.4 The fundamental solutions of the radial equation

problem is simple enough, so that particular solutions can be found more directly. One uses the ansatz $R(r) = r^{\alpha}$ and finds

$$\alpha\,(\alpha-1)\,r^{\alpha-2}r^2 + 2\alpha r^{\alpha-1}r - \kappa r^{\alpha} = 0$$

or properly sorted

$$\{\alpha\,(\alpha+1) - \kappa\} = 0\,.$$

This means, that r^{α} is a solution if the separation constant κ is equal to $\alpha(\alpha+1)$. In order to determine the allowed values of κ or α one regards the function

$$\kappa = \alpha(\alpha+1)\,.$$

This parabola (Fig. 3.4) has a minimum at $\alpha_{\text{min}} = -1/2$, so that the minimal value of κ_{min} is $-1/4$. In addition one can extract the following statements

(i) Only κ-values with $\kappa \geq -1/4$ lead to real exponents α. As the solution of the differential equation should be a real function (potential differences are measurable quantities), only these κ-values are possible.
(ii) Each κ-value (with $\kappa > -1/4$) leads to two α-values:

$$\alpha_1 = \alpha \qquad \kappa = \alpha_1(\alpha_1+1) = \alpha(\alpha+1)$$

$$\alpha_2 = -(\alpha+1) \qquad \kappa = \alpha_2(\alpha_2+1) = -(\alpha+1)(-\alpha) = \alpha(\alpha+1)\,.$$

If one restricts the range of α-values to $\alpha > -1/2$, one finds, that two linearly independent particular solutions exist for each value of α

$$R_1(r) = r^{\alpha} \qquad R_2(r) = r^{-(\alpha+1)}\,.$$

Both satisfy the differential equation

$$r^2 \frac{\mathrm{d}^2 R_i(r)}{\mathrm{d}r^2} + 2r \frac{\mathrm{d}R_i(r)}{\mathrm{d}r} - \alpha(\alpha+1) R_i(r) = 0 \qquad (i = 1,\ 2) \tag{3.13}$$

and the Wronski determinant does not vanish

$$W(R_1,\ R_2) = \begin{vmatrix} r^{\alpha} & r^{-(\alpha+1)} \\ \alpha\, r^{\alpha-1} & -(\alpha+1) r^{-(\alpha+2)} \end{vmatrix} = -(2\alpha+1)\frac{1}{r^2}\,.$$

The general solution of the radial equation is therefore

$$R_\alpha(r) = A_\alpha\, r^{-(\alpha+1)} + B_\alpha\, r^{\alpha}\,, \tag{3.14}$$

for all values of the alternative separation constant α with $\alpha > -1/2$.

The discussion of the angular part (3.12) requires the solution of the differential equation

$$\frac{1}{\sin\theta}\frac{\mathrm{d}}{\mathrm{d}\theta}\left(\sin\theta \frac{\mathrm{d}P(\theta)}{\mathrm{d}\theta}\right) + \alpha(\alpha+1) P(\theta) = 0 \qquad (\alpha > -1/2)\,. \tag{3.15}$$

The substitution

$$x = \cos\theta \qquad (-1 \le x \le 1)$$

leads to a differential equation with coefficient functions in the form of polynomials of the variable x. For the derivatives with respect to θ one applies the chain rule

$$\frac{\mathrm{d}}{\mathrm{d}\theta} = \frac{\mathrm{d}}{\mathrm{d}x}\frac{\mathrm{d}x}{\mathrm{d}\theta} = -\sin\theta\,\frac{\mathrm{d}}{\mathrm{d}x} = -\sqrt{1-x^2}\,\frac{\mathrm{d}}{\mathrm{d}x}$$

and obtains the differential equation

$$\frac{\mathrm{d}}{\mathrm{d}x}\left[\left(1-x^2\right)\frac{\mathrm{d}P(x)}{\mathrm{d}x}\right] + \alpha(\alpha+1) P(x) = 0\,,$$

or explicitly

$$\left(1-x^2\right)\frac{\mathrm{d}^2 P(x)}{\mathrm{d}x^2} - 2x\frac{\mathrm{d}P(x)}{\mathrm{d}x} + \alpha(\alpha+1) P(x) = 0\,. \tag{3.16}$$

This differential equation, which one meets in many branches of theoretical physics, is the **Legendre differential equation**. The solutions of this differential equation, the Legendre functions or the Legendre polynomials, are introduced and listed in Appendices C.6.1 to C.6.3.

The arguments, which lead to this listing can be summarised in the following manner: With an ansatz in the form of power series one can obtain two linearly independent solutions of the Legendre differential equation. The two solutions, $P_{\alpha,\,1}$ and $P_{\alpha,\,2}$, are initially defined for all values of the separation constant α with $\alpha > -1/2$. In order to be able to calculate the enclosed charge with the solution of the Laplace equation via the relation

$$\oiint \nabla V_a(r,\,\theta)\cdot \mathrm{d}\,\boldsymbol{f} = -4\pi k_e\; Q$$

only solutions of this equation, which do not have any singularities in the interval $-1 \leq \cos\theta \leq 1$ can be admitted. For values with $\alpha > -1/4$, which are not integers, this is not possible. It thus follows that α can only be a positive integer

$$\alpha \equiv l = 0,\; 1,\; 2,\; \ldots\; .$$

The power series for each l-value is then reduced to a polynomial in the variable x . The polynomial solutions, the **Legendre polynomials**

$$P_l(x) \equiv\; P_l(\cos\theta)\,,$$

and their properties can be found in Appendix C.6.1. One of their direct properties is the recursion relation

$$l\,P_l(x) - (2l-1)P_{(l-1)}(x) + (l-1)P_{(l-2)}(x) = 0\,,$$

which should be started for $l = 2$ with $P_0(x) = 1$ and $P_1(x) = x$.

The solutions of the original form of the Laplace equation (3.8) with relevance in physics are the polynomials

$$V_a(r,\theta) = k_e \sum_{l=0}^{\infty}\left(\frac{A_l}{r^{(l+1)}} + B_l r^l\right) P_l(\cos\theta))\,. \tag{3.17}$$

The proportionality factor of the Coulomb law k_e is usually not included in the constant of integration. The original partial differential equation does not depend on the separation constant $\kappa = l(l+1)$. This means, that the general solution can be obtained by summation over the individual solutions, eliminating in this fashion the dependence on l or κ.

The boundary condition $V(r,\theta) \longrightarrow 0$ for $r \longrightarrow \infty$ requires that $B_l = 0$ for all values of l, so that finally the solution of the potential problem posed, is

$$V_a(r,\theta) = k_e \sum_{l=0}^{\infty} \frac{A_l}{r^{(l+1)}} P_l(\cos\theta)\,. \tag{3.18}$$

The expansion (integration) constants A_l have to be determined by connecting this solution to the solution in the region $r \le R(\theta)$. The fact, that any number of integration constants can occur, is in line with the present situation. For a one-dimensional space one can have two boundary values, which determine the two possible integration constants. In a world with two space dimensions the border is a line, in the example discussed an ellipse $r = R(\theta)$. This means that there exist any number of border points, so that all points of the line can be used to determine all integration constants.

There exist two options for the determination of the constants A_l (for a density $\rho(r, \theta)$):

Option 1: Solve the Poisson equation $\Delta V(r, \theta) = -4\pi k_e \rho(r, \theta)$ in the interior region and use the condition for the connection to the exterior region in order to obtain the constants A_l.

Option 2: Provided knowledge of the potential in the interior region is not of interest, the connection can be achieved directly with the general formula

$$V(r, \theta) = k_e \iiint \frac{\rho(\boldsymbol{r}', \theta')\,\mathrm{d}V'}{|\boldsymbol{r} - \boldsymbol{r}'|} \qquad (r > r')$$

for the solution in the exterior region.

In order to apply the second option, it is necessary to expand the expression for the distance between the end points of the vectors $\boldsymbol{r}$ and $\boldsymbol{r}'$ in the exterior region in terms of Legendre polynomials. The function

$$|\boldsymbol{r} - \boldsymbol{r}'| = \left[(x - x')^2 + (y - y')^2 + (z - z')^2\right]^{1/2}$$

can be expressed either in spherical coordinates

$$|\boldsymbol{r} - \boldsymbol{r}'| = \left[r^2 - 2rr'(\cos(\varphi - \varphi')\sin\theta\sin\theta' + \cos\theta\cos\theta') + r'^2\right]^{1/2}$$

or in terms of the angle γ between the vectors ($\boldsymbol{r}$ and $\boldsymbol{r}'$)

$$|\boldsymbol{r} - \boldsymbol{r}'| = \left[r^2 - 2rr'\cos\gamma + r'^2\right]^{1/2}.$$

By comparison of the two expressions one can extract a useful relation between $\cos\gamma$ and the angles of the spherical coordinates

$$\cos\gamma = \cos(\varphi - \varphi')\sin\theta\sin\theta' + \cos\theta\cos\theta'. \tag{3.19}$$

The (multipole-) expansion, to be used, is

$$\frac{1}{|\boldsymbol{r} - \boldsymbol{r}'|} = \sum_{l=0}^{\infty} \frac{r_<^l}{r_>^{l+1}} P_l(\cos\gamma) , \tag{3.20}$$

where $r_>$ respectively $r_<$ denotes the larger or the smaller of the distances from the coordinate origin. The resulting expansion of the potential in the exterior region

$$V_a(r, \theta) = k_e \sum_{l=0}^{\infty} \iiint \rho(r', \theta') \frac{r'^l}{r^{l+1}} P_l(\cos\gamma)\, \mathrm{d}V' \tag{3.21}$$

can be compared with the explicit solution of the Laplace equation (3.18). As the two representations of the solution should agree for all values of the variable r, one finds the coefficients of the power $r^{-(l+1)}$ by direct comparison

$$\iiint \rho(r', \theta') r'^l P_l(\cos\gamma)\, \mathrm{d}V' = A_l P_l(\cos\theta) .$$

The somewhat indirect dependence of the angle γ on the angle θ, is the reason why this sorting is not a trivial task. It can be done directly for low l-values. For $l = 0$ one has $P_0(x) = 1$ and therefore

$$A_0 = \iiint \rho(r', \theta')\, \mathrm{d}V' = \iiint_{cd} \mathrm{d}V' \rho(r', \theta') = Q .$$

One has to integrate over the specified charge distribution (cd). The coefficient for $l = 1$ with $P_1(x) = x$ is

$$A_1 \cos\theta = \iiint_{cd} \mathrm{d}V' r' \rho(\boldsymbol{r}', \theta') \left(\cos(\varphi - \varphi') \sin\theta \sin\theta' + \cos\theta \cos\theta'\right) .$$

This expression for the cosine of the azimuthal angle θ can be treated with the addition theorem, followed by the integration

$$\int_0^{2\pi} \cos\varphi' \mathrm{d}\varphi' = \int_0^{2\pi} \sin\varphi' \mathrm{d}\varphi' = 0$$

to give

$$A_1 \cos\theta = \left[\iiint_{cd} r' \rho(\boldsymbol{r}', \theta') \cos\theta'\, \mathrm{d}V'\right] \cos\theta .$$

The coefficient A_1 is thus

$$A_1 = \iiint_{cd} z' \rho(\mathbf{r}', \theta')\, \mathrm{d}V' .$$

The quantity A_1 is the dipole moment of the charge distribution. This name is consistent with a dimensional check

$$\left[\frac{\text{length} \cdot \text{volume} \cdot \text{charge}}{\text{volume}}\right] = [\text{charge} \cdot \text{length}\ ] .$$

One obtains in particular for a point dipole with the charges on the z-axis (a charge distribution with azimuthal symmetry) because of

$$\rho(r', \theta) = q \left[\delta(z' - a) - \delta(z' + a)\right] \delta(x')\, \delta(y')$$

the old result for the dipole moment of the point dipole

$$A_1 = q(a - (-a)) = 2qa .$$

The attempt to obtain the subsequent coefficients in this elementary fashion quickly leads to rather involved calculations, as for instance for

$$A_2 \left(\frac{1}{2}\left(3\cos^2\theta - 1\right)\right) = \iiint_{cd} \rho(\mathbf{r}', \theta') r'^2 \left[\frac{1}{2}\left(3\cos^2\gamma - 1\right)\right] \mathrm{d}V' .$$

A more rational method is needed in order to separate the dependence on the angles of the spherical coordinates, which is entangled in $P_l(\cos\gamma)$. The tools for this task are found, if one considers the solution of the Laplace equation with spherical coordinates in the full three dimensional space.

3.3 General Problems: Solution with the Full Set of Spherical Coordinates

In a general situation with a charge distribution, which does not have a specific symmetry but is restricted to a finite region around the origin, one has to discuss the solution of the Laplace equation in three space dimensions. The partial differential equation in terms of all three spherical coordinates is then

$$\Delta V_a(r, \theta, \varphi) = 0 . \tag{3.22}$$

The method to be used is again the separation of variables with the ansatz

$$V_a(r, \theta, \varphi) = R(r) P(\theta) S(\varphi) .$$

The differential equation (3.22), using the standard form of the Laplace operator in spherical coordinates[1] yields directly

$$P(\theta)S(\varphi)\left\{\frac{1}{r^2}\frac{\mathrm{d}}{\mathrm{d}r}\left(r^2\frac{\mathrm{d}R(r)}{\mathrm{d}r}\right)\right\}+\frac{R(r)S(\varphi)}{r^2\sin\theta}\left\{\frac{\mathrm{d}}{\mathrm{d}\theta}\left(\sin\theta\frac{\mathrm{d}P(\theta)}{\mathrm{d}\theta}\right)\right\}$$

$$+\frac{R(r)P(\theta)}{r^2\sin^2\theta}\frac{\mathrm{d}^2S(\varphi)}{\mathrm{d}\varphi^2}=0\,.$$

The separation into three differential equations for $R(r)$, $P(\theta)$ and $S(\varphi)$ demands in this case two steps. In the first step one multiplies the differential equation with

$$\frac{r^2\sin^2\theta}{R(r)P(\theta)S(\varphi)}\,.$$

The result is

$$\sin^2\theta\left\{\frac{1}{R(r)}\frac{\mathrm{d}}{\mathrm{d}r}\left(r^2\frac{\mathrm{d}R(r)}{\mathrm{d}r}\right)+\frac{1}{P(\theta)\sin\theta}\frac{\mathrm{d}}{\mathrm{d}\theta}\left(\sin\theta\frac{\mathrm{d}P(\theta)}{\mathrm{d}\theta}\right)\right\}$$

$$+\frac{1}{S(\varphi)}\frac{\mathrm{d}^2S(\varphi)}{\mathrm{d}\varphi^2}=0\,,$$

that is an equation of the form

$$F_1(r,\,\theta)+F_2(\varphi)=0\,.$$

The standard argument, that each of the functions has to be a constant, allows to write

$$F_1(r,\,\theta)=\kappa^2\,,\qquad F_2(\varphi)=-\kappa^2\,.$$

The form of the separation constant is sufficiently general if κ can be a complex number. The ordinary differential equation for the function of the polar angle

$$\frac{\mathrm{d}^2S(\varphi)}{\mathrm{d}\varphi^2}+\kappa^2S(\varphi)=0 \tag{3.23}$$

is the differential equation of the harmonic oscillators. The fundamental solutions are

$$S(\varphi)=\mathrm{e}^{\pm\mathrm{i}\kappa\varphi}\,. \tag{3.24}$$

[1] This Laplace operator can be found in Appendix C.5.1.

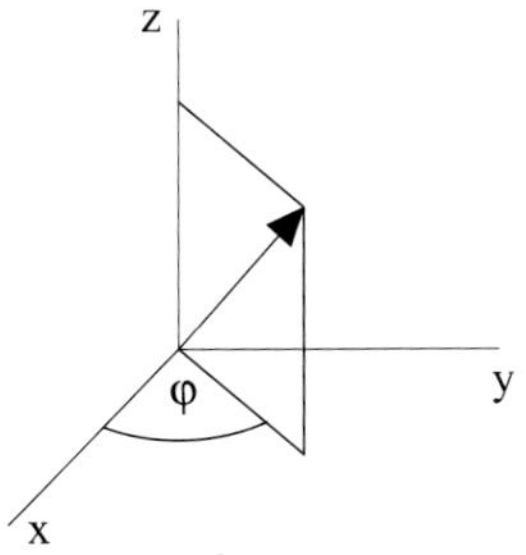

Fig. 3.5 The polar angle

A solution, which is relevant for physical problems, has to be

- finite, unique and continuous.

This condition can be met in the following manner: The end point of a vector in space is characterised by the polar angle φ, but also by the angles φ modulo 2π (Fig. 3.5). Any acceptable solution has to be independent of the (otherwise trivial) periodicity $S(\varphi) = S(\varphi + 2\pi)$. The condition

$$e^{i\kappa\varphi} = e^{i\kappa\varphi} \cdot e^{i2n\kappa\pi} \qquad \text{or} \qquad e^{i2n\kappa\pi} = 1$$

can only be satisfied, if κ is a whole number and real

$$\kappa = 0, \ \pm 1, \ \pm 2, \ \pm 3, \ldots .$$

In summary one can state: The differential equation for the polar angle is (using standard notation with m instead of κ)

$$\frac{d^2 S}{d\varphi^2} + m^2 S = 0 .$$

The separation constant can only have the values

$$m^2 = 0, \ 1, \ 4, \ 9, \ \ldots$$

and the general solution is therefore

$$S_m(\varphi) = C_1 \, e^{im\varphi} + C_2 \, e^{-im\varphi} \qquad (m = 0, \ \pm 1, \ \pm 2, \ \pm 3, \ \ldots) . \tag{3.25}$$

Additional remarks concerning this function are

(i) Instead of the functions $\exp(im\varphi)$ and $\exp(-im\varphi)$ one can use the real functions $\sin(m\varphi)$ and $\cos(m\varphi)$ as a fundamental system, the complex functions are, however, handled more easily.

(ii) The functions

$$\tilde{S}_m(\varphi) = \frac{1}{\sqrt{2\pi}}\, \mathrm{e}^{im\varphi} \qquad (m = -\infty,\, \ldots,\, 0,\, \ldots,\, \infty) \tag{3.26}$$

constitute an orthonormal system of functions in the interval $0 \le \varphi \le 2\pi$ if the scalar product is defined in the following fashion

$$\langle m_1|m_2\rangle = \int_0^{2\pi} \tilde{S}^*_{m_1}(\varphi)\tilde{S}_{m_2}(\varphi)\mathrm{d}\varphi = \delta_{m_1,\,m_2}\,. \tag{3.27}$$

One should keep in mind, that one of the functions in the integrand is a complex conjugate function, so that
for $m_1 \neq m_2$:

$$\int_0^{2\pi} \mathrm{e}^{\mathrm{i}(m_2-m_1)\varphi}\mathrm{d}\varphi = \frac{1}{\mathrm{i}(m_2-m_1)}\, \mathrm{e}^{\mathrm{i}(m_2-m_1)\varphi}\Big|_0^{2\pi} = 0$$

and for $m_1 = m_2$:

$$\int_0^{2\pi} \mathrm{d}\varphi = 2\pi\,.$$

(iii) The system of functions is complete, in the sense that every function $f(\varphi)$, which is defined in the interval $[0,\, 2\pi]$ can be represented by a Fourier series

$$f(\varphi) = \sum_{m=-\infty}^{\infty} a_m\, \mathrm{e}^{\mathrm{i}m\varphi}\,.$$

The second step in the separation process is composed of the following consecutive steps. One multiplies the equation

$$F_1(r,\, \theta) = m^2 \qquad \text{with} \quad \frac{1}{\sin^2\theta}$$

and finds

$$\left\{\frac{1}{R(r)}\frac{\mathrm{d}}{\mathrm{d}r}\left(r^2\frac{\mathrm{d}R(r)}{\mathrm{d}r}\right)\right\}$$

$$+\left\{\frac{1}{\sin\theta\, P(\theta)}\frac{\mathrm{d}}{\mathrm{d}\theta}\left(\sin\theta\frac{\mathrm{d}P(\theta)}{\mathrm{d}\theta}\right) - \frac{m^2}{\sin^2\theta}\right\} = 0\,.$$

(a) If the separation constant is called $\kappa' = l(l+1)$, as in the case of azimuthal symmetry (3.13), the differential equation for the radial part reads

$$\frac{\mathrm{d}}{\mathrm{d}r}\left(r^2\frac{\mathrm{d}R(r)}{\mathrm{d}r}\right) - \kappa' R(r) = 0\,. \tag{3.28}$$

It has the general solution (3.14)

$$R_l(r) = \frac{A_l}{r^{l+1}} + B_l r^l\,.$$

(b) For the angular part one obtains, again with the substitution $x = \cos\theta$, the differential equation

$$(1-x^2)\frac{\mathrm{d}^2 P(x)}{\mathrm{d}x^2} - 2x\frac{\mathrm{d}P(x)}{\mathrm{d}x} + \left[l(l+1) - \frac{m^2}{1-x^2}\right]P(x) = 0\,. \tag{3.29}$$

This extension of the Legendre differential equation is named associated Legendre differential equation.[2]

The **associated Legendre polynomials** can be obtained by differentiation (for positive of m) of the Legendre polynomials

$$P_l^m(x) = (-1)^m(1-x^2)^{(m/2)}\frac{\mathrm{d}^m}{\mathrm{d}x^m}P_l(x)\,, \tag{3.30}$$

where the integer values of m are restricted to the interval $0 \le m \le l$. Functions with negative values of the index m can be obtained by the symmetry relation

$$P_l^{(-m)}(x) = (-1)^m\frac{\Gamma(l-m+1)}{\Gamma(l+m+1)}P_l^m(x)\,. \tag{3.31}$$

The function $\Gamma(n)$ is a generalisation of the factorial.

It is convenient to combine the associated Legendre polynomials $P_l^m(\cos\theta)$ and the exponential functions of the differential equation for the polar angle to a complex valued function of two variables. The resulting functions are called spherical harmonics and are defined by

$$Y_{lm}(\theta,\varphi) = \left[\frac{(2l+1)(l-m)!}{4\pi(l+m)!}\right]^{1/2} P_l^m(\cos\theta)\,\mathrm{e}^{im\varphi} \tag{3.32}$$

[2] The solutions of the associated Legendre differential equation, which are of special interest in physics, are the associated Legendre polynomials. They are discussed in Appendix C.6.3.

on the surface of a sphere with radius 1. The indices can take the values

$$l = 0,\ 1,\ 2,\ \ldots \qquad -l \le m \le l\,.$$

The region of definition is, as expressed by the name,

$$0 \le \theta \le \pi \qquad 0 \le \varphi \le 2\pi\,.$$

This set of angular functions are introduced here in connection with the potential problem of electrostatics. They are used in many parts of physics, as e.g. in quantum mechanics for discussion of the solution of the Schrödinger equation, etc. Here they are used for the solution of the general potential problem with simple boundary conditions in spherical coordinates.

The problem can be stated as: Calculate the potential of a charge distribution $\rho(r,\ \theta,\ \varphi)$, which is located in the vicinity of the origin of a coordinate system, in the exterior region specified by

$$\rho(r,\ \theta,\ \varphi) = 0 \qquad \text{for} \qquad r > R(\theta, \varphi)\,.$$

The execution of this task includes the steps:

(1) Find the solution of the Laplace equation

$$\Delta V_a(r,\ \theta,\ \varphi) = 0$$

in spherical coordinates, which satisfies the conditions

$$V_a \xrightarrow{r\,\to\,\infty} 0 \qquad \text{and} \qquad \left|\int \nabla V_a \cdot \mathbf{d}\boldsymbol{f}\right| < \infty\,.$$

This solution has, in extension of (3.18), the form

$$V_a(r,\ \theta,\ \varphi) = k_e \sum_{l=0}^{\infty} \sum_{m=-l}^{l} \frac{A_{lm}}{r^{l+1}} \left[\frac{4\pi}{2l+1}\right]^{1/2} Y_{lm}(\theta,\ \varphi)\,. \tag{3.33}$$

The functions Y_{lm} are complex. This demands, that the integration constants A_{lm} also have to be complex in general, so that the total expression on the right hand side of (3.33) is real. Splitting off a factor $[4\pi/(2l+1)]^{1/2}$ is not necessary, but is quite useful.

(2) The general solution of the Poisson-/Laplace equations in the interior and the exterior regions for simple boundary conditions can also be written in the form

$$V(r,\ \theta,\ \varphi) = k_e \iiint \frac{\rho(r',\ \theta',\ \varphi')}{|\boldsymbol{r} - \boldsymbol{r}'|}\, \mathrm{d}V'\,.$$

If the distance of two points in the exterior region $r > R > r'$ is expanded in terms of Legendre polynomials, one obtains

$$V_a(r,\theta,\varphi) = k_e \sum_l \iiint \rho(\boldsymbol{r}') \frac{(r')^l}{r^{l+1}} P_l(\cos\gamma)\, \mathrm{d}V' .$$

In order to rewrite the Legendre polynomial $P_l(\cos\gamma)$, where γ is the angle between the vectors $\boldsymbol{r}$ and $\boldsymbol{r}'$, one uses the addition theorem of the spherical harmonics (see Appendix C.6.4) and obtains instead of the standard multipole expansion (3.20) a multipole expansion, in which all six spherical coordinates involved are separated

$$\frac{1}{|\boldsymbol{r}-\boldsymbol{r}'|} = \sum_{l=0}^{\infty}\sum_{m=-l}^{l} \frac{4\pi}{(2l+1)} \frac{r_<^l}{r_>^{l+1}} Y_{lm}(\theta,\varphi) Y_{lm}^*(\theta',\varphi') . \tag{3.34}$$

The potential in the exterior region is

$$V_a(r,\theta,\varphi) = k_e \sum_{l=0}^{\infty}\sum_{m=-l}^{l} \frac{4\pi}{(2l+1)} \frac{Y_{lm}(\theta,\varphi)}{r^{l+1}} \cdot \left[\iiint \mathrm{d}V' \rho(\boldsymbol{r}')(r')^l Y_{lm}^*(\theta',\varphi')\right] . \tag{3.35}$$

(3) By comparison of the expressions (3.33) and (3.35) for the exterior potential one can extract an expression for the integration constants A_{lm}

$$A_{lm} = \left[\frac{4\pi}{2l+1}\right]^{1/2} \iiint \mathrm{d}V' \rho(\boldsymbol{r}')(r')^l Y_{lm}^*(\theta',\varphi') . \tag{3.36}$$

As the spherical harmonics satisfy the symmetry relation

$$Y_{l,-m}(\theta,\varphi) = (-1)^m\, Y_{lm}^{\star}(\theta,\varphi) ,$$

one finds for the coefficients A_{lm} the relation

$$A_{l,-m} = (-1)^m A_{lm}^{\star} , \tag{3.37}$$

provided the charge distribution is real. The application of the formula (3.36) for different charge distributions is carried out in the next section. It is, however, convenient to indicate briefly the use of different sets of coordinates for the solution of the Laplace equation. The Laplace equation is separable in quite a number of sets of orthogonal coordinates.[3] A problem can arise, however, if one attempts to connect the solution of the Laplace equation in the exterior region with the solution of the Poisson equation in the interior.

As an example, one can look at the Laplace equation in cylindrical coordinates (ϱ, z, φ)

$$\left(\frac{\partial^2}{\partial\varrho^2} + \frac{1}{\varrho}\frac{\partial}{\partial\varrho} + \frac{1}{\varrho^2}\frac{\partial^2}{\partial\varphi^2} + \frac{\partial^2}{\partial z^2}\right) V_a(\varrho,\, z,\, \varphi) = 0\,. \tag{3.38}$$

The separation

$$V_a(\boldsymbol{r}) = R(\varrho)S(\varphi)Z(z)$$

leads to three ordinary differential equations

$$\frac{\mathrm{d}^2 Z(z)}{\mathrm{d}z^2} - \kappa^2 Z(z) = 0$$

$$\frac{\mathrm{d}^2 S(\varphi)}{\mathrm{d}\varphi^2} + m^2 S(\varphi) = 0$$

$$\frac{\mathrm{d}^2 R(\varrho)}{\mathrm{d}\varrho^2} + \frac{1}{\varrho}\frac{\mathrm{d}R(\varrho)}{\mathrm{d}\varrho} + \left(\kappa^2 - \frac{m^2}{\varrho^2}\right) R(\varrho) = 0\,.$$

The solutions of the first two equations are elementary functions

$$Z_\kappa(z) = a_+\, \mathrm{e}^{\kappa z} + a_-\, \mathrm{e}^{-\kappa z} \qquad (\kappa \geq 0 \rightarrow \text{real, arbitrary})$$

$$S_m(\varphi) = b_+\, \mathrm{e}^{\mathrm{i}m\varphi} + b_-\, \mathrm{e}^{-\mathrm{i}m\varphi} \qquad (m \geq 0,\ \text{integer})\,.$$

The third equation is a form of the Bessel differential equation. The solutions, the Bessel functions, also named cylinder functions.

As the general solution can not depend on the separation constants, the desired solution in cylindrical coordinates calls for the ansatz

$$V_a(\boldsymbol{r}) = k_e \sum_{\kappa,m} A_{\kappa,m} R_{\kappa,m}(\varrho) S_m(\varphi) Z_\kappa(z)\,.$$

[3] Further sets of orthogonal coordinates can be found, for example, in P. Moon and D.E. Spencer 'Field Theory Handbook' Springer Berlin 1961.

The solutions, which are useful for the characterisation of physical problems, still have to be selected. For the connection to the solution in the region of the charge distribution, the same options, as in the case of spherical coordinates are possible. They will not be addressed here.

3.4 Multipole Moments

The simplest class of charge distributions are the charge distributions with spherical symmetry $\rho(\boldsymbol{r}) = \rho(r)$. For such distributions one calculates the expansion coefficients according to

$$A_{lm} = \left[\frac{4\pi}{2l+1}\right]^{1/2} \int_0^\infty \rho(r')(r')^{l+2}\mathrm{d}r' \iint Y^*_{lm}(\theta', \varphi')\,\mathrm{d}\Omega' \,.$$

The surface integral

$$\iint \mathrm{d}\Omega'\; Y^*_{lm}(\Omega') \qquad (\Omega' \to (\theta', \varphi'))$$

can be evaluated by replacing the factor 1 with $1 = Y_{00}\sqrt{4\pi}$ and application of the orthogonality relation of the spherical harmonics

$$\iint \mathrm{d}\Omega' Y^*_{lm}(\Omega') = \sqrt{4\pi} \iint \mathrm{d}\Omega' Y^*_{lm}(\Omega') Y_{00}(\Omega') = \sqrt{4\pi}\,\delta_{l,0}\,\delta_{m,0}\,.$$

The radial integral

$$\int_0^\infty \rho(r')(r')^2\,\mathrm{d}r' = \frac{Q}{4\pi}$$

leads back to the previous result

$$A_{lm} = Q\,\delta_{l,0}\,\delta_{m,0} \quad \text{and} \quad \text{thus} \qquad V(\boldsymbol{r}) = \frac{Q}{r}\,.$$

The general result for a charge distribution with azimuthal symmetry

$$\rho(\boldsymbol{r}) = \rho(r, \theta)$$

is

$$A_{lm} = \left[\frac{4\pi}{2l+1}\right]^{1/2} \iiint \mathrm{d}r'\mathrm{d}\varphi'\mathrm{d}\cos\theta'\,\rho(r', \theta')\,(r')^{l+2}\,Y^*_{lm}(\theta', \varphi')\,.$$

The integration over φ can be executed readily (compare (3.27))

$$\int_0^{2\pi} \mathrm{d}\varphi' \, \mathrm{e}^{\pm im\varphi'} = 2\pi \, \delta_{m,0} \, .$$

The associated Legendre polynomial contained in the spherical harmonic P_l^m with $m = 0$ corresponds to the simple Legendre polynomial P_l, so that, including all factors, one arrives at the expression

$$A_{lm} = \delta_{m,0} \, 2\pi \iint \mathrm{d}r' \mathrm{d}\cos\theta' \, (r')^{l+2} \, \rho(r', \theta') \, P_l(\cos\theta') \, .$$

One defines for simplicity

$$A_{lm} = \delta_{m,0} \, A_l$$

and writes

$$A_l = \iiint \mathrm{d}V'(r')^l \, \rho(r', \theta') \, P_l(\cos\theta') \, . \tag{3.39}$$

This is the formula for the coefficients of charge distributions with azimuthal symmetry, which have already been found in Chap. 3.3 with elementary means.

The general result (3.33) for the potential in the exterior region corresponds also, because of

$$\left[\frac{4\pi}{2l+1}\right]^{1/2} Y_{l,0}(\theta, \varphi) = P_l(\cos\theta) \, ,$$

to the previous result (3.18)

$$V_a(r, \theta) = k_e \sum_l \frac{A_l}{r^{l+1}} \, P_l(\cos\theta) \, .$$

The lowest order results are the charge and the dipole moment

$$A_0 = \iiint \mathrm{d}V' \rho(r', \theta') = Q$$

$$A_1 = \iiint \mathrm{d}V' z' \, \rho(r', \theta') = D \, .$$

The coefficient with $l = 2$

$$A_2 = \iiint (r')^2\,\rho(r',\theta')\,P_2(\cos\theta')\,\mathrm{d}V'$$

$$= \frac{1}{2}\iiint (r')^2\,\rho(r',\theta')\,(3z'^2 - r'^2)\mathrm{d}V'$$

is called the quadrupole moment (of an azimuthal charge distribution).

The classification of the different terms with increasing index l are named 2^l-pole contributions, where each carries the corresponding name of the Greek numbers, as e.g.

$$\begin{aligned}&l = 3 \Longrightarrow 2^3\text{-pole} \Longrightarrow \text{octupole}\\ &l = 4 \Longrightarrow 2^4\text{-pole} \Longrightarrow \text{hexadecupole}\\ &\text{etc.}\end{aligned}$$

The exterior potential (3.18)

$$V_a(r,\theta) = k_e \sum_{l=0}^{\infty} \frac{A_l}{r^{(l+1)}}\,P_l(\cos\theta)$$

is a simple example of a multipole expansion. A multipole expansion of this form in the exterior region of a charge distribution would represent the exact solution of the Laplace equation, provided all the terms are included. As the contribution of higher multipoles decreases with $1/r^{l+1}$, usually only the first (non-vanishing) terms of the expansion are generally important.

The following explicit examples are supposed to illustrate the calculation of the multipole expansion. The first example is once more the point dipole, but this time from the point of view of the multipole expansion. The charge distribution can be represented in terms of spherical coordinates as a product of δ-functions

$$\rho(r,\theta) = \frac{q}{2\pi r^2}\delta(r-a)\left\{\delta(\cos\theta - 1) - \delta(\cos\theta + 1)\right\}\,.$$

The point charges are located at the intersection of a sphere with radius a and the z-axis. The result for the expansion coefficients A_l is therefore (using $x = \cos\theta$)

$$A_l = q\int_0^{\infty}\mathrm{d}r'\,(r')^l\delta(r-a)\int_{-1}^{1}\mathrm{d}x'\,P_l(x')\left\{\delta(x'-1) - \delta(x'+1)\right\}$$

$$= q\,a^l\left\{P_l(1) - P_l(-1)\right\}\,.$$

From $P_l(1) = 1$ and $P_l(-1) = (-1)^l$ follows

$$A_l = \begin{cases} 0 & \text{for} \quad l \text{ even} \\ 2qa^l & \text{for} \quad l \text{ odd}\,. \end{cases}$$

The first terms of the multipole expansion of the potential of a point dipole are

$$V_a(r,\,\theta) = k_e \left\{ \frac{2qa}{r^2}\, P_1(\cos\theta) + \frac{2q\,a^3}{r^4}\, P_3(\cos\theta) + \ldots \right.$$

$$\left. + \frac{2q\,a^{2l+1}}{r^{2l+2}}\, P_{2l+1}(\cos\theta) + \ldots \right\} \qquad (r > a)\,.$$

The potential in the vicinity of the dipole is rather complicated, but the first terms of the expansion dominate for larger distances. The multipole expansion agrees with

$$V(r,\,\theta) = \frac{k_e q}{\left[r^2 - 2ar\cos\theta + a^2\right]^{1/2}} - \frac{k_e q}{\left[r^2 + 2ar\cos\theta + a^2\right]^{1/2}}$$

as the expansion of this function has the form

$$V_a = k_e q \left\{ \sum_l \frac{a^l}{r^{l+1}} P_l(\cos\theta) - \sum_l \frac{a^l}{r^{l+1}} P_l(-\cos\theta) \right\}$$

for $r > a$. The symmetry of the Legendre polynomials $P_l(-x) = (-)^l P_l(x)$ is the reason, why terms with odd values of l do not contribute.

The second example is a distribution with three charges along the z-axis, which add up to the value zero

$$\sum_i q_i = 0\,.$$

This arrangement (Fig. 3.6) is characterised by

$$\rho(r,\,\theta) = \frac{q}{2\pi r^2}\Big[\delta(r-a)\,\delta(\cos\theta - 1) - \delta(r) + \delta(r-a)\,\delta(\cos\theta + 1)\Big]\,.$$

The total charge has indeed the value zero

$$\int \mathrm{d}V'\,\rho(r',\,\theta') = 0\,,$$

Fig. 3.6 Charge distribution: Point quadrupole

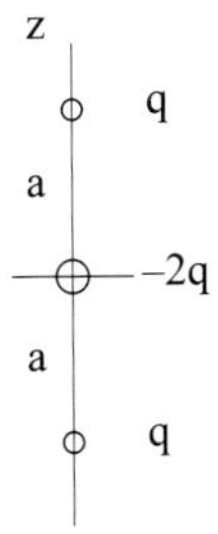

as

$$\iiint \frac{1}{r'^2}\,\delta(r'-a)\,\delta(\cos\theta' \pm 1)\,\mathrm{d}V' = 2\pi$$

$$\iiint \frac{1}{r'^2}\,\delta(r')\,\mathrm{d}V' = 4\pi\,.$$

For the expansion coefficients A_l one obtains the following results

$$A_l = q\int_0^\infty \mathrm{d}r' \int_{-1}^{1} \mathrm{d}x' (r')^l$$

$$\cdot\big(\delta(r'-a)\,\delta(x'-1) - \delta(r') + \delta(r'-a)\,\delta(x'+1)\big)\,P_l(x')$$

$$= q\left(a^l P_l(1) - 2\delta_{l,0} + a^l P_l(-1)\right),$$

or in detail

$$A_0 = 0$$

$$A_l = 0 \qquad \text{for } l \text{ odd}$$

$$A_l = 2qa^l \text{ for } l \text{ even}, l \geq 2\,.$$

There exist only multipoles with even l, beginning with $l = 2$. This is the reason, why this charge distribution is referred to as a stretched quadrupole. Usually, the lowest contribution, which is different from zero, determines the name. The multipole expansion of the potential of this charge distribution is

$$V(r,\theta) = k_e\left\{\frac{2qa^2}{r^3}\,P_2(\cos\theta) + \frac{2qa^4}{r^5}\,P_4(\cos\theta) + \ldots\right\}.$$

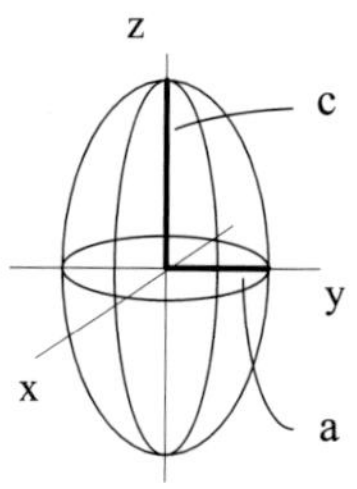

Fig. 3.7 Charge distribution: Ellipsoid of rotation

An example for an extended charge distribution is a uniformly charged ellipsoid of rotation (Fig. 3.7).

$$\rho(r,\theta) = \begin{cases} \rho_0 & \text{for } r \le R(\theta) \\ 0 & \text{for } r > R(\theta)\,. \end{cases}$$

The surface is given by the relation (Chap. 3.1)

$$\frac{R^2\sin^2\theta}{a^2} + \frac{R^2\cos^2\theta}{c^2} = 1\,.$$

It can be resolved with respect to $R(\theta)$, so that one can distinguish the shapes

$a \ge c$	oblate ellipsoid (form of a pan cake)
$c \ge a$	prolate ellipsoid (form of a cigar).

The equation of the boundary can be solved as

$$R(\cos\theta) = \frac{a}{\left[1-\kappa^2\cos^2\theta\right]^{1/2}} \qquad \text{with} \qquad 0 \le \kappa^2 = \frac{c^2-a^2}{c^2} \le 1$$

for a prolate ellipsoid. The calculation of the multipole moments calls for the evaluation of the integral

$$A_l = 2\pi\,\rho_0 \int_{-1}^{1} \mathrm{d}x'\,P_l(x') \int_0^{R(x')} (r')^{l+2}\mathrm{d}r'$$

with the result

$$A_l = \frac{2\pi\,\rho_0\,a^{l+3}}{(l+3)} \int_{-1}^{1} \mathrm{d}x'\,P_l(x')\left[1-\kappa^2 x'^2\right]^{-(l+3)/2}\,.$$

The integrand is an odd function of x for odd l-values, so that the integral has the value zero

$$A_l = 0 \qquad \text{for } l \text{ odd}.$$

For even l-values ($l = 2n$) a short side calculation is required, which is found in Detail 3.5.2. The individual steps are

(i) Expand the root in the integrand.
(ii) Calculate a standard integral with Legendre polynomials.
(iii) Summarise this expansion.

The result is

$$A_l = \frac{4\pi\, \rho_0\, a^{l+3}}{(l+1)(l+3)} \frac{\kappa^l}{\left[1-\kappa^2\right]^{(l+1)/2}} \qquad (l = \text{even}).$$

The parameters of the ellipse can be introduced again using

$$1-\kappa^2 = 1 - \frac{c^2-a^2}{c^2} = \frac{a^2}{c^2}$$

$$\kappa^l = (\kappa^2)^{l/2} = \left(\frac{c^2-a^2}{c^2}\right)^{l/2},$$

so that one obtains

$$A_l = \frac{4\pi\, \rho_0\, a^{l+3}}{(l+1)(l+3)} \frac{\left(c^2-a^2\right)^{l/2}}{c^l} \frac{c^{l+1}}{a^{l+1}}$$

$$= \frac{4\pi\, \rho_0 a^2 c}{(l+1)(l+3)} \left[c^2-a^2\right]^{l/2}.$$

As the total charge of the ellipsoid is

$$Q = \rho_0 V = \frac{4}{3}\pi\, \rho_0 a^2 c,$$

one finally arrives at the result

$$A_l = \frac{3Q}{(l+1)(l+3)} \left[c^2-a^2\right]^{l/2} \qquad (l = \text{even}).$$

The first three coefficients are e.g.

$$A_0 = Q\,, \quad A_2 = \frac{1}{5}Q(c^2 - a^2)\,, \quad A_4 = \frac{3}{35}Q(c^2 - a^2)^2\,.$$

The multipole expansion of the potential starts with

$$V(\boldsymbol{r}) = \frac{k_e Q}{r}\left\{1 + \frac{1}{5}\frac{(c^2 - a^2)}{r^2}P_2(\cos\theta) + \ldots\right\}\,.$$

On the basis of this result one may calculate the multipole moments with any accuracy desired. Re summation into a simple closed expression is, however, not possible. An alternative evaluation with elliptical coordinates does not produce a simple result either.

The multipole expansion for the general case with $\rho(\boldsymbol{r}) = \rho(r,\,\theta,\,\varphi)$ will only be considered briefly. The formulae for the potential (3.33) and the coefficients (3.36) are

$$V_a(\boldsymbol{r}) = k_e \sum_{lm}\left[\frac{4\pi}{2l+1}\right]^{1/2}\frac{A_{lm}}{r^{l+1}}\,Y_{l,m}(\theta,\,\varphi)$$

$$A_{lm} = \left[\frac{4\pi}{2l+1}\right]^{1/2}\iiint \mathrm{d}V'\rho(\boldsymbol{r}')(r')^l\,Y^*_{l,m}(\theta',\,\varphi')\,.$$

The monopole moment ($l = 0$) is, as expected, the total charge

$$A_{0,0} = \iiint \mathrm{d}V'\rho(\boldsymbol{r}') = Q\,.$$

For the higher l-contributions one finds, that for each value of l $(2l + 1)$ moments exist, e.g. for the dipole ($l = 1$)

$$A_{1,1} = -\sqrt{\frac{1}{2}}\iiint \rho(\boldsymbol{r}')r'\sin\theta'\,\mathrm{e}^{-\mathrm{i}\varphi'}\mathrm{d}V'$$

$$A_{1,0} = \iiint \rho(\boldsymbol{r}')r'\cos\theta'\mathrm{d}V'$$

$$A_{1,-1} = +\sqrt{\frac{1}{2}}\iiint \rho(\boldsymbol{r}')r'\sin\theta'\,\mathrm{e}^{\mathrm{i}\varphi'}\mathrm{d}V'\,.$$

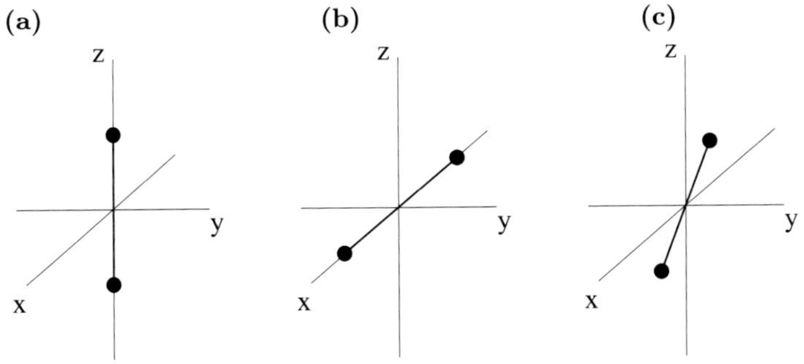

Fig. 3.8 Point dipoles with different orientations

These three quantities represent the vectorial components of the dipole moment

$$\boldsymbol{D} = \iiint \rho(\boldsymbol{r}')\boldsymbol{r}'\mathrm{d}V' .$$

This statement can be justified with a look at the Cartesian components:

$$D_x = \iiint \mathrm{d}V'\, x'\, \rho(\boldsymbol{r}') = \langle x \rangle$$

for the x-component and similar expressions for the y- and z-components. These quantities D_i with ($i = x, y, z$ or $i = 1, 2, 3$) can be interpreted in a simple fashion. For example, for a dipole oriented along the z-axis (Fig. 3.8a) one has use the charge density

$$\rho(\boldsymbol{r}') = q\ \delta(x')\,\delta(y')\,\big(\delta(z' - a) - \delta(z' + a)\big)$$

and obtains

$$D_x = D_y = 0 \qquad D_z \neq 0 .$$

For a point dipole along the x-axis (Fig. 3.8b) with

$$\rho(\boldsymbol{r}') = q\ \delta(y')\,\delta(z')\,\big(\delta(x' - a) - \delta(x' + a)\big)$$

the result is

$$D_x \neq 0 \qquad D_y = D_z = 0 .$$

Thus one may conclude: The quantities D_i describe the projections of the vector components of a dipole moment of an arbitrary charge distribution on the coordinate axis i (Fig. 3.8c)

$$D_i = D\cos\alpha_i \qquad i = x,\ y,\ z\,,$$

where $\cos\alpha_i$ is the directional cosine with respect to the direction i.

An interpretation of the quantities A_{1m} can be obtained, if the Cartesian components are expressed in terms of spherical coordinates (e.g. for the combination $(x - \mathrm{i}y) = r(\sin\theta\,\mathrm{e}^{-\mathrm{i}\varphi})$ as linear combinations of the three components D_i

$$A_{1,1} = -\frac{1}{\sqrt{2}}\left(D_x - \mathrm{i}D_y\right)$$

$$A_{1,0} = D_z$$

$$A_{1,-1} = \frac{1}{\sqrt{2}}\left(D_x + \mathrm{i}D_y\right)\,.$$

The quantities A_{1m} are the **spherical components** of the vector $\boldsymbol{D}$. They characterise in a different manner the magnitude

$$D = \left[\sum_i D_i^2\right]^{1/2} = \left[\sum_m A{*}_{1m}A_{1m}\right]^{1/2}$$

and the orientation of the dipole vectors in space, e.g.

$$\cos\alpha_x = \frac{A_{1,-1} - A_{1,1}}{\sqrt{2}\,D}\,.$$

The five components of a quadrupole ($l = 2$) A_{2m}, as

$$A_{2,-2} = \sqrt{\frac{3}{8}}\iiint \rho(\boldsymbol{r}')r'^2\sin^2\theta'\,\mathrm{e}^{2\mathrm{i}\varphi'}\,\mathrm{d}V'$$

can be interpreted in a similar manner. One begins by looking at the possible mean values of the Cartesian components in the form

$$Q_{ik} = \iiint \mathrm{d}V'\rho(\boldsymbol{r}')\left(3x_i'x_k' - r'^2\delta_{i,k}\right)\,,$$

and finds

(i) The situation is reminiscent of the matrix of the moments of inertia of mechanics. Therefore one can say, that the elements of the quadrupole matrix Q_{ik} are the elements of a tensor of second rank.
(ii) The matrix is symmetric

$$Q_{ik} = Q_{ki} \, .$$

(iii) The trace of the matrix has the value zero

$$\mathrm{trace}(Q) = \sum_i Q_{ii} = 0 \, .$$

The statements (ii) and (iii) show, that there exist five independent elements of the quadrupole tensors. The consequence is

(iv) The five spherical components can be related to five Cartesian elements as e.g.

$$\begin{aligned}
A_{2,-2} &= \sqrt{\frac{3}{8}} \iiint \rho(\boldsymbol{r}') r'^2 \sin^2\theta' \left(\cos 2\varphi' + \mathrm{i} \sin 2\varphi'\right) \\
&= \sqrt{\frac{3}{8}} \iiint \rho(\boldsymbol{r}') r'^2 \sin^2\theta' \left(\cos^2\varphi' - \sin^2\varphi' + 2\mathrm{i} \cos\varphi' \sin\varphi'\right) \\
&= \sqrt{\frac{3}{8}} \left(\langle x^2\rangle - \langle y^2\rangle + 2\mathrm{i}\langle xy\rangle\right) \\
&= \frac{1}{\sqrt{24}} \left(Q_{11} - Q_{22} + 2\mathrm{i}\, Q_{12}\right)
\end{aligned}$$

and in the same manner

$$\begin{aligned}
A_{2,-1} &= \frac{1}{\sqrt{6}} \left(Q_{13} + \mathrm{i} Q_{23}\right) \\
A_{2,0} &= \frac{1}{2} Q_{33} \\
A_{2,1} &= -\frac{1}{\sqrt{6}} \left(Q_{13} - \mathrm{i} Q_{23}\right) \\
A_{2,2} &= -\frac{1}{\sqrt{24}} \left(Q_{11} - Q_{22} - \mathrm{i} Q_{12}\right) \, .
\end{aligned}$$

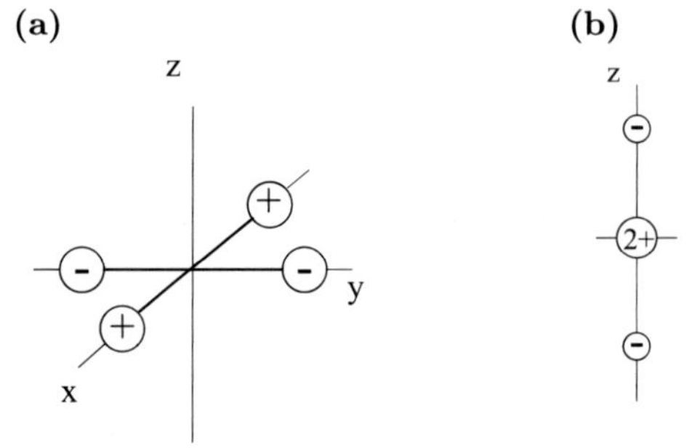

$A_{2,2} = A_{2,-2} \neq 0 \quad A_{2,0} = 0$ $\qquad$ $A_{2,2},\ A_{2,-2},\ A_{2,0} \neq 0$

Fig. 3.9 Basic point quadrupoles

(v) The five Cartesian components of the quadruple matrix can (in analogy to the matrix of inertia) interpreted as:

- Three of the components describe the orientation of an elementary quadrupole in space.
- Two of the components characterise two basic forms with $\sum_i q_i = 0$ (Fig. 3.9).

The discussion of the multipoles with $l > 2$ is more involved but follows the same pattern. E.g. an octupole ($l = 3$) consists of 8 point charges in the basic arrangement and there are seven independent quantities A_{3m}, which are associated with the seven linearly independent orientations of the 8 point charges. The Cartesian components Q_{ikl} are elements of a tensor of rank 3 with corresponding symmetry properties.

The next chapter addresses the more complicated potential problem with metallic and/or polarisable materials distributed in space, where the separation of the charges in the materials induced by external charges plays a role.

3.5 Details

3.5.1 The Potential of a Uniformly Charged Sphere in a Uniformly Charged Sphere

The potential of an assembly of spherical charge distributions can be obtained with the superposition principle. The potential problem of a uniformly charged sphere with the charge density ρ_2 in a larger uniformly charged sphere with the charge density ρ_1 is discussed here in more detail.

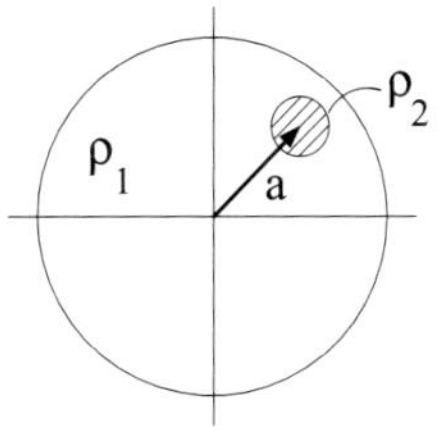

Fig. 3.10 Geometry of the charged spheres

The origin of the coordinate system is chosen as the centre of the larger sphere with radius R_1 (see Chap. 3.1). The potential of this sphere is then

$$V_1^{(i)}(\boldsymbol{r}) = 2\pi k_e\, \rho_1 \left(R_1^2 - \frac{1}{3} r^2 \right) \qquad r \le R_1$$

$$V_1^{(a)}(\boldsymbol{r}) = \frac{4}{3}\pi k_e \rho_1\, R_1^3\, \frac{1}{r} \qquad r > R_1 \,.$$

For the potential, which is produced by the smaller sphere (Fig. 3.10) with the radius R_2 and the additional charge distribution $\Delta\rho = \rho_2 - \rho_1$ around the position $\boldsymbol{a}$, one finds

$$V_2^{(i)}(\boldsymbol{r}) = 2\pi k_e\, \Delta\rho \left(R_2^2 - \frac{1}{3} |\boldsymbol{r} - \boldsymbol{a}|^2 \right) \qquad |\boldsymbol{r} - \boldsymbol{a}| \le R_2$$

$$V_2^{(a)}(\boldsymbol{r}) = \frac{4}{3}\pi k_e \Delta\rho\, R_2^3\, \frac{1}{|\boldsymbol{r} - \boldsymbol{a}|} \qquad |\boldsymbol{r} - \boldsymbol{a}| > R_2 \,.$$

The potential of the total charge distributions in the different domains is therefore:

within the small sphere:

$$|\boldsymbol{r} - \boldsymbol{a}| \le R_2 \qquad V_A(\boldsymbol{r}) = V_1^{(i)} + V_2^{(i)} \,,$$

within the big, but on the outside of the small sphere:

$$r \le R_1,\ |\boldsymbol{r} - \boldsymbol{a}| \le R_2 \qquad V_B(\boldsymbol{r}) = V_1^{(i)} + V_2^{(a)} \,,$$

outside of the bigger sphere:

$$r \ge R_1 \qquad V_C(\boldsymbol{r}) = V_1^{(a)} + V_2^{(a)} \,.$$

If the centre of the second sphere is placed on the z-axis, one has $\boldsymbol{a} = (0, 0, a)$. The potential observed for points on a straight line through the centres of the two spheres $\boldsymbol{r} = (0, 0, z)$ with $-\infty \le z \le \infty$ are given by

$$V_A(z) = 2\pi k_e \left\{\rho_1 \left(R_1^2 - \frac{1}{3}z^2\right) + \Delta\rho \left(R_2^2 - \frac{1}{3}(z-a)^2\right)\right\}$$

$$V_B(z) = 2\pi k_e \left\{\rho_1 \left(R_1^2 - \frac{1}{3}z^2\right) + \frac{2}{3}\Delta\rho \frac{R_2^3}{|z-a|}\right\}$$

$$V_C(z) = \frac{4}{3}\pi k_e \left\{\rho_1 \frac{R_1^3}{z} + \Delta\rho \frac{R_2^3}{|z-a|}\right\}.$$

Other choices of the relative position of the two spheres are possible, e.g. a situation with the centre of the small sphere at

$$\boldsymbol{a} = (a\cos\varphi\sin\theta,\ a\sin\varphi\sin\theta,\ a\cos\theta)\,.$$

In this case a straight line through the centres of the two spheres is given by

$$\boldsymbol{r} = (r\cos\varphi\sin\theta,\ r\sin\varphi\sin\theta,\ r\cos\theta)\,.$$

The difference between the centres is now $|\boldsymbol{r} - \boldsymbol{a}| = |r - a|$, where the coordinate $\boldsymbol{r}$ has been used instead of z.

The value of the total potential along the straight line through the centres is plotted in the Figs. 3.11–3.13 (in CGS units).

In Fig. 3.11 the position of the small sphere is varied, in Figs. 3.12 and 3.13 the charge density of the small sphere for different values of the radius.

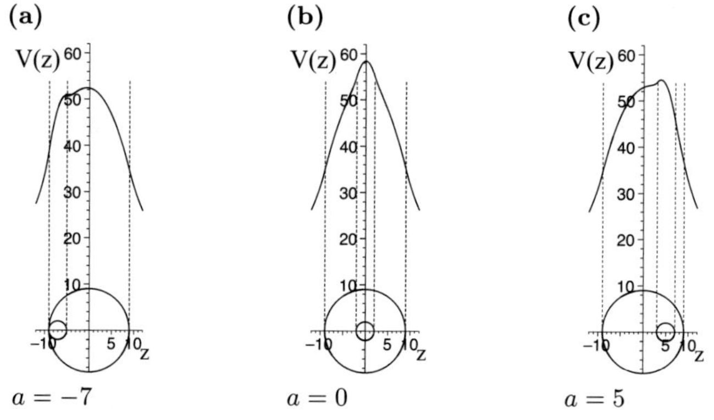

Fig. 3.11 Fixed density, variable position $R_1 = 9$, $R_2 = 2$, $\rho_1 = 0.1$, $\rho_2 = 0.4$

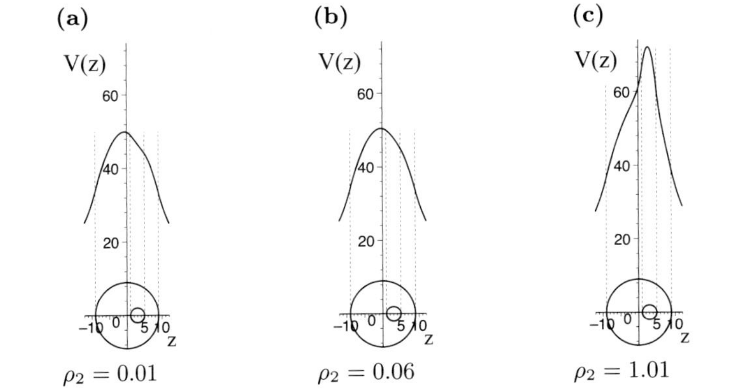

Fig. 3.12 Variation of the density ρ_2 for fixed position $R_1 = 9$, $R_2 = 2$, $\rho_1 = 0.1$, $a_x = 3$

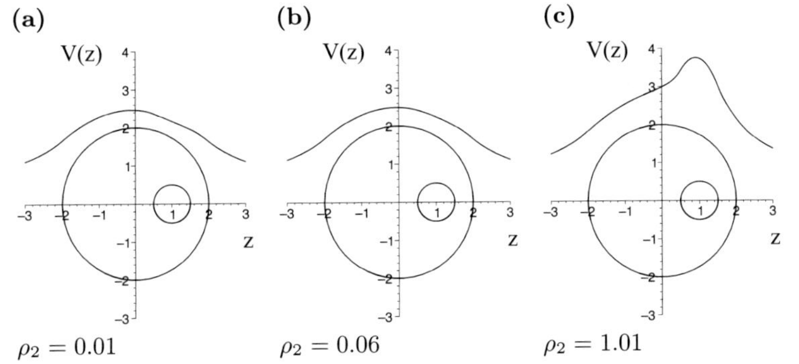

Fig. 3.13 Variation of the density ρ_2 for fixed position $R_1 = 2$, $R_2 = 0.5$, $\rho_1 = 0.1$, $a_x = 1$

3.5.2 The Electric Potential of a Rotational Ellipsoid

The calculation of the electric potential for a charge distribution with a more complicated geometry is in general not possible in analytic form. Options are then the multipole expansion (often only in the region outside the charge distribution) or a numerical solution of the Poisson/Laplace equation (with optimal use of the remaining symmetry).

The multipole expansion of a charge distribution with azimuthal symmetry in the exterior region has the form (Chap. 3.2, (3.18))

$$V_A(r,\theta) = k_e \sum_{l=0}^{\infty} \frac{A_l}{r^{(l+1)}} P_l(\cos\theta)\,. \tag{3.40}$$

The expansion coefficients are given by

$$A_l = \iiint \mathrm{d}V' (r')^l \, \rho(r', \theta') \, P_l(\cos\theta') \,,$$

in the case of a uniformly charged rotational ellipsoid by

$$A_l = \frac{2\pi \, \rho_0 \, a^{l+3}}{(l+3)} I_l \,.$$

The following integrals over the angle θ are needed (use $\cos\theta' = x'$)

$$I_l = \int_{-1}^{1} \mathrm{d}x' \; P_l(x') \left[1 - \kappa^2 x'^2\right]^{-(l+3)/2} \,. \tag{3.41}$$

The parameter κ is determined by the semi-axes of the (here prolate) ellipsoid

$$0 \le \kappa^2 = \frac{c^2 - a^2}{c^2} \le 1 \,. \tag{3.42}$$

Some formulae and preparative measures are required for the ensuing calculation

- $$(1+x)^\alpha = \sum_{m=0}^{\infty} \binom{\alpha}{m} x^m \,. \tag{3.43}$$

  Here α can be any positive or negative number.
- The binomial symbol is initially only defined for integer values of $\alpha \to n$

  $$\binom{n}{k} = \frac{n!}{k!(n-k)!} = \frac{\Gamma(n+1)}{\Gamma(k+1)\,\Gamma(n-k+1)} \,,$$

  where the factorial can be represented by the Γ-function (Gamma function). With the Γ-function one is able to extend the binomial symbol to arbitrary values of $n \to \alpha$

  $$\binom{\alpha}{k} = \frac{\Gamma(\alpha+1)}{\Gamma(k+1)\,\Gamma(\alpha-k+1)} \,.$$

- One also needs the relation

  $$\binom{-\alpha}{k} = (-1)^k \binom{k+\alpha-1}{k} = (-1)^k \frac{\Gamma(k+\alpha)}{\Gamma(k+1)\,\Gamma(\alpha)} \,, \tag{3.44}$$

  which follows from the properties of the Γ-function.

- Further properties of the Γ-function, that are used, are

$$\Gamma(z+1) = z\,\Gamma(z) \tag{3.45}$$

and

$$\Gamma(2z) = (2\pi)^{-1/2}\, 2^{2z-1/2}\, \Gamma(z)\, \Gamma(z+1/2)\,. \tag{3.46}$$

- A basic integral with Legendre polynomials is

$$\int_0^1 P_\nu x^\rho \mathrm{d}x = \frac{2^{-\rho-1}\sqrt{\pi}\,\Gamma(1+\rho)}{\Gamma(1+\rho/2-\nu/2)\,\Gamma(\rho/2+\nu/2+3/2)}\,. \tag{3.47}$$

It can be found in suitable literature on special functions

The evaluation of the integral (3.41) is different for even and odd values of the parameter l. If l is **odd** ($l = 2n+1$), one has to note, that the integrand in the integral (3.41)

$$I_{2n+1} = \int_{-1}^{1} \mathrm{d}x' \frac{P_{2n+1}(x')}{\left[1-\kappa^2 x'^2\right]^{n+2}}$$

is a product of an even function of the variable x'

$$f(x') = \left[1-\kappa^2 x'^2\right]^{-(n+2)}$$

with an odd polynomial $P_{2n+1}(x')$. The integral over the interval $[-1, 1]$ has the value zero

$$I_{2n+1} = 0\,.$$

If l is **even** ($l = 2n$), one obtains with the expansion of the square root of the integrand

$$I_{2n} = \int_{-1}^{1} \mathrm{d}x' \frac{P_{2n}(x')}{\left[1-\kappa^2 x'^2\right]^{n+3/2}}$$

with (3.43) on exchanging the sequence of the operations (summation and integration)

$$I_{2n} = \sum_{m=0}^{\infty} \binom{-n-\frac{3}{2}}{m} (-1)^m \kappa^{2m} \int_{-1}^{1} (x')^{2m} P_{2n}(x')\mathrm{d}x'\,. \tag{3.48}$$

The first task is to show that the part integrals in (3.48) are only different from zero, if $m \geq n$ is valid. For a proof of this statement one can use the fact, that every power (or polynomial) can be represented by a series of Legendre polynomials

$$x^{2m} = \sum_{j=0}^{m} c_j P_{2j}(x) .$$

The coefficients c_j could be determined, though this is not of interest at this stage. If this relation is inserted in (3.48), then one finds on the basis of the orthogonality of the Legendre polynomials a reduction of the sum

$$I_{2n} = \sum_{m=0}^{\infty} \binom{-n-\frac{3}{2}}{m} (-1)^m \kappa^{2m} \int_{-1}^{1} \sum_{j=0}^{m} c_j P_{2j}(x') P_{2n}(x') \mathrm{d}x'$$

$$= \sum_{m=0}^{\infty} \binom{-n-\frac{3}{2}}{m} (-1)^m \kappa^{2m} \sum_{j=0}^{m} c_j \frac{2}{2j+1} \delta_{2j,2n} .$$

If l is even and $m < n$, one finds, that the integral I_{2n} is equal to zero as a result of the requirement $j = n$. One can restrict the calculation to values of m with $m \geq n$.

The interval of integration can be reduced to [0, 1], as P_{2n} and x^{2m} are both even functions

$$\int_{-1}^{1} (x')^{2m} P_{2n}(x') \mathrm{d}x' = 2 \int_{0}^{1} (x')^{2m} P_{2n}(x') \mathrm{d}x' .$$

Using (3.47) one obtains for this integral the result

$$2 \int_{0}^{1} (x')^{2m} P_{2n}(x') \mathrm{d}x' = 2 \, \frac{2^{-2m-1} \sqrt{\pi}\, \Gamma(1+2m)}{\Gamma(1+m-n)\Gamma(m+n+3/2)}$$

respectively with (3.46)

$$= \frac{\Gamma(m+1/2)\, \Gamma(m+1)}{\Gamma(1+m-n)\Gamma(m+n+3/2)} \qquad (m \geq n) .$$

Insertion in (3.48) with (3.44) leads to

$$I_{2n} = \sum_{m=n}^{\infty} \binom{-n-\frac{3}{2}}{m} (-1)^m \kappa^{2m} \frac{\Gamma(m+1/2)\,\Gamma(m+1)}{\Gamma(1+m-n)\Gamma(m+n+3/2)}$$

$$= \sum_{m=n}^{\infty} (-1)^{2m} \kappa^{2m} \frac{\Gamma(n+3/2+m)\Gamma(m+1/2)}{\Gamma(m+1)\,\Gamma(n+3/2)}$$

$$\cdot \frac{\Gamma(m+1)}{\Gamma(1+m-n)\,\Gamma(m+n+3/2)}$$

$$= \sum_{m=n}^{\infty} (-1)^{2m} \kappa^{2m} \frac{\Gamma(m+1/2)}{\Gamma(m-n+1)\,\Gamma(n+3/2)} .$$

The substitution $m = n + M$ allows a re-summation with (3.44) and (3.43)

$$I_{2n} = \sum_{M=0}^{\infty} (-1)^{2n+2M} \kappa^{2M+2n} \frac{\Gamma(n+M+1/2)}{\Gamma(M+1)\,\Gamma(n+1/2)}$$

$$= \frac{\kappa^{2n}}{(n+1/2)} (-1)^{2n} \sum_{M=0}^{\infty} (-1)^M \binom{-n-1/2}{M} \kappa^{2M}$$

$$= \frac{\kappa^{2n}}{(n+1/2)} (-1)^{2n} (1-\kappa^2)^{-n-1/2} .$$

The result for even values of l ($(-1)^l = 1$) is therefore

$$I_{2n} = \frac{2\kappa^l}{(l+1)(1-\kappa^2)^{(l+1)/2}} .$$

The convergence of the resulting expansion of the potential (3.40) in the region outside the charge distribution (replace κ^2 by the parameters of the ellipsoid (3.42) and use the total charge $Q = 4\pi\rho_0 a^2 c/3$)

$$V_A(\boldsymbol{r}) = \frac{k_e Q}{r} \sum_{n=0}^{\infty} \frac{3}{(2n+1)(2n+3)} \frac{\left[c^2 - a^2\right]^n}{r^{2n}} P_{2n}(\cos\theta)$$

$$= \frac{k_e Q}{r} \sum_{n=0}^{\infty} \frac{3}{(2n+1)(2n+3)} \kappa^{2n} \left(\frac{c}{r}\right)^{2n} P_{2n}(\cos\theta)$$

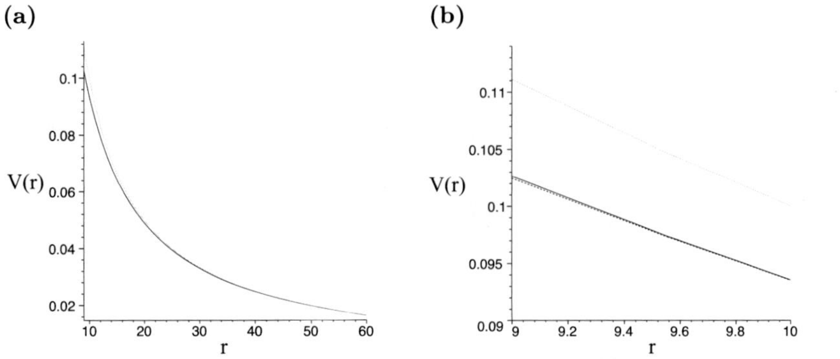

Expansion up to order $V_A(\boldsymbol{r})^{(8)}$

Clipping: $V_A(\boldsymbol{r})^{(0)}$ (grey) , $V_A(\boldsymbol{r})^{(6)}$ (broken) and $V_A(\boldsymbol{r})^{(8)}$ (black)

Fig. 3.14 Potential $V_A(\boldsymbol{r})$ with $\kappa = 0.943, \theta = 0.5\pi$

with the first terms

$$V_A(\boldsymbol{r}) = \frac{k_e Q}{r}\left\{1 + \frac{1}{5}\frac{(c^2 - a^2)}{r^2}P_2(\cos\theta) + \frac{3}{35}\frac{(c^2 - a^2)^2}{r^4}P_4(\cos\theta)\ldots\right\}$$

is illustrated in Figs. 3.14 and 3.15 for different values of the eccentricity

$$\kappa = \sqrt{1 - \frac{a^2}{c^2}}$$

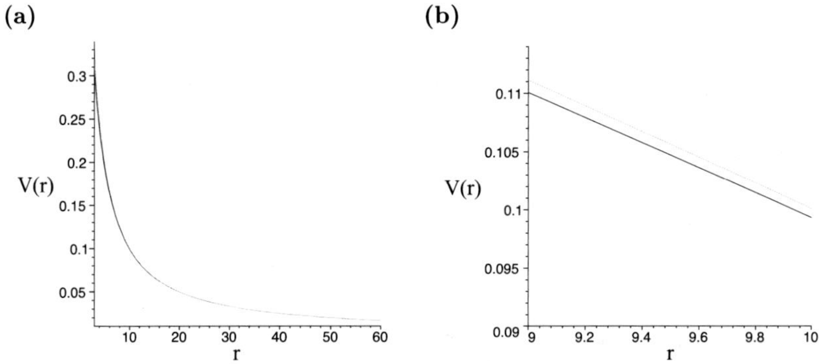

Expansion up to order $V_A(\boldsymbol{r})^{(8)}$

Clipping: $V_A(\boldsymbol{r})^{(0)}$ (grey) , $V_A(\boldsymbol{r})^{(6)}$ (broken) and $V_A(\boldsymbol{r})^{(8)}$ (black)

Fig. 3.15 Potential $V_A(\boldsymbol{r})$ with $\kappa = 0.866, \theta = 0.5\pi$

and $k_e Q = 1$ and different orders of the expansion

$$V_A(\boldsymbol{r})^{(j)} = \frac{k_e Q}{r} \sum_{n=0}^{j} \frac{3}{(2n+1)(2n+3)} \kappa^{2n} \left(\frac{c}{r}\right)^{2n} P_{2n}(\cos\theta) \, .$$

The limiting values correspond to a sphere ($c = a = R$) and to a linear charge distribution ($a = 0$) of the length $2c$ along the z-axis.

Solution of the Poisson Equation: General Boundary Conditions

4

The discussion of the solution of the Poisson equation is slightly more involved, if more general boundary conditions are considered. The boundary value problems named after Dirichlet as well as the one named after von Neumann will be introduced in this chapter. The two boundary value problems differ in the sense, that in the first case values for the potential on (metallic) surfaces are specified, in the second problem one specifies the values for the normal derivative of the potential (the electric field).

The method of mirror images is a method, which is used for the treatment of Dirichlet problems as the separation of charges, which occurs e.g. if a metallic object (e.g. a metallic sphere) is placed into a (uniform) electric field.

A practical method, which is used for the solution of potential problems is the method of Green's functions. These functions are characterised by a partial differential equation, in which the boundary conditions and therefore the geometry of the problem are included. This is achieved by a reformulation of the integral theorems of Green, which are an extension of the divergence theorem, in order to arrive at general formulae for the solution with a specific geometry. The solution of the differential equation for the Green's function of a Dirichlet problem with spherical symmetry will be discussed as a sample and used to treat several specific problems.

If charge distributions and charged surfaces are not placed in a vacuum but are embedded in a dielectric material, one must include the possibility of the polarisation of the material. This discussion will be initiated by a brief excursion concerning capacitors. It is convenient for the discussion of polarisation to introduce the concepts of the *dielectric displacement* and the *polarisation field*. The electric field, the dielectric displacement and the polarisation field are determined (in this order) by the complete set of charges, by the free charges (that is the freely movable charges) and by the polarisation charges. A dielectric sphere in a uniform external field will be used for the explicit illustration of the phenomenon of polarisation.

© The Author(s), under exclusive license to Springer-Verlag GmbH, DE, part of Springer Nature 2024

R. M. Dreizler, C. S. Lüdde, *Electrostatics and Magnetostatics*,
https://doi.org/10.1007/978-3-662-69933-1_4

The comparison of the behaviour of a metallic and a dielectric object in a uniform electric field demonstrates the similarities of but also the differences between the two situations.

An alternative method for the treatment of potential problems relies on the use of complex functions. It can be applied for problems with translational symmetry, so that projection onto the plane of complex numbers is possible and useful. The final pages of this chapter offer a glimpse of this interesting, but not globally applicable technique.

4.1 Remarks on the Classification of Boundary Conditions

In order to introduce the two types of boundary conditions, that are used in electrostatics, it is convenient to consider first a simple situation. Assume that there is just one metallic object with a surface charge in a region of space. One can then argue: As the charge is distributed over the surface, the Laplace equation is the equation, which is responsible for the situation in the interior and the exterior of the object

$$\boldsymbol{\Delta} V(\boldsymbol{r}) = 0\,.$$

As no charge is moved in a stationary situation, the potential in the *interior and on the surface of the object* is given by the trivial solution, a constant electric potential $V_{\text{int}}(\boldsymbol{r}) = V_0$. The next step is the calculation of the potential *in the exterior region* by solution of the Laplace equation for a given value of the potential on the surface S of the metallic object, that is

$$V(\boldsymbol{r})\Big|_{\text{S}} = V(\boldsymbol{R}_S) = V_0\,. \tag{4.1}$$

This is a **Dirichlet** boundary value problem with the boundary condition (4.1). The surface S can be open or closed. The electric field on the surface can be calculated by the normal derivative of the potential. With this field one can calculate the distribution of the surface charges in the final equilibrium situation using the discontinuity relation (see Fig. 4.1a and compare Chap. 2.2)

$$\frac{\partial V_{ext}(\boldsymbol{r})}{\partial n}\Big|_S - \frac{\partial V_{\text{int}}(\boldsymbol{r})}{\partial n}\Big|_S = -4\pi k_e\,\sigma(\boldsymbol{r})\Big|_S\,,$$

or as the potential in the interior of the metal object is constant and the potential in the exterior can just be called $V(\boldsymbol{r})$

$$\sigma(\boldsymbol{R}_S) = -\frac{1}{4\pi k_e}\frac{\partial V(\boldsymbol{r})}{\partial n}\Big|_S\,.$$

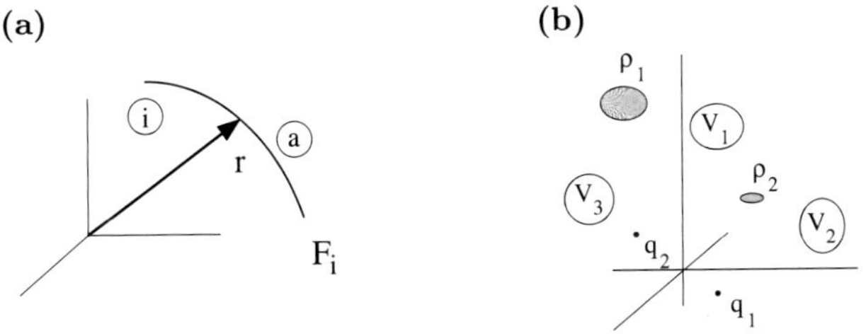

(a) The discontinuity condition

(b) Charge situation for the Dirichlet problem

Fig. 4.1 The boundary value problem

Knowledge of the potential on the outside of a metallic body allows the determination of the distribution of the charges on its surface.

The general Dirichlet problem (Fig. 4.1b), which is a direct extension of the simple variant, can be formulated as:

> Given are the values $V(\boldsymbol{R}_i)$ of the potential on the surfaces S_i of a set of metallic objects $i = 1, 2, \ldots, N$ and a distribution of space and point charges in the space outside of the metallic objects. The task is: Calculate the solution of the Poisson or the Laplace equations in the outside region of the metal objects, with the boundary conditions $V(\boldsymbol{R}_i) = V_{i,0}$.
> The values of the potentials are specified on the surfaces of charged metal objects. If some additional uncharged metal surfaces and point and space charges are introduced in the space, induction will be observed on all the metal surfaces. The task is the calculation of the potential of the final charge distribution and the distribution of the surface charges of the system in the final equilibrium situation. The values of the fields on each surface S_i can be found with the discontinuity condition as in the simpler case

$$\sigma(\boldsymbol{r})\Big|_{S_i} = -\frac{1}{4\pi k_e}\frac{\partial V(\boldsymbol{r})}{\partial n}\Big|_{S_i} .$$

The **von Neumann** boundary value problem addresses the following variant of the simple boundary value problem: A distribution of surface charges $\sigma(\boldsymbol{r})$ is specified on a (closed) surface S. Calculate the potential which is generated by the surface charges in the space not covered by the objects which produce the surface charges. The process of solution involves the following steps:

1. Determine the general solution of the Laplace equation (or the Poisson equation, if there are additional point or space charges) in the space exterior to the metal objects.

Fig. 4.2 The role of geometry in boundary value problems

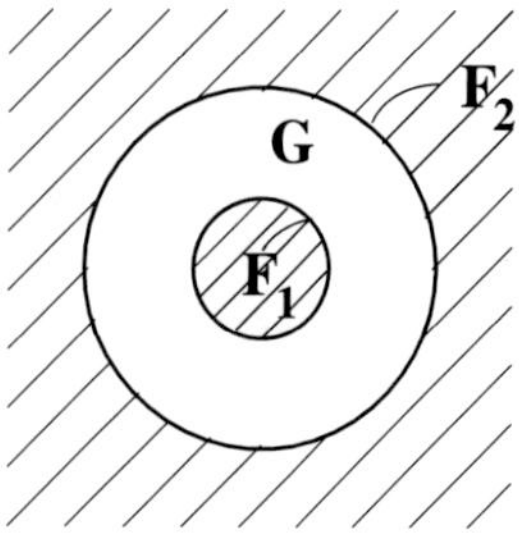

2. Use the boundary condition

$$\left.\frac{\partial V(\boldsymbol{r})}{\partial n}\right|_S = -4\pi k_e\,\sigma(\boldsymbol{r})\Big|_S \tag{4.2}$$

in order to select the solution required.

The two problems are distinguished by the following properties: In the Dirichlet-problem the potential V is specified on surfaces, in the von Neumann problem it is the electric field $\boldsymbol{E}$. The general von Neumann problem can be formulated with a similar extension as for the Dirichlet problem.

A third type of a boundary value problem, which is possible, is the Cauchy boundary value problem, in which both the potential V and the field $\partial V \partial n$ are specified on surfaces. Such a problem can not be discussed in electrostatics, as it would be over determined. Such specification are, however, possible in partial differential equations describing other phenomena as e.g. wave equations or diffusion equations.

One question, that turns up, if one is dealing with boundary value problems in electrostatics (or in different contexts) is the question, whether the solutions obtained are unique. In order to answer this question, one can use the following argument: Consider a region of space G, which is enclosed by two surfaces S_1 and S_2 as e.g. in Fig. 4.2. The outer surface can include infinitely distant points. Consider a solution of the Laplace equation $\Delta(V\boldsymbol{r}) = 0$ or the Poisson equation $\Delta(V\boldsymbol{r}) = -4\pi\rho(\boldsymbol{r})$ with corresponding boundary conditions on the surfaces S_i $(i = 1, 2)$

$$\text{(i)} \qquad V(\boldsymbol{R}_i) = f(\boldsymbol{R}_i) \qquad \text{(Dirichlet)} \tag{4.3}$$

$$\text{(ii)} \qquad \left.\frac{\partial V(\boldsymbol{r})}{\partial n}\right|_{S_i} = g(\boldsymbol{R}_i) \qquad \text{(von Neumann)}\,. \tag{4.4}$$

The vectors $\boldsymbol{R}_i$, $(i = 1, 2)$ represent the surfaces from the point of view of a chosen coordinate system. If one assumes, that two solutions $V_1(\boldsymbol{r})$ and $V_2(\boldsymbol{r})$ exist, which satisfy the corresponding differential equation and boundary condition, then the difference of the two solutions

$$\phi(\boldsymbol{r}) = V_1(\boldsymbol{r}) - V_2(\boldsymbol{r})$$

satisfies the Laplace equation

$$\Delta\phi(\boldsymbol{r}) = 0$$

and, depending on the type of problem under consideration, one of the boundary conditions

$$\text{(a)}\quad \phi(\boldsymbol{R}_i) = 0 \qquad \text{or} \qquad \text{(b)}\quad \left.\frac{\partial\phi(\boldsymbol{r})}{\partial n}\right|_{\boldsymbol{R}_i} = 0\,.$$

One can show, that the statement $\phi(\boldsymbol{r}) = const.$ is valid for the difference of two solutions. Solutions of the Laplace or Poisson equations with either Dirichlet or von Neumann boundary conditions can only differ by a constant. In the case of Dirichlet conditions one finds, because of

$$\phi(\boldsymbol{R}_1) = \phi(\boldsymbol{R}_2) = 0$$

even the result $\phi(\boldsymbol{r}) = 0$. The constant has the value zero. The solution $V(\boldsymbol{r})$ is unique. In the case of von Neumann conditions one finds

$$\left.\frac{\partial\phi(\boldsymbol{r})}{\partial n}\right|_{\boldsymbol{R}_1,\boldsymbol{R}_2} = 0\,.$$

The constant is not determined by the boundary condition.

In order to prove these statements concerning the solution, one has to assume that the domains S_i are simply connected and closed.[1]

A method, which is often used to treat Dirichlet problems is the method of mirror charges. This method will be discussed in the next section for some selected problems.

4.2 Solution of Dirichlet Problems

A standard example for the application of the method of **mirror charges** is the following: A metal sphere with radius R is connected to ground (this means, that the potential of the surface S has the value zero) (Fig. 4.3a). The boundary condition

[1] A domain is called simply connected, if a closed curve in the domain without double points can be contracted to a point.

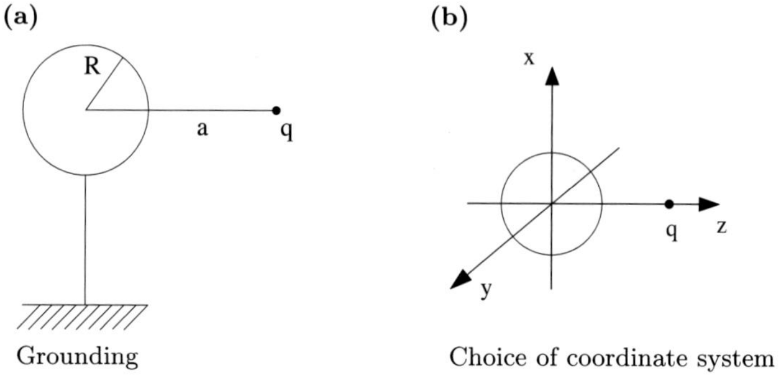

Fig. 4.3 Metal sphere and point charge

on the surface is therefore

$$V_S(\boldsymbol{r}) = 0\,.$$

In a distance a from the centre of the sphere one places a point charge q. For $a > R$ the charge is outside of the sphere. It produces an electric field in each point of surface. Charges are flowing along the line from the ground to the surface (or from the surface to the ground) until the potential on the surface has reached the value zero again. In the final equilibrium situation, one finds a distribution of induced charges on the surface, which are to be calculated.

The solution of the corresponding Dirichlet problem requires the following steps

1. Calculate the potential in the exterior (the region with $R < a < \infty$) of the sphere for the boundary value $V_S = 0$ on the surface of the metal sphere.
2. Determine the distribution of the induced surface charges.

A useful choice of a coordinate system for the task at hand is the following: Place the origin in the centre of the sphere and the z-axis through the point charge (Fig. 4.3b). The coordinates of the points $\boldsymbol{r}$ and $\boldsymbol{a}$ are then

$$\boldsymbol{r} = (r\sin(\theta)\cos(\phi),\ r\sin(\theta)\sin(\phi)\ r\cos(\theta))$$

and

$$\boldsymbol{a} = (0,\ 0,\ a)$$

with $r \geq R$ and $a \geq R$. For this choice with azimuthal symmetry one first attempts to determine the general solution of the Poisson equation

$$\Delta V(\boldsymbol{r}) = -4\pi k_e\, q\, \delta(\boldsymbol{r} - \boldsymbol{a}) \tag{4.5}$$

in the exterior region. A general (and physically acceptable) solution can be found with the superposition principle in the form of a power series for the contribution of the surface charges plus the contribution of the point charge to the potential

$$V(\boldsymbol{r}) = k_e \sum_{l=0}^{\infty} \frac{A_l}{r^{l+1}} P_l(\cos\theta) + \frac{k_e q}{|\boldsymbol{r} - \boldsymbol{a}|} \qquad r \geq R\,. \tag{4.6}$$

If the Laplace operator is applied to this ansatz, one obtains a formal Poisson equation $\Delta V(\boldsymbol{r})$. The first term on the right hand side represents an arbitrary modification of the potential generated by the surface charges of the sphere in agreement with the azimuthal symmetry. The ansatz is also in accord with the asymptotic boundary condition

$$V(\boldsymbol{r}) \xrightarrow{r \to \infty} 0\,.$$

The determination of the integration constants A_l with the aid of the boundary condition

$$V(\boldsymbol{R}) = 0$$

requires a multipole expansion of the contribution of the point charge to the potential at the surface of the sphere

$$\frac{1}{|\boldsymbol{R} - \boldsymbol{a}|} = \frac{1}{a} \frac{1}{\left[1 - 2\frac{R}{a}\cos(\theta) + \frac{R^2}{a^2}\right]^{1/2}}\,.$$

This expansion with $R/a < 1$ is

$$\frac{1}{|\boldsymbol{R} - \boldsymbol{a}|} = \frac{1}{a} \sum_l \left(\frac{R}{a}\right)^l P_l(x)\,.$$

The boundary condition can now be included by considering

$$k_e \sum_l \left\{ \frac{A_l}{R^{l+1}} + \frac{q}{a}\left(\frac{R}{a}\right)^l \right\} P_l(x) = 0\,.$$

This Legendre series of a function $f(x) = 0$ is satisfied, if the coefficient of every Legendre polynomial has the value zero. For a proof, one can multiply the series with a Legendre polynomial $P_{l'}(x)$, where l' is arbitrary, and integrate over the interval $[-1 < x < 1]$. Due to the orthogonality of the Legendre polynomials one finds

$$\left\{ \frac{A_{l'}}{R^{l'+1}} + \frac{q}{a} \left(\frac{R}{a} \right)^{l'} \right\} = 0 \,,$$

so that the result for the coefficients is

$$A_l = -q \, \frac{R^{2l+1}}{a^{l+1}} \,. \tag{4.7}$$

For the potential in the region outside of the sphere one obtains

$$V(\boldsymbol{r}) = V(r, \theta) = k_e q \left\{ \frac{1}{|\boldsymbol{r} - \boldsymbol{a}|} - \frac{R}{ar} \sum_l \left(\frac{R^2}{ar} \right)^l P_l(x) \right\} .$$

In order to use this result one has to write the second term in a more compact form

$$\frac{1}{r} \sum_l \left(\frac{R^2}{ar} \right)^l P_l(x) = \left[r^2 - 2 \left(\frac{R^2}{a} \right) rx + \left(\frac{R^2}{a} \right)^2 \right]^{-1/2}$$

and define a vector

$$\boldsymbol{a}' = (0,\ 0,\ R^2/a) \,, \tag{4.8}$$

with a length, which is smaller than R (because of $R < a$) and points in the direction of the z-axis. The second term of the intermediate result can be written as (Fig. 4.4a)

$$\frac{1}{r} \sum_l \left(\frac{R^2}{ar} \right)^l P_l(x) = \frac{1}{|\boldsymbol{r} - \boldsymbol{a}'|} \,.$$

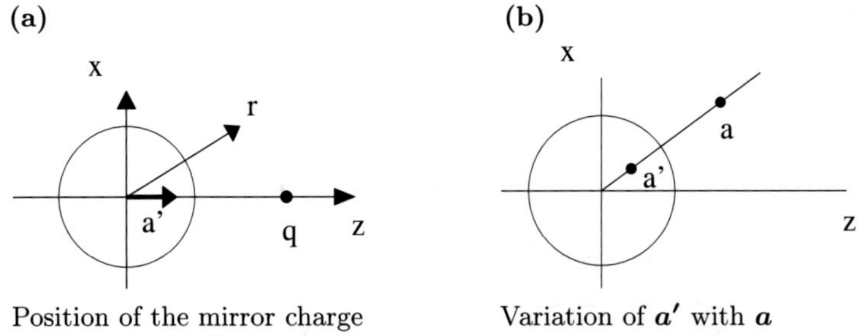

Fig. 4.4 The method of mirror charges

Define now the mirror (or image) charge

$$q' = -q\frac{R}{a}\,, \tag{4.9}$$

so that $|q'| < |q|$ and note the final result as

$$V(\boldsymbol{r}) = k_e\left\{\frac{q}{|\boldsymbol{r}-\boldsymbol{a}|} + \frac{q'}{|\boldsymbol{r}-\boldsymbol{a}'|}\right\} \qquad r \geq R\,. \tag{4.10}$$

This compact result allows the comments:

1. The point $\boldsymbol{a}'$ is referred to as the mirror point with respect to the point $\boldsymbol{a}$. It can be obtained by reflection on the surface of the sphere. This operation of reflection is defined in the following fashion: Each point with the distance a from the centre of the sphere (Fig. 4.4b) is transformed into a point with the distance a' $(a' \leq a)$ on the same ray as a with the aid of the relation $aa' = R^2$, as for instance

$$a = \ R\ 2R\ \ 3R\ \ \ldots \infty$$

$$a' = R\ R/2\ R/3 \ldots 0\,.$$

2. The result states: The potential of a point charge plus the potential of a grounded sphere with an induced surface charge, so that $V(\boldsymbol{R}) = 0$, is identical in the exterior region with the potential of two point charges. The second point charge is found at the (mirror) point $\boldsymbol{a}'$ and has the value $q' = -qR/a$. The potential of the two charges satisfies the boundary condition $V(R) = 0$ on the surface of the sphere. The mirror charge q' is, because of $R/a \leq 1$, always smaller than the original charge and has the opposite sign. It grows in magnitude, if the original charge is moved closer to the sphere.

   It is worthwhile emphasising once more the statement: The solution discussed above is only valid for the exterior of the metal sphere. The potential has the value zero in the interior of the metal sphere and on its surface. The field lines of the grounded sphere and the charge q are compared in Fig. 4.5 with the field lines of

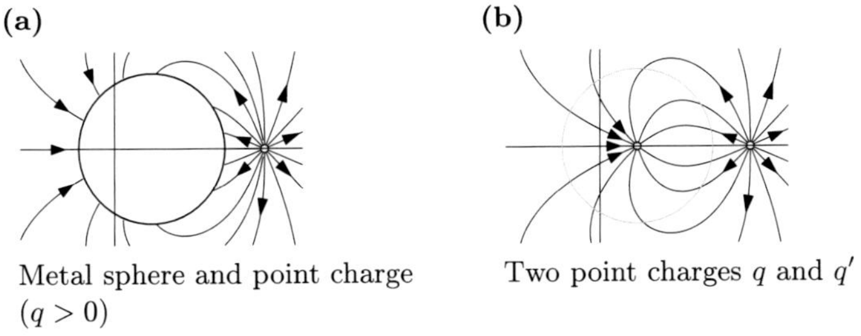

Fig. 4.5 Field lines and the method of mirror charges

the point charges q and q'. The two sets of field lines are identical in the exterior region of the sphere but differ in the interior region. The lines are perpendicular to the surface of the sphere for both the charge and the mirror charge.

3. The distribution of the induced surface charges can be calculated with the expression

$$\sigma(R,\,\theta) = -\frac{1}{4\pi k_e}\,\frac{\partial V(\boldsymbol{r})}{\partial n}\bigg|_{r=R} = -\frac{1}{4\pi k_e}\,\frac{\partial V(\boldsymbol{r})}{\partial r}\bigg|_{r=R}\,.$$

The differentiation is elementary and results in (Detail 4.7.1.1)

$$\sigma(R,\,\theta) = -\frac{q}{4\pi\,R}\,\frac{(a^2-R^2)}{\left[a^2-2aRx+R^2\right]^{3/2}} \qquad (x=\cos\theta)\,. \tag{4.11}$$

The variation of the surface charge with the distance a is the following: The surface charge is zero ($\sigma \to 0$), if the point charge is very far from the sphere ($a \to \infty$). There is no induced charge in this limit. If the original charge is closer to the sphere, one finds e.g. for $a = 4R$

$$\sigma(R,\,\theta) = -\frac{q}{4\pi\,R^2}\,\frac{15}{[17-8\cos(\theta)]^{3/2}}\,.$$

This function of the angle θ increases quite slowly with the angle of the observation point (Fig. 4.6a), There is, however, some excess of (negative) induced charge at the position directly opposite the point charge. For $a = 2R$ the function $\sigma(R,\,\theta)$ is

$$\sigma(R,\,\theta) = -\frac{q}{4\pi\,R^2}\,\frac{3}{[5-4\cos(\theta)]^{3/2}}\,.$$

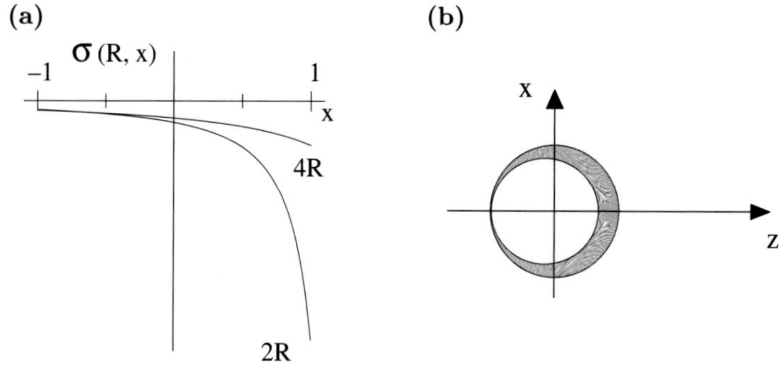

With $x = \cos\theta$ for $a = 2R$ and $4R$ Qualitative polar diagram

Fig. 4.6 Variation of the induced surface charge

This function shows the largest increment of negative charge in the vicinity of the point, where the z-axis cuts through the sphere. An indication of this function for $a = 2R$ is seen in Fig. 4.6b. In the limit, that the original point charge is on the sphere $a \to R$, one obtains

$$x \neq 1 \qquad \sigma = \lim_{\Delta R \to 0} \left\{ -\frac{q}{4\pi\, R} \frac{\Delta R}{\sqrt{2} R^2\, [1 - x]^{3/2}} \right\} = 0$$

$$x = 1 \qquad \sigma = \lim_{a \to R} \left\{ -\frac{q}{4\pi\, R} \frac{a^2 - R^2}{(a - R)^3} \right\} \to -\infty\,,$$

which can be expressed as

$$\sigma \xrightarrow{a \to R} -\frac{q\, \delta(\cos\theta)}{2\pi\, R^2}\,.$$

The point charge $q' = -q$ is found at the same point as the charge $+q$. For this case one has not only $V(\boldsymbol{R}) = 0$, but also $V(\boldsymbol{r}) = 0$. The potential has the value zero in the complete space if the charge q is on the surface of the sphere.

The integral over this distribution of the induced charge q_{ind} (see Detail 4.7.1.2)

$$q_{\text{ind}} = \iint \sigma(R,\, \theta) \mathrm{d}f = \iint \sigma(R,\, \theta) R^2\, \mathrm{d}\,\Omega = -\frac{R}{a} q = q'$$

is (as expected) equal to the (fictive) mirror charge.

A comparison of the explicit solution and the construction with the mirror charge in the example shows (Fig. 4.7), that the agreement of the results is a consequence of the uniqueness of the solution of Dirichlet problems. In both situations one demands, that the potential has the values

$$V(\boldsymbol{R}) = V(\boldsymbol{R}_{\infty}) = 0$$

on the surface of the sphere with radius R and an infinite sphere. The potential for all points of the domain between the two spheres should be the same because of the

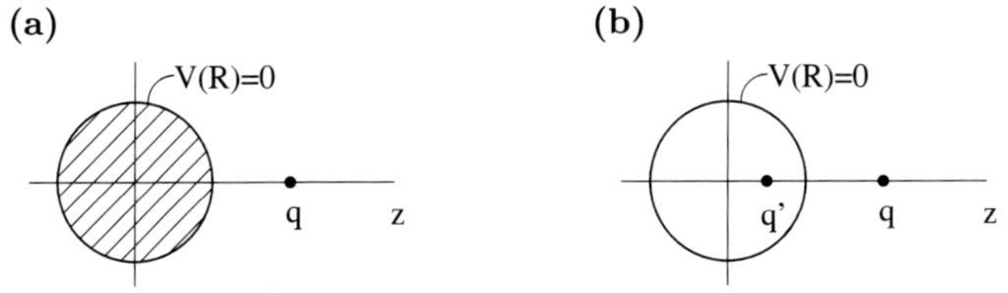

Fig. 4.7 Uniqueness of the solution with the method of mirror charges

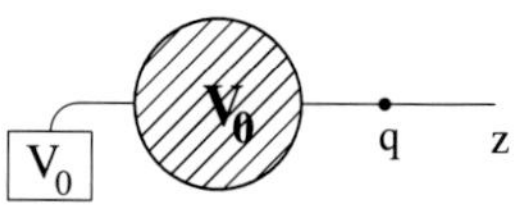

Fig. 4.8 Metal sphere at potential V_0 and a point charge

uniqueness of the solution of the Dirichlet problem

$$V_{\text{problem}}(\boldsymbol{r}) = V_{\text{image problem}}(\boldsymbol{r}) \qquad \text{for } r \geq R\,.$$

There are variants of the problem with a sphere and a point charge. In the first variant one considers the same arrangement, but the metal sphere is not grounded. It has been connected to a battery with the potential $V_0 \neq 0$ (Fig. 4.8), so that the surface of the sphere is now at $V(\boldsymbol{R}) = V_0$. The general solution of the Poisson equation in the exterior is the same as in the case of a grounded sphere (4.6), but the boundary condition on the surface of the sphere is

$$k_e \sum_l \left(\frac{A_l}{R^{l+1}} + \frac{q}{a}\left(\frac{R}{a}\right)^l \right) P_l(x) = V_0\,.$$

In order to apply this boundary condition, one writes for the potential V_0

$$V_0 = V_0\,\delta_{l,0}\,P_l(x)$$

(as $P_0(x)$ is 1) and rearranges the series

$$k_e \sum_l \left(\frac{A_l}{R^{l+1}} + \frac{q}{a}\left(\frac{R}{a}\right)^l - \frac{V_0}{k_e}\,\delta_{l,0} \right) P_l(x) = 0\,.$$

The Legendre series in this function $f(x) = 0$ allows to extract the coefficients

$$A_0 = -q\,\frac{R}{a} + \frac{V_0 R}{k_e}$$

$$A_l = -q\,\frac{R^{2l+1}}{a^{l+1}} \qquad l \geq 1\,,$$

so that the solution of the Dirichlet problem is now

$$V(r) = \begin{cases} V_0 & r = R \\[2ex] \dfrac{k_e q}{|\boldsymbol{r}-\boldsymbol{a}|} + \dfrac{k_e q'}{|\boldsymbol{r}-\boldsymbol{a}'|} + \dfrac{V_0 R}{r} & r \geq R \end{cases}\,. \tag{4.12}$$

The distance $\boldsymbol{a}'$ and the mirror charge q' are, as before, given by Eqs. (4.8) and (4.9). The potential in the exterior differs from the case of the grounded sphere by

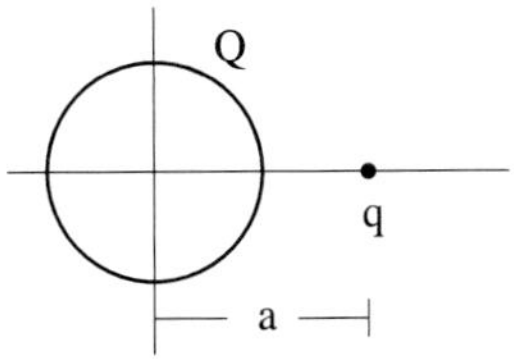

Fig. 4.9 Uniformly charged metal sphere and point charge

an additional potential of a (fictive) point charge at the origin with the magnitude $Q = V_0 R$. The mirror charge q' vanishes, if the point charge has the value $q = 0$, that is, if it is removed, and only the potential of a sphere with a uniform surface charge remains. The distribution of the induced charge is indeed

$$\sigma(R, \theta) = \sigma_{(V_0=0)}(R, \theta) + \frac{V_0}{4\pi k_e R} . \tag{4.13}$$

That is the charge distribution of the example (4.11) plus a uniform charge distribution (independent of θ). The total surface charge of the sphere is

$$q_{\text{sphere}} = \iint \sigma \, \mathrm{d}f = -\frac{R}{a} q + \frac{V_0 R}{k_e} .$$

If a positive (negative) charge q is moved in the direction of the sphere, then the original charge $Q = V_0 R$ of the sphere connected with the battery is reduced (increased) by an increasing negative (positive) charge. Induced charges move from the battery onto the sphere (or vice versa).

The charge Q is distributed uniformly over an isolated metal sphere in the second variant. The sphere was connected to the battery in order to charge it, afterwards the battery was removed. For this initial arrangement one places the point charge q again to the position $\boldsymbol{a}$ (Fig. 4.9). This problem is a hidden Dirichlet problem. The potential is not specified directly on the surface of the sphere. It is, however, possible to find a solution in the following way. Solve the problem for every distance a of the point charge from the centre of the sphere and find for every value of a the constant value of the potential on the surface of the sphere

$$V(\boldsymbol{R}) = V_a .$$

If one uses for each distance the solution of the first variant (4.12) with an unknown potential V_a, then one is faced with the equations

$$V(\boldsymbol{r}) = \begin{cases} V_a & r = R \\ \dfrac{k_e q}{|\boldsymbol{r} - \boldsymbol{a}|} + \dfrac{k_e q'}{|\boldsymbol{r} - \boldsymbol{a}'|} + \dfrac{V_a R}{r} & r \geq R . \end{cases}$$

In order to obtain the value of the potential V_a one considers

(a) charge distribution on the sphere (4.13)

$$\sigma(R, \theta) = \sigma_{(V_0=0)}(R, \theta) + \frac{V_a}{4\pi k_e R},$$

(b) and the total charge of the sphere

$$\iint \sigma(R, \theta)\mathrm{d}f = -\frac{R}{a}q + \frac{V_a R}{k_e}.$$

As no additional charges have been placed on the sphere in this case and the original total charge Q has only been redistributed, the charge Q must still be

$$Q = -\frac{R}{a}q + \frac{V_a R}{k_e}.$$

Thus V_a can be found from

$$V_a = k_e\left(\frac{Q}{R} + \frac{q}{a}\right) = \frac{k_e}{R}(Q - q').$$

The final solution of this potential problem is therefore

$$V(\boldsymbol{r}) = \begin{cases} \dfrac{k_e}{R}(Q - q') & r = R \\[2ex] \dfrac{k_e q}{|\boldsymbol{r} - \boldsymbol{a}|} + \dfrac{k_e q'}{|\boldsymbol{r} - \boldsymbol{a}'|} + \dfrac{k_e(Q - q')}{r} & r \geq R\,. \end{cases} \tag{4.14}$$

The potential in the exterior seems to have been generated by a charge q at the position $\boldsymbol{a}$, a mirror charge q' at the position $\boldsymbol{a}'$ and a point charge $Q - q'$ in the centre of the sphere. The surface of the sphere is an equipotential surface, so that the field lines are perpendicular to it. The boundary condition is arranged, so that the total charge of the surface always has the value Q. The distribution of this charge on the surface changes with the distance a of the exterior charge according to

$$\sigma(a, x) = \frac{q}{4\pi R}\left\{\left(\frac{Q}{qR} + \frac{1}{a}\right) - \frac{(a^2 - R^2)}{[a^2 - 2aRx + R^2]^{3/2}}\right\} \tag{4.15}$$

$$(x = \cos\theta)\,.$$

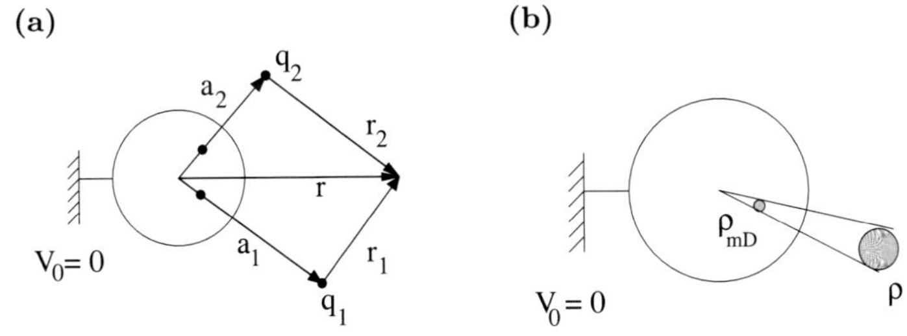

Fig. 4.10 Extension of the method of mirror charges

If the distance a is infinite, the charge Q is uniformly distributed over the sphere ($\sigma(\infty, x) = Q/(4\pi R^2)$). For $a = R$ the surface charge density is

$$\sigma(R, x) = \frac{(Q+q)}{(4\pi R^2)}$$

if $x \neq 1$. The charge $(Q+q)$ is distributed uniformly over the surface.

The method of mirror charges can be extended for application in different situations. In the following problem one considers a metal sphere (radius R at potential $V_0 = 0$) with two point charges q_i at the positions $\boldsymbol{a}_i$, $i = 1, 2$ (Fig. 4.10a). In this case the ansatz is

$$V(\boldsymbol{r}) = k_e \left(\frac{q_1}{r_1} + \frac{q_1'}{r_1'} \right) + k_e \left(\frac{q_2}{r_2} + \frac{q_2'}{r_2'} \right)$$

$$(\boldsymbol{r}_i = \boldsymbol{r} - \boldsymbol{a}_i, \quad \boldsymbol{r}_i' = \boldsymbol{r} - \boldsymbol{a}_i' \qquad i = 1, 2)$$

for the potential in the exterior. The task is the calculation of the quantities q_1', q_2', $\boldsymbol{a}_1'$ and $\boldsymbol{a}_2'$, so that $V(\boldsymbol{R}) = 0$. The corresponding calculation (see Detail 4.7.2) follows the same pattern indicated above. The result for the mirror charges q_i' and the distances from the centre of the sphere is

$$q_i' = -\frac{R}{a_i} q_i \qquad \boldsymbol{a}_i' = \frac{R^2}{a_i^2} \boldsymbol{a}_1' \qquad (i = 1, 2) .$$

The case, that a distribution of space charges (Fig. 4.10b) is present outside of a metal sphere can be treated by dividing the distribution into infinitely small elements and reflecting these into the interior. The potential of the metal sphere and the external charge distribution is then

$$V(\boldsymbol{r}) = k_e \iiint_{\mathrm{D}} \frac{\rho(\boldsymbol{r}')}{|\boldsymbol{r} - \boldsymbol{r}'|} \mathrm{d}V' + k_e \iiint_{\mathrm{mD}} \frac{\rho_{\mathrm{mD}}(\boldsymbol{r}')}{|\boldsymbol{r} - \boldsymbol{r}'|} \mathrm{d}V' \qquad (r > R) . \tag{4.16}$$

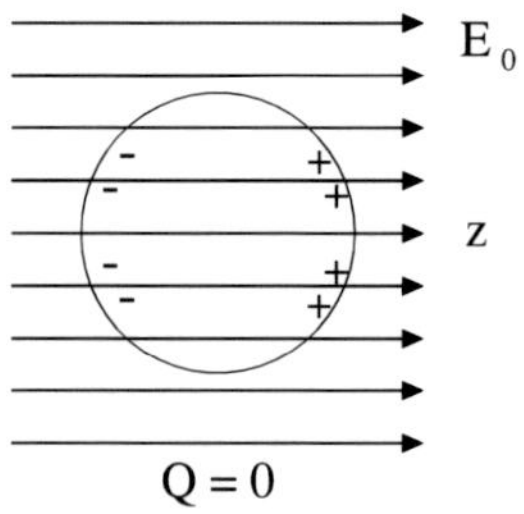

Fig. 4.11 Uncharged metal sphere in a uniform electric field

The notation indicates the original domain of the charges D and the mirror image of the charge distribution ρ_{mD}, which have to be constructed point by point. One should note, that the mirror image of a homogeneous charge distribution is not homogeneous due to the distortion by the reflection.

The method can also be used in the case of more general charge distributions (e.g. ellipsoids), but at the cost of more complicated equations.

Another classic example for a Dirichlet boundary value problem is the following problem: An uncharged metal sphere (radius R) is placed in a uniform electric field (Fig. 4.11). The task is: Calculate the change of the field, caused by the charge separation on the surface of the sphere, and the distribution of the induced charges on the surface of the sphere.

The first step is always the choice of a coordinate system, in the present case:

(a) The origin is the centre of the sphere and the z-direction is the direction of the field. The potential associated with the field is

$$V_0(\boldsymbol{r}) = -E_0\, z$$

or

$$V_0(\boldsymbol{r}) = -E_0\, r\, P_1(\cos\theta)\,.$$

(b) The exterior of the sphere is characterised by the Laplace equation

$$\Delta V(\boldsymbol{r}) = 0\,,$$

for which one has to use the boundary condition

(i) $V(\boldsymbol{R}) = V_R$. The potential on the surface of the sphere is constant. The boundary value is initially not known. It can be obtained in retrospect from the statement that the total charge of the sphere is zero ($Q_{\mathrm{sphere}} = 0$).

(ii) For large distances from the sphere the solution of the Laplace equation should become the potential of the uniform field

$$V(\boldsymbol{r}) \xrightarrow{r\to\infty} -E_0\, r\, P_1(\cos\theta)\,.$$

The symmetry of the present problem is azimuthal. The solution of the Laplace equation should for this reason have the form

$$V(\boldsymbol{r}) = V(r,\,\theta) = k_e \sum_{l=0}^{\infty} \left[\frac{A_l}{r^{l+1}} + B_l r^l \right] P_l(\cos\theta)\,.$$

It is necessary to use the complete radial part of the solution as the standard boundary condition

$$V(r,\,\theta) \to 0 \qquad \text{for} \to \infty$$

can not be applied in this problem. The constants A_l and B_l are to be determined by the boundary conditions.

(a) On the surface of the sphere one should have

$$k_e \sum_{l=0}^{\infty} \left[\frac{A_l}{R^{l+1}} + B_l R^l \right] P_l(\cos\theta) = V_0 = V_0 P_0(\cos\theta)\,.$$

Comparison of the coefficients (see (4.7)) leads to the statements

$$\frac{A_0}{R} + B_0 - \frac{V_0}{k_e} = 0 \qquad \text{for} \quad l = 0\,.$$

For $l \geq 1$, the relation is

$$\frac{A_l}{R^{l+1}} + B_l R^l = 0\,.$$

(b) For infinitely distant points it is

$$\sum_{l=0}^{\infty} B_l r^l \; P_l(\cos\theta)\Big|_{r\to\infty} = -\frac{E_0}{k_e} r\, P_1(\cos\theta)\Big|_{r\to\infty}\,.$$

The terms in $1/r$ do not contribute. Comparison of the coefficients gives

$$B_0 = 0, \qquad B_1 = -\frac{E_0}{k_e}, \qquad B_l = 0 \qquad \text{for} \qquad l \geq 2\,.$$

Evaluation of the two sets of conditions yields the following results for the coefficients

$$
\begin{aligned}
l = 0: &\quad B_0 = 0 &\quad A_0 &= \frac{RV_0}{k_e} \\
l = 1: &\quad B_1 = -\frac{E_0}{k_e} &\quad A_1 &= E_0 R^3 \\
l \geq 2: &\quad B_l = 0 &\quad A_l &= 0\,.
\end{aligned}
$$

The potential on the outside of the sphere is

$$
V(r,\,\theta) = \frac{RV_0}{r} - E_0\left[r - \frac{R^3}{r^2}\right] P_1(\cos\theta)\,.
$$

In order to find the value of the constant V_0 one has to calculate first the distribution of the induced surface charges

$$
\sigma(R,\,\theta) = -\frac{1}{4\pi k_e}\,\frac{\partial V}{\partial r}\bigg|_{r=R}\,.
$$

A simple calculation leads to

$$
\sigma(R,\,\theta) = \frac{1}{4\pi k_e}\left(\frac{V_0}{R} + 3E_0 P_1(\cos\theta)\right)\,.
$$

Integration over the sphere with this surface density give the total charge

$$
\begin{aligned}
Q &= \iint \sigma(R,\,\theta) R^2 \,\mathrm{d}\Omega \\
&= \frac{RV_0}{4\pi k_e}\iint \mathrm{d}\Omega + \frac{3E_0 R^2}{4\pi k_e}\iint P_1(\cos\theta)\mathrm{d}\Omega \\
&= \frac{RV_0}{k_e}\,.
\end{aligned}
$$

As the sphere is not charged in the present problem, one obtains $V_0 = 0$, so that the final result is

$$
V(\boldsymbol{r}) = \begin{cases} 0 & r \leq R \\ -E_0\left[r - \dfrac{R^3}{r^2}\right]\cos\theta & r \geq R \end{cases} \tag{4.17}
$$

and

$$\sigma(R, \theta) = \frac{3}{4\pi k_e} E_0 \cos\theta \,. \tag{4.18}$$

In order to have a better idea of the modification, one has to calculate the electric field in the exterior region. It is preferable to use spherical coordinates for this purpose

$$\begin{aligned} \boldsymbol{E}(r, \theta) &= \left(-\frac{\partial V}{\partial r},\ -\frac{1}{r}\frac{\partial V}{\partial \theta},\ 0\right) \\ &= E_0 \left(\left(1 + 2\frac{R^3}{r^3}\right)\cos\theta,\ -\left(1 - \frac{R^3}{r^3}\right)\sin\theta,\ 0\right) \qquad (r \geq R)\,. \end{aligned} \tag{4.19}$$

The vector $(\cos\theta, -\sin\theta, 0)$ describes the decomposition of the original constant field, the terms with $1/r^3$ represent the modifications of the field by the charge separation on the surface of the sphere, for which one finds

$$E_r(R, \theta) = 3E_0 \cos\theta \qquad E_\theta(R, \theta) = 0\,.$$

The field lines are (as expected for a metallic sphere) perpendicular to the surface (Fig. 4.12a). The strength of the field is maximal in the z-direction, it decreases towards the value zero in the direction of the equator ($\theta = \pi/2$) with the cosine function. The field is negative on the back of the sphere. The field lines end on the negative induced charges. The modification of the field decreases quite rapidly (with $1/r^3$). The modified field lines are displayed (note the cylindrical symmetry) in Fig. 4.12b.

The charge distribution changes in a similar manner with the cosine function. The charge density is maximal at the poles and decreases to the value zero at the equator. The hemisphere in the front (corresponding to the positive z-axis) carries a positive charge, the hemisphere in the back a negative charge.

The discussion of the electric potential offers an access to the topic Green's functions, which constitute a very important tool in theoretical physics. The aspects

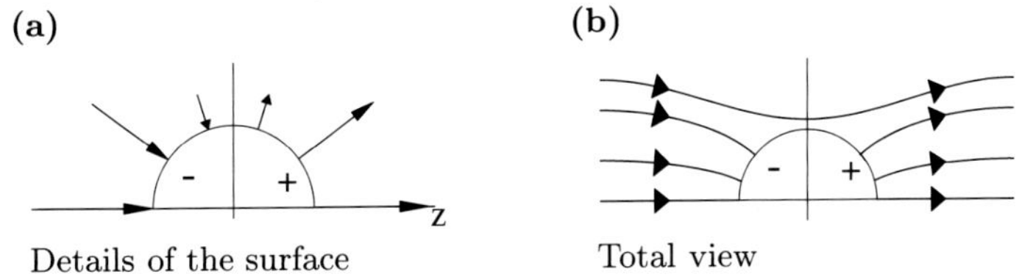

Fig. 4.12 The field lines of a metal sphere polarised by induction

relevant for the calculation of the electric potential problem are collected in the next section.

4.3 Green's Functions

Green's functions are used for the discussion of stationary as well as dynamic problems. One example for the application of such functions in electrodynamics is the calculation of the electric field of a dipole antenna (see Dreizler and Lüdde, 2024, Electrodynamics and Special Theory of Relativity (Springer Berlin Heidelberg), Chap. 2.3). Other areas, in which Green's functions play a major role (often under the name propagators) are collision problems, the many body problem of quantum mechanics and the corresponding extensions in quantum field theories. The solutions of the Laplace/Poisson equation

$$\Delta V(\boldsymbol{r}) = -4\pi k_e \, \rho(\boldsymbol{r}) \tag{4.20}$$

with the help of Green's functions will be investigated in this section. The following types of boundary conditions will be considered:

(a) Simple boundary conditions with the requirement that the potential vanishes for infinitely distant points

$$V(\boldsymbol{r}) \xrightarrow{r \to \infty} 0 \, .$$

(b) Dirichlet boundary conditions with $V(\boldsymbol{r})$ specified on closed metallic surfaces.
(c) von Neumann boundary conditions with $\partial V(\boldsymbol{r})/\partial n$ specified on closed surfaces.

The question, that will be answered in this section, is:

Is it possible to find a compact formula for the potential in electrostatics, in which the boundary conditions described above, are fully included?
The short answer is: This is indeed possible on the basis of Green's functions and the ansatz

$$V(\boldsymbol{r}) = V_{\text{hom}}(\boldsymbol{r}) + \iiint G(\boldsymbol{r}, \boldsymbol{r}')\rho(\boldsymbol{r}')\mathrm{d}V' \, . \tag{4.21}$$

This ansatz is in accord with the statement, that the solution of the potential problem in the form of an inhomogeneous partial differential equation of second order must be the sum of the general solution of the homogeneous differential equation and a particular solution of the inhomogeneous differential equation.
The function $V_{\text{hom}}(\boldsymbol{r})$ is the general solution of the homogeneous differential equation, the Laplace equation. The particular integral of the inhomogeneous

differential equation is represented by the Green's function $G(\boldsymbol{r}, \boldsymbol{r}')$ multiplied the charge distribution $\rho(\boldsymbol{r}')$.

If one acts with the Laplace operator $\Delta_{\boldsymbol{r}}$ on the ansatz (4.21), one finds

$$\Delta V(\boldsymbol{r}) = \Delta V_{\text{hom}}(\boldsymbol{r}) + \iiint \Delta_{\boldsymbol{r}} G(\boldsymbol{r}, \boldsymbol{r}')\rho(\boldsymbol{r}')\, \mathrm{d}V' .$$

As V is a solution of the Poisson equation (4.20) and V_{hom} a solution of the Laplace equation, the result of this operation can be written as

$$-4\pi k_e\, \rho(\boldsymbol{r}) = \iiint \mathrm{d}V' \Delta_{\boldsymbol{r}} G(\boldsymbol{r}, \boldsymbol{r}')\rho(\boldsymbol{r}') .$$

It can be used to extract a differential equation for a Green's function

$$\Delta_{\boldsymbol{r}} G(\boldsymbol{r}, \boldsymbol{r}') = -4\pi\, k_e\, \delta(\boldsymbol{r} - \boldsymbol{r}') . \tag{4.22}$$

This Green's function can be interpreted, according to (4.21) and (4.22), as the Green's function of a point charge with the magnitude 1, which is situated at the position $\boldsymbol{r}'$, for the potential at the position $\boldsymbol{r}$. As a consequence of the property

$$\delta(\boldsymbol{r} - \boldsymbol{r}') = \delta(\boldsymbol{r}' - \boldsymbol{r})$$

of the delta function, one can infer, that this Green's function, a function of six variables, is symmetric with respect to the interchange[2] of $\boldsymbol{r}$ and $\boldsymbol{r}'$

$$G(\boldsymbol{r}, \boldsymbol{r}') = G(\boldsymbol{r}', \boldsymbol{r}) .$$

The characterisation of this Green's function can be stated as well in the form: contribution of a point charge at the position $\boldsymbol{r}$ to the potential at the position $\boldsymbol{r}'$. The total potential due to an assembly of point charges is given by the sum of the contributions of all point charges weighted with a specified charge distribution $\rho(\boldsymbol{r})$.

The proof, that this argumentation is correct, can be demonstrated with the integral theorems of Green, which follow directly from the divergence theorem (see Vol. 1, Math. Chap. 5.3.3), which says

$$\iiint_{\mathrm{B}} \nabla \cdot \boldsymbol{F}(\boldsymbol{r})\mathrm{d}V = \oiint_{\mathrm{O(B)}} \boldsymbol{F}(\boldsymbol{r}) \cdot \mathbf{d}\boldsymbol{f} .$$

---

[2] The proof of the symmetry and a more detailed discussion of Green's functions can be found in Detail 4.7.3.1.

The domain B in space is enclosed by a surface $O(B)$. This relation is valid for arbitrary vector fields $\boldsymbol{F}$, in particular also for vector fields with the structure

$$\boldsymbol{F}(\boldsymbol{r}) = \phi(\boldsymbol{r})[\nabla\psi(\boldsymbol{r})]\,.$$

The functions $\phi(\boldsymbol{r})$ and $\psi(\boldsymbol{r})$ are two arbitrary, differentiable functions characterising scalar fields. With the vector function $\boldsymbol{F}(\boldsymbol{r})$ one can calculate the scalar product

$$\nabla\cdot\boldsymbol{F}(\boldsymbol{r}) = \phi(\boldsymbol{r})\Delta\psi(\boldsymbol{r}) + \nabla\phi(\boldsymbol{r})\cdot\nabla\psi(\boldsymbol{r})$$

and the infinitesimal surface element

$$\boldsymbol{F}(\boldsymbol{r})\cdot\mathbf{d}\boldsymbol{f} = \boldsymbol{F}(\boldsymbol{r})\cdot\boldsymbol{e}_{\mathrm{n}}\,\mathrm{d}f = \phi(\boldsymbol{r})\,\frac{\partial\psi(\boldsymbol{r})}{\partial n}\,\mathrm{d}f\,.$$

The derivative $\partial\psi(\boldsymbol{r})/\partial n = \boldsymbol{e}_n\cdot\nabla\psi(\boldsymbol{r})$ is the normal derivative of the scalar function ψ on the surface $O(B)$. The direction of the normal vector is towards the outside of the domain B. Insertion of these relations into the divergence theorem yields

$$\iiint_{\mathrm{B}}[\phi(\boldsymbol{r})\Delta\psi(\boldsymbol{r}) + \nabla\phi(\boldsymbol{r})\cdot\nabla\psi(\boldsymbol{r})]\,\mathrm{d}V = \oiint_{\mathrm{O(B)}}\phi(\boldsymbol{r})\,\frac{\partial\psi(\boldsymbol{r})}{\partial n}\,\mathrm{d}f\,. \tag{4.23}$$

This relation, which connects a volume integral with a surface integral of two scalar functions is the **first integral theorem of Green**. The argumentation can be repeated with the vector function

$$\boldsymbol{F}(\boldsymbol{r}) = \psi(\boldsymbol{r})\cdot\nabla\phi(\boldsymbol{r})\,,$$

leading to the relation

$$\iiint_{\mathrm{B}}[\psi(\boldsymbol{r})\Delta\phi(\boldsymbol{r}) + \nabla\psi(\boldsymbol{r})\cdot\nabla\phi(\boldsymbol{r})]\,\mathrm{d}V = \oiint_{\mathrm{O(B)}}\psi(\boldsymbol{r})\,\frac{\partial\phi(\boldsymbol{r})}{\partial n}\,\mathrm{d}f\,. \tag{4.24}$$

Subtraction of the equations (4.23) and (4.24) yields **the second integral theorem of Green**, which is usually just called the **Green's Theorem**.

$$\iiint_{\mathrm{B}}[\phi(\boldsymbol{r})\Delta\psi(\boldsymbol{r}) - \psi(\boldsymbol{r})\Delta\phi(\boldsymbol{r})]\,\mathrm{d}V =$$

$$\oiint_{\mathrm{O(B)}}\left[\phi(\boldsymbol{r})\,\frac{\partial\psi(\boldsymbol{r})}{\partial n} - \psi(\boldsymbol{r})\,\frac{\partial\phi(\boldsymbol{r})}{\partial n}\right]\mathrm{d}f\,. \tag{4.25}$$

This is the point to introduce the Green's function and the potential into Green's theorem in order to find a compact formula for the solution of the potential problem with a specific geometry by choosing

$$\phi(\boldsymbol{r}) = V(\boldsymbol{r}) \qquad\qquad \psi(\boldsymbol{r}) = G(\boldsymbol{r},\, \boldsymbol{r}')$$

and extracting the general formula in a few steps. With the differential equations

$$\Delta V(\boldsymbol{r}) = -4\pi k_e\, \rho(\boldsymbol{r}) \qquad \text{and} \qquad \Delta_{\boldsymbol{r}} G(\boldsymbol{r},\, \boldsymbol{r}') = -4\pi k_e\, \delta(\boldsymbol{r} - \boldsymbol{r}')$$

one obtains with Green's theorem (4.25)

$$-4\pi k_e \iiint_{\mathrm{B}} \left[V(\boldsymbol{r})\, \delta(\boldsymbol{r} - \boldsymbol{r}') - G(\boldsymbol{r},\, \boldsymbol{r}')\rho(\boldsymbol{r})\right] \mathrm{d}V$$

$$= \oiint_{\mathrm{O(B)}} \left[V(\boldsymbol{r})\, \frac{\partial G(\boldsymbol{r},\, \boldsymbol{r}')}{\partial n} - G(\boldsymbol{r},\, \boldsymbol{r}')\, \frac{\partial V(\boldsymbol{r})}{\partial n}\right] \mathrm{d}f\, .$$

If one resolves this equation with respect to $V(\boldsymbol{r}')$, renames the variables $\boldsymbol{r} \longleftrightarrow \boldsymbol{r}'$, one obtains the formula for the solution of the electric potential problem

$$V(\boldsymbol{r}) = \iiint_{\mathrm{B}} G(\boldsymbol{r}',\, \boldsymbol{r})\rho(\boldsymbol{r}')\mathrm{d}V' \tag{4.26}$$

$$+ \frac{1}{4\pi k_e} \oiint_{\mathrm{O(B)}} \left[G(\boldsymbol{r}',\, \boldsymbol{r})\, \frac{\partial V(\boldsymbol{r}')}{\partial n'} - V(\boldsymbol{r}')\, \frac{\partial G(\boldsymbol{r}',\, \boldsymbol{r})}{\partial n'}\right] \mathrm{d}f'\, .$$

This equation establishes the connection between the potential $V(\boldsymbol{r})$ and the Green's function $G(\boldsymbol{r}',\, \boldsymbol{r})$. It states: If the charge distribution and the Green's function are known and if the quantities $V(\boldsymbol{R})$ and $\partial V(\boldsymbol{R})/\partial n$ on the relevant surfaces, which are borders of the volume B, are specified, then one is able to calculate the potential $V(\boldsymbol{r})$ in every point of the domain B. However as both $V(\boldsymbol{R})$ *and* $\partial V(\boldsymbol{R})/\partial n$ can not be specified at the same time on the surfaces, marking the border of B, the Green's function has to satisfy certain constraints, which have to correspond to the boundary conditions of the problem at hand. So, the next task is the discussion of these constraints.

Simple Boundary Conditions

The charge distribution $\rho(\boldsymbol{r})$ is concentrated around the origin. There exist no metal surface in the domain B, which encompasses the complete space. The border $O(B)$ is the infinite sphere. The requirement, that the potential vanishes on $O(B)$

$$V(\boldsymbol{r}) \xrightarrow{r \to \infty} 0\, ,$$

reduces the central equation (4.26) for $r \to \infty$ to

$$0 = \lim_{r\to\infty}\left[\iiint_B G(\boldsymbol{r}', \boldsymbol{r})\rho(\boldsymbol{r}')\mathrm{d}V' + \oiint_{\infty\ \text{sphere}} \frac{G(\boldsymbol{r}', \boldsymbol{r})}{4\pi k_e}\frac{\partial V(\boldsymbol{r}')}{\partial n'}\mathrm{d}f'\right].$$

This relation leads to a unique potential, if the Green's function satisfies the boundary condition

$$G(\boldsymbol{r}', \boldsymbol{r}) \xrightarrow{r\to\infty} 0\,.$$

The Green's function, which satisfies this boundary condition, is the potential of a point charge

$$G(\boldsymbol{r}', \boldsymbol{r}) = \frac{k_e}{|\boldsymbol{r} - \boldsymbol{r}'|} = G(\boldsymbol{r}, \boldsymbol{r}')\,. \tag{4.27}$$

The symmetry of the Green's function is clearly apparent. The expression for the general potential is the well known result

$$V(\boldsymbol{r}) = \iiint_B G(\boldsymbol{r}, \boldsymbol{r}')\rho(\boldsymbol{r}')\,\mathrm{d}V' = k_e \iiint_B \frac{\rho(\boldsymbol{r}')}{|\boldsymbol{r} - \boldsymbol{r}'|}\mathrm{d}V'\,. \tag{4.28}$$

Dirichlet Boundary Conditions

In this case, the domain B is an arbitrary domain, enclosed by metal surfaces (Fig. 4.13). The potential is specified on the surfaces of the metallic border

$$V(\boldsymbol{R}) = f(\boldsymbol{R}) \qquad\qquad \boldsymbol{R} \in \mathrm{O(B)}\,.$$

The directional derivative $\partial V(\boldsymbol{R})/\partial n$ in the formula (4.26) should not be included, so that the Green's function on the surfaces has to be

$$G(\boldsymbol{r}', \boldsymbol{r}) = 0 \qquad \text{for} \quad \boldsymbol{r}' = \boldsymbol{R} \in O(B) \tag{4.29}$$

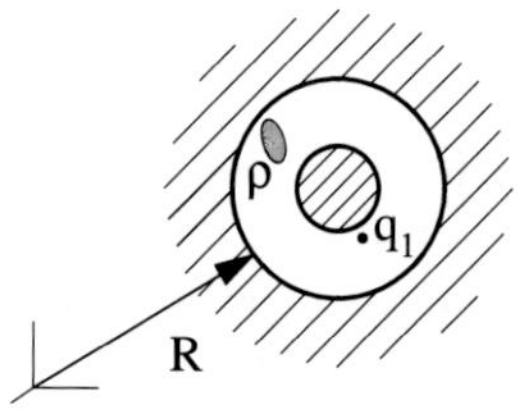

Fig. 4.13 The Dirichlet problem

in order to avoid a contradiction The Green's function should vanish on the boundary surface. The solution is therefore reduced to

$$V(\boldsymbol{r}) = \iiint_{\mathrm{B}} G(\boldsymbol{r}', \boldsymbol{r})\rho(\boldsymbol{r}')\mathrm{d}V' - \frac{1}{4\pi k_e} \oiint_{\mathrm{O(B)}} V(\boldsymbol{r}') \frac{\partial G(\boldsymbol{r}', \boldsymbol{r})}{\partial n'} \mathrm{d}f' .$$

With the Green's function one can obtain the solution of the Poisson/Laplace problem, if the charge distribution and the values of the potential on the border are given.

Von Neumann Boundary Conditions

The derivative $\partial V(\boldsymbol{R})/\partial n$ has to be specified in the case of von Neumann boundary conditions. The general solution could be used, if the Green's function does satisfy the boundary condition

$$\frac{\partial}{\partial n'} G(\boldsymbol{r}', \boldsymbol{r}) = 0 \qquad \text{for} \quad \boldsymbol{R} \in O(B) .$$

This condition can, however, not be used. It contradicts the differential equation for the Green's function as a consequence of the fact, that application of the divergence theorem would lead to

$$\begin{aligned} \oiint_{\mathrm{O(B)}} \frac{\partial G(\boldsymbol{r}', \boldsymbol{r})}{\partial n'} \mathrm{d}f' &= \oiint_{\mathrm{O(B)}} \nabla_{\boldsymbol{r}'} G(\boldsymbol{r}', \boldsymbol{r}) \cdot \mathrm{d}\boldsymbol{f}' \\ &= \iiint_{\mathrm{B}} \Delta_{\boldsymbol{r}'} G(\boldsymbol{r}', \boldsymbol{r})\, \mathrm{d}V' \\ &= -4\pi k_e \iiint_{\mathrm{B}} \delta(\boldsymbol{r} - \boldsymbol{r}')\, \mathrm{d}V' = -4\pi k_e \quad (\boldsymbol{r} \in \mathrm{B}) . \end{aligned}$$

A possible (but still simple) choice for the boundary condition is

$$\frac{\partial G(\boldsymbol{r}', \boldsymbol{r})}{\partial n'} = \text{const.} \qquad \text{for} \quad \boldsymbol{r}' = \boldsymbol{R} \in O(B) .$$

The divergence theorem demands for the constant

$$\text{const.} = -\frac{4\pi k_e}{F(\mathrm{O(B)})}$$

and the formula for the solution takes in the present case the form

$$V(\boldsymbol{r}) = \iiint_B G(\boldsymbol{r}', \boldsymbol{r})\rho(\boldsymbol{r}')\mathrm{d}V' \tag{4.30}$$

$$+\frac{1}{F_{O(B)}} \oiint_{O(B)} V(\boldsymbol{r}')\mathrm{d}f' + \frac{1}{4\pi k_e} \oiint_{O(B)} G(\boldsymbol{r}', \boldsymbol{r}) \frac{\partial V(\boldsymbol{r}')}{\partial n'} \mathrm{d}f' .$$

The second term is the mean value of the potential on the total surface of the problem. The fact, that it appears at this stage, is related to the statement, that the von Neumann problem does not have a completely unique solution. If on the other hand the surface is an infinite sphere with an infinite area $F_{O(B)}$, this term can be set equal to zero. The other terms represent then the contribution of the specified charge distribution and a contribution of the surface, which arises from the distribution of the field on the surface. Both terms can be obtained, if $G(\boldsymbol{r}, \boldsymbol{r}')$ is known.

The answer to the question concerning a compact formula is therefore: Such a formula can be found. The Green's function, which features in this formula can be calculated with the differential equation (4.26) and the boundary conditions of the potential problem. As the Green's function is only determined by the geometry of the surfaces of the problem, it is possible to obtain the solution of the Poisson-/Laplace problem for arbitrary charge distributions in the area enclosed by the surfaces and the potential values on the border by direct integration.

The application of the method of Green's functions is illustrated with an example of the Dirichlet type: The potential is specified on two concentric spherical surfaces with radius $R_1 = R$ and $R_2 \to \infty$. In the domain B between these surfaces there is an (arbitrary) distribution of point and space charges (Fig. 4.14). It is necessary to use the following steps in order to find the appropriate Green's function (a more detailed discussion can be found in (Detail 4.7.3)): One first has to obtain a solution of the differential equation (4.26), which satisfies the symmetry and the boundary conditions for the Green's function of this Dirichlet problem. The geometry calls for a solution in spherical coordinates. As in the case of the function of the distance of two points an expansion is called for, this time in view of the more general situation

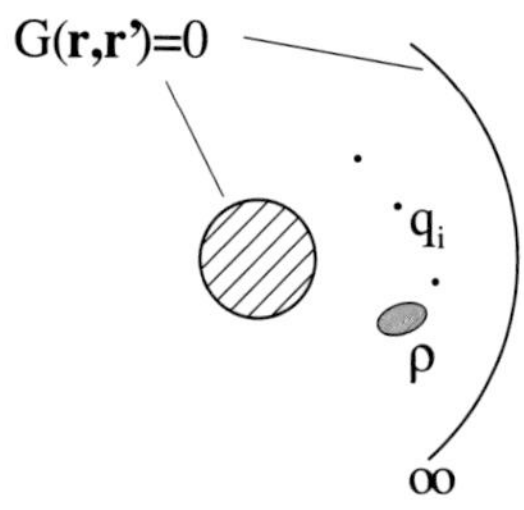

Fig. 4.14 Specification of a Dirichlet problem

in terms of spherical harmonics

$$G(\boldsymbol{r},\ \boldsymbol{r}') = \sum_{l',m'} g_{l'}(r, r')Y_{l',m'}(\Omega)Y^{*}_{l',m'}(\Omega') . \tag{4.31}$$

Insertion into the differential equation for $G(\boldsymbol{r},\ \boldsymbol{r}')$ leads to, after elimination of the angular parts (integration after multiplication with appropriate spherical harmonics), a differential equation for the radial parts

$$\frac{\mathrm{d}}{\mathrm{d}r}\left(r^2\frac{\mathrm{d}}{\mathrm{d}r}g_l(r, r')\right) - l(l+1)g_l(r, r') = -4\pi k_e \delta(r - r') \tag{4.32}$$

$$(l = 0, 1, 2, \ldots) .$$

The solution of one boundary value problem has been replaced by an infinite number of boundary value problems (for each l) with an ordinary differential equation. The solution of these differential equations can be found with the following argument: In the domain with $r > r'$ and also in the domain with $r < r'$ one has arrived at a homogeneous differential equation, which is similar to the differential equation for the radial part of the Laplace problem (3.28). The solution in each of the domains can therefore be written in the form

$$g_l(r, r') = a_l(r')r^l + \frac{b_l(r')}{r^{l+1}} .$$

The dependence of the solution on the variable r' is not determined by the equation (4.32). As the solutions in the two domains are connected by the symmetry relation, a dependence on r' must exist. It can be exposed by using a different ansatz in the two possible domains

$$g_{1,l}(r, r') = a_{1,l}(r')r^l + b_{1,l}(r')\frac{1}{r^{l+1}} \qquad r < r'$$

$$g_{2,l}(r, r') = a_{2,l}(r')r^l + b_{2,l}(r')\frac{1}{r^{l+1}} \qquad r' < r .$$

The functions for the coefficients $a_{il},\ b_{il}$ are determined by the conditions

- The boundary condition on the inner spherical shell

  $$g_{1,l}(R_1, r') = 0 ,$$

  with $r = R_1 < r'$ gives

  $$b_{1,l}(r') = -a_{1,l}(r')R_1^{2l+1} .$$

The corresponding condition on the outer shell

$$\lim_{R_2 \to \infty} g_{2l}(R_2, r') = 0$$

with $r' < r = R_2 \longrightarrow \infty$ gives

$$a_{2l}(r') = 0\,.$$

The result of this first step is (change the notation from R_1 to R)

$$g_{1,l}(r, r') = a_{1,l}(r')\left[r^l - \frac{R^{2l+1}}{r^{l+1}}\right] \qquad r < r'$$

$$g_{2,l}(r, r') = b_{2,l}(r')\frac{1}{r^{l+1}} \qquad r' < r\,.$$

- The symmetry relation for the Green's function leads to the condition $g_{1,l}(r, r') = g_{2,l}(r', r)$ for the radial parts. This condition, which connects the functions in the two domains, requires

$$a_{1,l}(r')\left[r^l - \frac{R^{2l+1}}{r^{l+1}}\right] = b_{2,l}(r)\frac{1}{(r')^{l+1}}\,.$$

Comparison of the r- and the r'- dependence gives

$$a_{1,l}(r') = c_l\frac{1}{(r')^{l+1}} \qquad b_{2,l}(r) = c_l\left[r^l - \frac{R^{2l+1}}{r^{l+1}}\right],$$

so that the result for the two radial functions is

$$g_{1,l}(r, r') = c_l\left[\frac{r^l}{(r')^{l+1}} - \frac{R^{2l+1}}{(rr')^{l+1}}\right] \qquad r < r'$$

$$g_{2,l}(r, r') = c_l\left[\frac{r'^l}{r^{l+1}} - \frac{R^{2l+1}}{(rr')^{l+1}}\right] \qquad r' < r\,.$$

The symmetry condition also proofs, that the functions $g_{1,l}$ and $g_{2,l}$ are continuous at the position $r = r'$

$$g_{1l}(r, r) = g_{2l}(r, r) \qquad \text{for} \quad r' = r\,.$$

- The remaining factor c_l can be determined by the last condition, the behaviour of the first derivative of the functions at the position $r = r'$. This behaviour can be extracted directly from the differential equation of second order (4.32), by writing

$$\frac{\mathrm{d}^2}{\mathrm{d}r^2}\big(r g_l(r, r')\big) - \frac{l(l+1)}{r} g_l(r, r') = -\frac{4\pi k_e}{r}\delta(r - r')$$

and calculating the integral

$$\lim_{\varepsilon \to 0} \int_{r=r'-\varepsilon}^{r=r'+\varepsilon} \mathrm{d}r$$

for both sides of this equation. The terms with g_l do not contribute because of the continuity of the functions for $r = r'$, so that

$$\lim_{\varepsilon \to 0} \left\{ \frac{\mathrm{d}}{\mathrm{d}r}\big(r g_{2l}(r, r')\big)_{r=r'+\varepsilon} - \frac{\mathrm{d}}{\mathrm{d}r}\big(r g_{1l}(r, r')\big)_{r=r'-\varepsilon} \right\} = -\frac{4\pi k_e}{r'}$$

remains. The singularity of the differential equation at $r = r'$ demands, that first derivative of the radial part is discontinuous at the critical position. Exploitation of this property yields

$$\left\{ \left[-\frac{l}{r'} + \frac{l R^{2l+1}}{(r')^{2l+2}} \right] - \left[\frac{l+1}{r'} + \frac{l R^{2l+1}}{(r')^{2l+2}} \right] \right\} c_l = -\frac{4\pi k_e}{r'}$$

respectively

$$c_l = \frac{4\pi k_e}{2l+1} \,.$$

The final result for the coefficient functions is

$$g_{1,l}(r, r') = \frac{4\pi k_e}{(2l+1)} \left[\frac{r^l}{(r')^{l+1}} - \frac{R^{2l+1}}{(r r')^{l+1}} \right] \qquad r < r'$$

$$g_{2,l}(r, r') = \frac{4\pi k_e}{(2l+1)} \left[\frac{r'^l}{r^{l+1}} - \frac{R^{2l+1}}{(r r')^{l+1}} \right] \qquad r' < r \,.$$

It is possible to derive the complete Green's function, using $g_{1,l}$ or $g_{2,l}$ and the symmetry relation

$$G(\boldsymbol{r}, \boldsymbol{r}') = \sum_{lm} \frac{4\pi k_e}{(2l+1)} \left[\frac{r^l}{(r')^{l+1}} - \frac{R^{2l+1}}{(rr')^{l+1}} \right] Y_{lm}(\Omega) Y^*_{lm}(\Omega')$$

$$= k_e \sum_l \left[\frac{r^l}{(r')^{l+1}} - \frac{R^{2l+1}}{(rr')^{l+1}} \right] P_l(\cos\alpha) , \qquad (4.33)$$

if $g_{1,l}$ is used. The second line follows with the addition theorem of the spherical harmonics (see Appendix C.5). If one uses $g_{2,l}$ one finds

$$G(\boldsymbol{r}', \boldsymbol{r}) = k_e \sum_l \left[\frac{(r')^l}{r^{l+1}} - \frac{R^{2l+1}}{(rr')^{l+1}} \right] P_l(\cos\alpha) .$$

As can be expected, the same result is found with the re-summation of the expansion using the generating function of the Legendre polynomials

$$G(\boldsymbol{r}, \boldsymbol{r}') = \frac{k_e}{|\boldsymbol{r} - \boldsymbol{r}'|} - \frac{k_e R}{r'} \frac{1}{\left| \boldsymbol{r} - \dfrac{R^2}{r'^2} \boldsymbol{r}' \right|} . \qquad (4.34)$$

This result agrees with the result of the calculation with the simpler mirror charge problem. The advantage of the method of Green's functions is the fact, that one gains the solution of a whole class of problems for the same geometry of the boundary conditions.

One still needs to insert the Green's function into the formula for the solution (4.26) and evaluate this expression. For the present problem one has to calculate the normal derivative of the Green's function on the surfaces. The definition of this derivative is towards the exterior of the enclosed volume, that is opposite to the direction of $\boldsymbol{r}$

$$\frac{\partial G(\boldsymbol{r}', \boldsymbol{r})}{\partial n'} = - \left. \frac{\partial G(\boldsymbol{r}', \boldsymbol{r})}{\partial r'} \right|_{r'=R} = - \frac{k_e (r^2 - R^2)}{R \left[r^2 + R^2 - 2rR\cos\alpha \right]^{3/2}} ,$$

where α is the angle between the vectors $\boldsymbol{r}$ and $\boldsymbol{R}$. The result in vector form is

$$\frac{\partial G(\boldsymbol{r}, \boldsymbol{r}')}{\partial n'} = -k_e \frac{(\boldsymbol{r} - \boldsymbol{R}) \cdot (\boldsymbol{r} + \boldsymbol{R})}{R\, |\boldsymbol{r} - \boldsymbol{R}|^3} .$$

The outer (infinite) spherical shell does not contribute. The complete result for the present boundary value problem is (use $\mathrm{d}\,f' = R^2\,\mathrm{d}\,\Omega'$)

$$V(\boldsymbol{r}) = k_e \iiint \frac{\rho(\boldsymbol{r}')}{|\boldsymbol{r} - \boldsymbol{r}'|}\,\mathrm{d}V' - k_e \iiint \frac{R\rho(\boldsymbol{r}')}{r'|\boldsymbol{r} - \frac{R^2}{r'^2}\boldsymbol{r}'|}\,\mathrm{d}V' \tag{4.35}$$

$$+\frac{1}{4\pi}\iint V(R, \theta', \varphi')\,\frac{R(r^2 - R^2)}{\left[r^2 + R^2 - 2rR\cos\alpha(\theta', \varphi')\right]^{3/2}}\,\mathrm{d}\Omega' .$$

It is possible to calculate the potential in the space between the two spheres, if the potential $V(\boldsymbol{R})$ on the inner sphere and the distribution of the charges between the spheres is given. The evaluation of the integrals is not necessarily trivial, but this does not subtract from value of the global character of the formula for the specified geometry of the surfaces. Simple examples are:

1. A grounded sphere with a point charge in the space between the two spheres. With

$$V(\boldsymbol{R}) = 0 \qquad \rho(\boldsymbol{r}) = q\,\delta(\boldsymbol{r} - \boldsymbol{a})$$

one can recover the previous result

$$V(\boldsymbol{r}) = k_e \frac{q}{|\boldsymbol{r} - \boldsymbol{a}|} - k_e \frac{R}{a}\,\frac{q}{|\boldsymbol{r} - \frac{R^2}{a^2}\boldsymbol{a}|} .$$

2. If the potential of the inner sphere is arbitrary and there are no charges in the space between the spheres

$$V(\boldsymbol{R}) = V_0 \qquad \rho(\boldsymbol{r}) = 0$$

it is only necessary to evaluate the surface term. The result is much simpler than the general formula suggests (see Detail 4.7.4)

$$V(\boldsymbol{r}) = \frac{V_0 R}{r} = k_e \frac{Q}{r} .$$

3. A point charge in the space between the spheres and a sphere, which is divided by an infinitesimal isolating tape, so that the potential on the two half spheres can be different

$$V(R, \theta > \pi/2) = V_o \qquad V(R, \theta < \pi/2) = V_u$$

$$\rho(\boldsymbol{r}) = q\,\delta(\boldsymbol{r} - \boldsymbol{a}) .$$

If the point charge is located on the z-axis one uses the input

$$\rho(\boldsymbol{r}) = q\,\delta(x)\delta(y)\delta(z-a)\,.$$

The modification of the calculation of the electric potential for a given charge distribution in the presence of metallic objects by induction in the metal has been discussed in this section. The continuation of this topic is the question: How can one deal with the modification, if the charge and the metal constellation is not located in a vacuum, but completely or partially in a polarisable medium? The modification of the potential problem due to the presence of insulators (dielectric materials) will be addressed in the next section after some remarks on 'capacitors'.

4.4 Capacitors

The discussion has so far assumed that the charges and metallic objects are located in a vacuum. An extension of the formulation of electrostatics is, however necessary, if non-conducting materials are also found in the space.

As an introduction to this extension some more general remarks on capacitors, besides the spherical capacitor, which was introduced in Chap. 1.4.1, will be useful. This capacitor consists of two concentric metal spheres. The inner one (radius R_1, see Fig. 4.15a) carries the charge $Q = 4\pi\,\sigma_1\,R_1^2$. The outer one (inner radius R_2) is kept at a constant potential V_0. The calculation of the potential in the space between the two spheres is a Dirichlet problem (see Detail 4.7.5.1 and 4.7.5.3) with the result

$$V(\boldsymbol{r}) = k_e\frac{Q}{r} - k_e\frac{Q}{R_2} + V_0\,.$$

The voltage (potential difference) between the two spherical surfaces is

$$U = V(\boldsymbol{R_1}) - V(\boldsymbol{R_2}) = k_e\left(\frac{Q}{R_1} - \frac{Q}{R_2}\right).$$

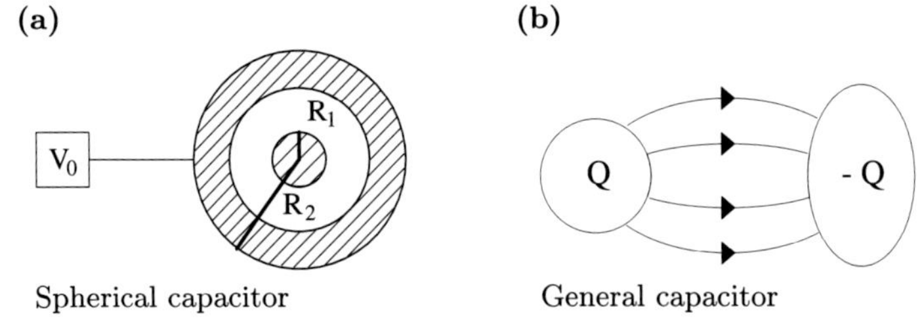

Fig. 4.15 Capacitors

The ratio of charge to voltage of such an arrangement is the **capacity**, which is generally denoted by C

$$C = \frac{Q}{U} = \left[\frac{k_e}{R_1} - \frac{k_e}{R_2}\right]^{-1} = \frac{R_1 R_2}{k_e(R_2 - R_1)} .$$

The capacity of the spherical capacitor is

$$C = \frac{R_1 R_2}{(R_2 - R_1)} \qquad \text{in the CGS-system}$$

$$C = 4\pi\,\varepsilon_0 \frac{R_1 R_2}{(R_2 - R_1)} \qquad \text{in the SI-system} .$$

The associated units are

$$[C] = \frac{\text{statcoul}}{\text{statvolt}} = \text{cm} \qquad \text{in the CGS-system}$$

and

$$[C] = \frac{\text{Coulomb}}{\text{Volt}} = \text{Farad} \qquad \text{in the SI-system} .$$

The general statement is: Every arrangement of two metal surfaces, which carry an equal charge of opposite sign represent a capacitor (Fig. 4.15b). The ratio Q/U for each arrangement depends only on the geometry.

The correctness of this statement can be demonstrated with the following argument: The relation between charge and potential on the two metal surfaces is

$$Q_1 = \oiint_1 \sigma_1 \,\mathrm{d}f_1 = -\frac{1}{4\pi k_e} \oiint_1 \left(\frac{\partial V}{\partial n}\right)_1 \mathrm{d}f_1 = Q$$

$$Q_2 = \oiint_2 \sigma_2 \,\mathrm{d}f_2 = -\frac{1}{4\pi k_e} \oiint_2 \left(\frac{\partial V}{\partial n}\right)_2 \mathrm{d}f_2 = -Q .$$

If the charge on surface 1 is changed by the factor a, one finds

$$Q_1' = a Q_1 = -\frac{1}{4\pi k_e} \oiint_1 \left(\frac{\partial a V}{\partial n}\right)_1 \mathrm{d}f_1 .$$

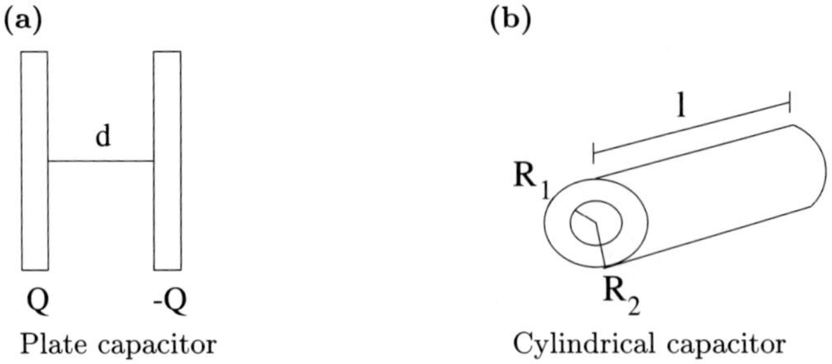

Fig. 4.16 Capacitors

An increase (decrease) of the charge by a factor a causes an increase (decrease) of the potential by the same factor, as the potential in the expression on the right hand side is the only quantity, which can vary. The second surface reacts to the increase with

$$-\frac{1}{4\pi k_e}\oiint_2 \left(\frac{\partial aV}{\partial n}\right)_2 \mathrm{d}f_2 = Q_2' = a\,Q_2\,.$$

The charge on the second surface is changed (by induction) by the same factor, so that one has

$$C = \frac{Q_1}{U_1} = \frac{Q_1'}{U_1'}\,.$$

The capacity C is independent of the actual value of the charge and the voltage. It can therefore depend only on the geometry.

The calculation of the capacity of capacitors with a simple geometry is not difficult:

A plate capacitor consists of two parallel plane plates, each with a surface F and a separation of the plates d (Fig. 4.16a). If one neglects the influence of the edges the capacity is

$$C = \frac{Q}{U} = \frac{F}{4\pi k_e\, d}\,.$$

This result can be extracted from the formula for the capacity of the spherical capacitor. For

$$R_2 - R_1 = d \ll R_1 \approx R_2 = R$$

one finds

$$C = \frac{R^2}{k_e d} = \frac{F}{4\pi k_e d}\,.$$

For a small space between large spheres one can neglect the curvature of the plates and the effect of the edges.

A cylindrical capacitor is constructed from two concentric metallic cylinders, where the radius of the inner one is R_1, the length is l and the inner radius of the second one is R_2 at the same length (Fig. 4.16b). If one neglects edge effects, one finds (see Detail 4.7.5.2)

$$C = \frac{1}{2k_e} \frac{l}{\ln(R_2/R_1)}\,.$$

The discussion can be generalised by looking at a system of n conductors, each with the potential V_i and with the total charge Q_i. According to the Coulomb law, the potential of the k-th conductor is linearly dependent on all the charges

$$V_k = \sum_{i=1}^{n} a_{ki} Q_i\,.$$

The coefficients a_{ki} depend only on the spatial arrangement and the form of the conductors. If one resolves this system of equations with respect to the charges one obtains

$$Q_i = \sum_{k=1}^{n} b_{ik} V_k\,.$$

Each of the charges is a linear function of all the potentials on the conductors. The coefficients b_{ik}, which depend also only the geometry, are referred to as **coefficients of capacity**. According to this statement one would refer to the coefficient of capacity b_{ii}, which is equal to the total charge on the i-th conductor, provided the conductor is kept on the potential with value 1 and all other conductors on the potential zero.

It is possible to express the potential energy of a system of conductors (see Chap. 2.5) with the aid of the coefficients of capacity by the potentials of all conductors

$$W = \frac{1}{2} \sum_i Q_i V_i = \frac{1}{2} \sum_{ik} b_{ik} V_i V_k\,.$$

This formula is the starting point for approximate calculations or estimations of the capacities in electrotechnical systems with variational methods.

An alternative characterisation of the situation, which is more readily adapted to experimental needs, is based on the consideration of the voltage U between the conductors

$$U_{k_1k_2} = V_{k_1} - V_{k_2} = \sum_{i=1}^{n} (a_{k_1i} - a_{k_2i}) Q_i \,.$$

If one considers the statement

$$\sum_{i=1}^{n} Q_i = 0 \,,$$

which says, that the system of conductors is closed electrically—or more directly, that each field line ends on one of the conductors—then one finds for the resolution of the $n(n-1)/2$ equations with respect to the charges, as e.g.

$$Q_i = \sum_{k \neq i} C_{ik} U_{ik} \,. \tag{4.36}$$

The coefficients in (4.36) are called **partial capacities**. They also depend only on the geometry of the conductors.

A relation, between these two possibilities to discuss the situation, can be demonstrated by looking at a system of two conductors. For the partial capacities the following statements are valid

$$Q_1 = C_{12}\, U_{12} \qquad\qquad Q_2 = C_{21}\, U_{21} \,.$$

Because of $U_{12} = U_{21}$ and $Q_1 + Q_2 = 0$ there follows

$$C_{12} = C_{21} \equiv C \,.$$

If the system is discussed from the point of view of the coefficients of capacity, one resolves the system of equations

$$Q_1 = b_{11} V_1 + b_{12} V_2 \qquad\qquad Q_2 = b_{21} V_1 + b_{22} V_2$$

with respect to the potentials, and finds

$$V_1 = \frac{(b_{22} Q_1 - b_{12} Q_2)}{(b_{11} b_{22} - b_{12} b_{21})} \qquad V_2 = \frac{(-b_{21} Q_1 + b_{22} Q_2)}{(b_{11} b_{22} - b_{12} b_{21})} \,.$$

If one uses

$$Q_1 = -Q_2 = Q \qquad \text{and} \qquad C = Q/(V_1 - V_2) \,,$$

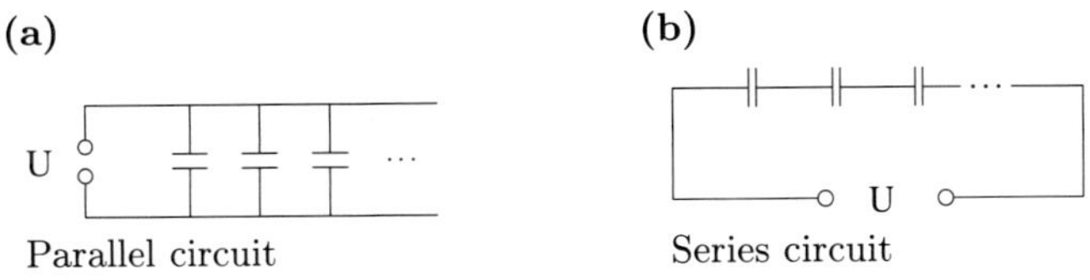

Fig. 4.17 Circuits with capacitors

one obtains

$$C = \frac{(b_{11}b_{22} - b_{12}b_{21})}{(b_{11} + b_{12} + b_{21} + b_{22})} .$$

The capacity can be calculated from the four coefficients of capacity.

Capacitors can be combined in electric circuits with the following rules for combining the capacitors in parallel or in series. For a combination in parallel (Fig. 4.17a) one adds the capacities

$$C = \sum_n C_n \tag{4.37}$$

as one voltage U produces the charges

$$Q_1 = C_1 U, \quad Q_2 = C_2 U \quad \ldots$$

on each of the capacitors. The total charge of the system is

$$Q = \sum_n Q_n .$$

The total capacity is

$$C = \frac{Q}{U} = \sum_n C_n .$$

For a circuit with capacitors in series (Fig. 4.17b) one adds the inverse values of the capacities in order to obtain the inverse of the total capacity

$$\frac{1}{C} = \sum_n \frac{1}{C_n} . \tag{4.38}$$

The total voltage of the system is the sum of the individual voltages

$$U = \sum_n U_n .$$

The voltage on each of the capacitors is $U_n = Q/C_n$. The total capacity is then

$$C = \frac{Q}{U} = \sum_n \frac{1}{\left(\frac{1}{C_1} + \frac{1}{C_2} + \ldots\right)} \qquad \text{or} \qquad \frac{1}{C} = \sum_n \frac{1}{C_n} .$$

4.5 Polarisation of Dielectric Materials

The following simple experiment concerning the effect of a dielectric material in the space between the plates of a plate capacitor initiates the pattern for the extension of the theory. The plates of the capacitor are charged with $+Q_0$ respectively $-Q_0$ by connection to a battery (Fig. 4.18a). The battery is removed and the voltage U_0 between the plates is measured

$$U_0 = \frac{C_0}{Q_0} \qquad \left(C_0 = \frac{F}{4\pi k_e\, d}\right) .$$

A glass plate or a different non-conducting material is then introduced into the space between the plates (Fig. 4.18b). The glass fills the space completely. The charges on the plates of the capacitor have not been changed. If the voltage is measured after the insertion of the glass, a drop of the voltage will be observed

$$U = \frac{U_0}{\varepsilon}, \qquad \varepsilon > 1 .$$

If the experiment is repeated with a different capacitor (e.g. larger plates or larger distance between the plates), but the same material as before, one finds that the voltage decreases with the same factor, as long as the material fills the space between the plates completely. The constant ε depends only on the type of material, that has been inserted. This quantity is called the dielectric constant. An impression of the range of values of this material constant can be gleaned from the Table 4.1.

The experiment will now be modified in the following fashion. The plates are connected to a source with voltage U_0 (Fig. 4.19). If one inserts the material this time, the voltage will not change, but the charge on the plates will change. The charge with the material is $Q = \varepsilon Q_0$, if it is Q_0 without the material. The charge is

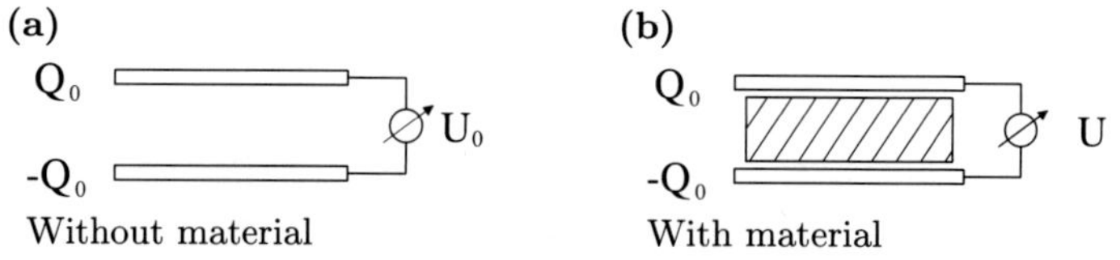

Fig. 4.18 Experiments with dielectric material (constant charge)

Table 4.1 Dielectric constants of selected materials

material	dielectric constant	
vacuum	$\varepsilon = 1$	for comparison
air	$\varepsilon = 1.00054$	
glass	$\varepsilon = 4.5$	typical insulator
paper	$\varepsilon = 3.5$	"
porcelain	$\varepsilon = 6.5$	"
$(H_2O)_{dist}$	$\varepsilon = 78.0$	distilled water
$Ti\,O_2$	$\varepsilon = 100.0$	titanium dioxide (rutile)

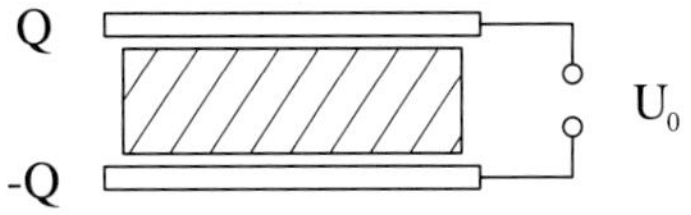

Fig. 4.19 Experiments with dielectric materials (constant voltage)

increased by the factor ε. The results of the two experiments can be summed up in the following fashion. The capacity of the capacitor filled with an insulator is

$$C = \frac{Q}{U} = \left.\frac{(\varepsilon Q_0)}{U_0}\right|_{\text{exp. 2}} = \left.\left(\frac{\varepsilon}{U_0}\right) Q_0\right|_{\text{exp. 1}} = \varepsilon\, C_0 \,.$$

The capacity of the capacitor has increased by the factor ε as a consequence of the insertion of the material. A quantitative understanding of these experiments requires an excursion into solid state physics. A qualitative interpretation is possible with simpler means e.g. after an explanation of the difference between a normal and a polar dielectric material.

The molecules of some materials (as $(H_2O)_{dist.}$) have a permanent dipole moment $\boldsymbol{p}$. Normally the dipoles are distributed in a statistical way, so that their effect is not detected. They can, however, be oriented by the electric field of a capacitor, at least partially (Fig. 4.20a). This produces a field in the opposite direction, which degrades the original field (decreases the voltage). Materials with this property are called **polar dielectric materials**. The degree of the orientation depends on the strength of the external field and the temperature (the thermal motion of the molecules).

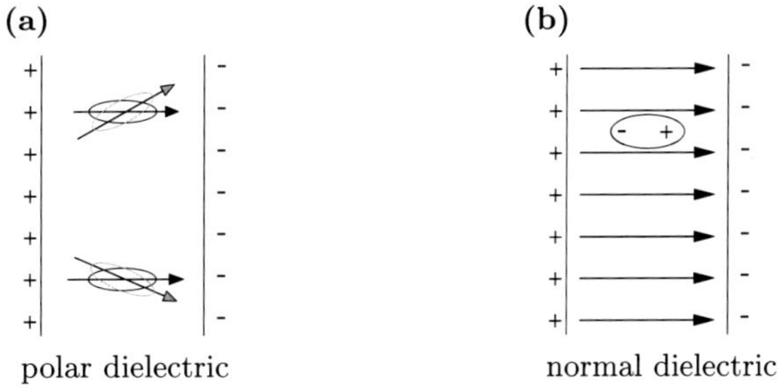

Fig. 4.20 The classification of dielectric materials

Substances without an atomic or molecular dipole moment can experience charge separation in an external field. The electron cloud around the nucleus (which can be moved more easily) is displaced by the external field (Fig. 4.20b).

The shift is small in comparison with the atomic radius, so that no macroscopic transport of charge takes place. It is proportional to the external field, provided this is not too strong. The dipole moments, which are induced in this fashion are (generally) oriented in the direction of the external field. They reduce, as in the case of polar substances the external field. Substances, in which dipole moments are created by the action of external fields are called **normal dielectric materials**.

A few numbers might indicate the size of the shifts on the atomic level. External fields in capacitors are typically smaller than $E_{\text{ext}} < 10^6$ Volt/cm. If they are stronger, a spark-over can occur. The strength of interatomic fields is typically

$$E_{\text{atomic}} \approx k_e \frac{e_0}{R^2} = \frac{4.8 \cdot 10^{-10}}{10^{-16}} \text{ esu} \approx 10^{10} \text{ Volt/cm},$$

that is definitely much stronger. The electron cloud is still mainly controlled by the interatomic forces, so that the shift is small.

The induced dipoles in some insulators (e.g. quartz) are not oriented in the direction of the external field. This is caused by the crystal structure, a much more complex effect. In the discussion from here on, the following model will be used to describe polarisation (be it polar or normal): The superposition of the induced or oriented dipoles can be imagined to create an induced surface charge (negative polarisation charge opposite the positive plate of the capacitor, positive polarisation charge in front of the negative plate of the capacitor (Fig. 4.20)). In the interior of material the effects of the shift or the orientation cancel from a macroscopic point of view.

One can imagine, that a plate capacitor filled with a dielectric material can be discussed in terms of two types of charges and three electric fields, if one accepts this simplified macroscopic picture. The two types of charge are referred to as free charges and polarisation charges. The **free** or **true charges** (Q_{tr}) are the charges on the plates of the capacitor. They are freely movable and have been added by the action of a battery. The surface charges on the dielectric, which can not be moved freely, are the **polarisation** charges (Q_{pol}, Fig. 4.21a).

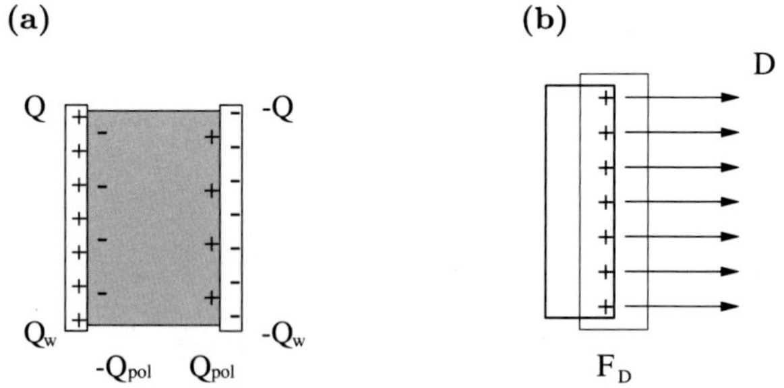

Fig. 4.21 Polarisation charges Q_{pol} and dielectric displacement $\boldsymbol{D}$

The three electric fields are characterised by the statements

(1) The electric field, which is the field produced by the true charges, is the **dielectric displacement *D***. It is represented by the Gaussian surface ***D*** in Fig. 4.21b and is defined by the relation

$$\oiint_{F_D} \boldsymbol{D} \cdot \mathbf{d}\boldsymbol{f} = 4\pi k_d\, Q_{\text{tr}}\,. \tag{4.39}$$

A new constant has to be introduced at this point in order to accommodate the different systems of units of electrodynamics.[3] For the two systems, used here, it is

$$k_{d,\text{SI}} = \frac{1}{4\pi} \qquad k_{d,\text{CGS}} = k_{e,\text{CGS}} = 1\,. \tag{4.40}$$

The direction of the vector of the dielectric displacement ***D*** starts at the positive charge and ends on the negative charge. The strength of the field of an ideal plate capacitor without filling (neglecting edge effects) is

$$|D| = 4\pi k_d \frac{|Q_{\text{tr}}|}{F}\,.$$

(2) The field, which is produced by the polarisation charges, is called the polarisation ***P***. By convention it points from the negative to the positive polarisation charges (!!). This field is determined by the surface integral over the Gaussian surface shown in Fig. 4.22a and defined by the equation

$$\oiint_{F_P} \boldsymbol{P} \cdot \mathbf{d}\boldsymbol{f} = -\,Q_{\text{pol}}\,. \tag{4.41}$$

The minus sign characterises the special direction of this field. For a plate capacitor one has

$$|P| = \frac{|Q_{\text{pol}}|}{F}\,.$$

(3) The combination of these two fields is the electric field ***E***

$$\boldsymbol{E} = \frac{k_e}{k_d}\,\boldsymbol{D} - 4\pi k_e\,\boldsymbol{P} = 4\pi k_e \left(\frac{1}{4\pi k_d}\,\boldsymbol{D} - \boldsymbol{P}\right)\,.$$

[3] A full discussion of the systems of units can be found in Appendix.

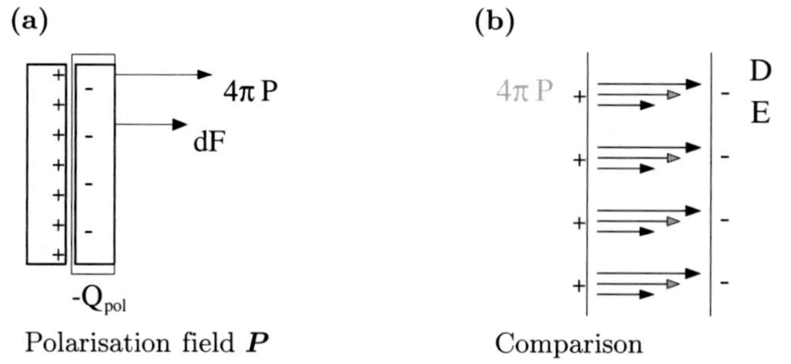

Fig. 4.22 The three electric fields

It is the field, which is measured in experiments. It is produced by the true **and** the polarisation charges

$$\oiint \boldsymbol{E} \cdot \mathbf{d}\boldsymbol{f} = 4\pi k_e \left(Q_{\text{tr}} + Q_{\text{pol}}\right) . \tag{4.42}$$

The strength is given by

$$|E| = \frac{4\pi k_e}{F} \left|Q_{\text{tr}} + Q_{\text{pol}}\right| .$$

The direction of all three fields is the same (by definition), provided one has $Q_{\text{tr}} > Q_{\text{pol}}$ (see (Fig. 4.22b)).

The three electric fields, which are produced by three types of different charges, can be related in the case of a uniform isotropic dielectric material with the aid of the dielectric constant. Instead of the preliminary notation, which has been used in the discussion of the simple experiments, the correct quantities will now be presented. The true charge on the plates was kept constant ($Q_0 = Q_{\text{tr}} = \text{const.}$) in the first experiment. This leads, according to (4.39), to a constant value of the dielectric displacement, so that the following statements are possible

$$\text{without material:} \quad P = 0 \longrightarrow D_0 = \frac{k_e}{k_d} E_0$$

$$\text{with material:} \qquad D_0 = \varepsilon \frac{k_e}{k_d} E$$

and there follows the relation

$$\varepsilon = \frac{E_0}{E} = \frac{U_0}{U} \qquad \left(U = \int_1^2 \boldsymbol{E} \cdot \mathbf{d}\boldsymbol{s} = E\, d\right),$$

which is again independent of the system of units.

The voltage between the plates of the capacitor was kept constant in the second experiment. In this case the electric field does not change but the dielectric displacement changes because of the reloading the plates, so that one finds in this case

$$\text{without material:}\quad P = 0 \longrightarrow D_0 = \frac{k_e}{k_d}\, E_0$$

$$\text{with material:}\qquad D = \varepsilon\, \frac{k_e}{k_d}\, E_0 \,.$$

The statement is in this case (again independent of the system of units)

$$\varepsilon = \frac{D}{D_0} = \frac{Q_{tr}}{Q_{tr0}} \,.$$

The experimental finding for the situation with filling is

$$\boldsymbol{D} = \varepsilon\, \frac{k_d}{k_e}\, \boldsymbol{E} \,,$$

or explicitly, addressing the two main systems of units,

$$\boldsymbol{D}_{CGS} = \varepsilon\, \boldsymbol{E}_{CGS} \qquad \boldsymbol{D}_{SI} = \varepsilon\, (\varepsilon_0\, \boldsymbol{E}_{SI}) \,.$$

In the SI-system the material constant ε is often referred to as ε_r (relative permittivity). The actual number is, as shown, independent of the system of units.

The polarisation field is connected with the two other fields by the relations

$$\boldsymbol{P} = \frac{1}{4\pi k_e}\left(\frac{k_e}{k_d}\boldsymbol{D} - \boldsymbol{E}\right) = \frac{1}{4\pi k_e}(\varepsilon - 1)\, \boldsymbol{E} = \frac{1}{4\pi k_d}\frac{(\varepsilon - 1)}{\varepsilon}\, \boldsymbol{D} \,.$$

The quantity

$$\kappa = \frac{1}{4\pi k_d}\,(\varepsilon - 1) \geq 0$$

is named **electric susceptibility**. It is a direct measure of the polarisation of the medium as it corresponds to the dielectric constant of the material in comparison with the dielectric constant of the vacuum ($\varepsilon = 1$). The polarisation charge of a capacitor with plates is

$$Q_{\text{pol}} = -F\, |\boldsymbol{P}| = \frac{F(\varepsilon - 1)}{4\pi k_d \varepsilon}\, |\boldsymbol{D}| = \frac{(\varepsilon - 1)}{\varepsilon}\, Q_{\text{tr}} \,.$$

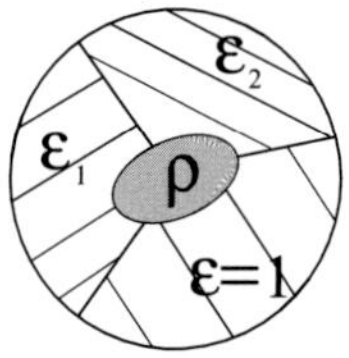

Fig. 4.23 General potential problem

The potential problem with other types of capacitors can be formulated and discussed in analogy to the case of a plate capacitor: In a domain of space, which is filled totally or in part with dielectric materials there exist point and space charges (Fig. 4.23). Dielectric materials are polarised by the fields produced by the true charges. The task is: Calculate (for given boundary conditions, given distribution of the dielectric materials and the distribution of the true charges) the electric field, which can be measured and the corresponding potential for every point of a domain. One has to use all three fields, which are created by the three kinds of charges, if one attempts to tackle this problem.

(1) The field, which is produced by the true charges is the $\boldsymbol{D}$-field, the dielectric displacement. This field is characterised by the differential equation

$$\operatorname{div} \boldsymbol{D}(\boldsymbol{r}) = 4\pi k_d \, \rho_{\text{tr}}(\boldsymbol{r}) \tag{4.43}$$

and boundary conditions. This field is not a measurable field. From the point of theory it is a useful aid.

(2) The measurable electric field, which is denoted by $\boldsymbol{E}$, is produced by the true and the polarisation charges

$$\operatorname{div} \boldsymbol{E}(\boldsymbol{r}) = 4\pi k_e \left(\rho_{\text{tr}}(\boldsymbol{r}) + \rho_{\text{pol}}(\boldsymbol{r})\right) . \tag{4.44}$$

The density of the polarisation charges ρ_{pol} has to be represented in the form

$$\rho_{\text{pol}}(\boldsymbol{r}) = \sigma(\boldsymbol{r}) \, \delta(\text{area}) , \tag{4.45}$$

as the polarisation charges are surface charges. These charges are, in general, not known, so that the differential equation for $\boldsymbol{E}$ can not be solved directly. One can, however, use the fact, that this field is always irrotational

$$\operatorname{rot} \boldsymbol{E}(\boldsymbol{r}) = \boldsymbol{0} .$$

It can be represented even in the general situation in terms of a potential

$$\boldsymbol{E}(\boldsymbol{r}) = -\nabla V(\boldsymbol{r}) .$$

(3) The polarisation field is defined by the relation

$$\boldsymbol{P}(\boldsymbol{r}) = \frac{1}{4\pi k_d}\boldsymbol{D}(\boldsymbol{r}) - \frac{1}{4\pi k_e}\boldsymbol{E}(\boldsymbol{r})\,. \tag{4.46}$$

This leads to the statement

$$\operatorname{div} \boldsymbol{P}(\boldsymbol{r}) = -\rho_{\text{pol}}(\boldsymbol{r})\,. \tag{4.47}$$

The sources of the polarisation field are the negative (!) polarisation charges.

With a theory founded on a microscopic basis (e.g. solid state theory) one would

(a) calculate the response of the material to the $\boldsymbol{D}$-field and determine in this fashion ρ_{pol}.
(b) In a second step one can with the knowledge of the distributions of both kinds of charges obtain the $\boldsymbol{E}$-field.

In a macroscopic approach one uses (similar as in the treatment of the plate capacitor) heuristic relations between the $\boldsymbol{D}$-field and the $\boldsymbol{E}$-field. These relations are known as **material equations**. The simplest option is: In the interior of a dielectric the relation

$$\boldsymbol{D} = \varepsilon\,\frac{k_d}{k_e}\,\boldsymbol{E} \tag{4.48}$$

is valid. This statement assumes, that the polarisation is uniform and isotropic, as well as proportional to the original field.

For crystalline dielectric materials one often has to use relations as

$$\begin{pmatrix} D_{\mathrm{x}} \\ D_{\mathrm{y}} \\ D_{\mathrm{z}} \end{pmatrix} = \frac{k_d}{k_e}\begin{pmatrix} \varepsilon_{\mathrm{xx}} & \varepsilon_{\mathrm{xy}} & \varepsilon_{\mathrm{xz}} \\ \varepsilon_{\mathrm{yx}} & \varepsilon_{\mathrm{yy}} & \varepsilon_{\mathrm{yz}} \\ \varepsilon_{\mathrm{zx}} & \varepsilon_{\mathrm{zy}} & \varepsilon_{\mathrm{zz}} \end{pmatrix}\begin{pmatrix} E_{\mathrm{x}} \\ E_{\mathrm{y}} \\ E_{\mathrm{z}} \end{pmatrix}$$

or short in matrix form

$$\boldsymbol{D} = \frac{k_d}{k_e}\,\hat{\varepsilon}\,\boldsymbol{E}\,. \tag{4.49}$$

The dielectric constant is replaced by a dielectric tensor (matrix). If the material is (in addition) inhomogeneous, one has to use a tensor with space dependent elements.

$$\varepsilon \longrightarrow \varepsilon(\boldsymbol{r})\,.$$

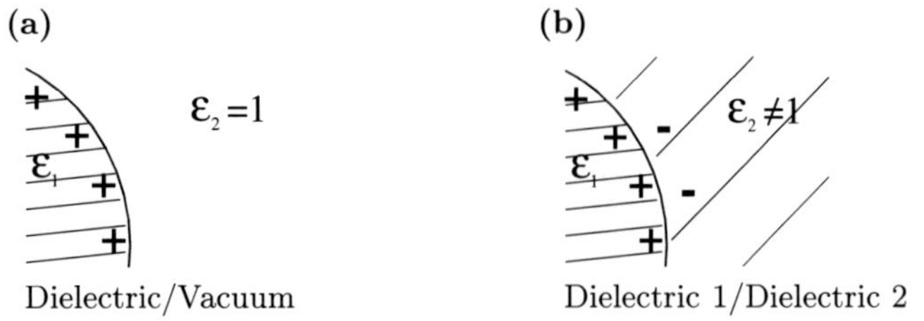

Fig. 4.24 General potential problem: situation on interlayers

It is, however, sufficient to use the simple relation for most materials and most strength of the fields. This relation will therefore be used in the following.

It is possible to extract statements concerning the behaviour of the three fields at surfaces separating different materials on the basis of the characteristics equations for the fields and the properties of the materials. Such interlayers can consist of

(a) simple layers of polarisation charges, if the surface divides a dielectric from the vacuum (Fig. 4.24a) or
(b) a layer of dipoles, if the surface divides two dielectric media (Fig. 4.24b).

It is assumed in the following discussion, that there are no true charges on the interlayers.

Polarisation charges do not contribute to the production of the $\boldsymbol{D}$-field. For this reason one finds (see discussion of the Gauss' theorem (1.4))

$$D_{a,\mathrm{n}} - D_{i,\mathrm{n}} = 0\,.$$

The normal component of the $\boldsymbol{D}$-field is continuous at the interlayer between two dielectric materials. Polarisation charges contribute to the creation of the $\boldsymbol{E}$-field. The normal component of this field has a discontinuity on the surface between two dielectric materials, which can be described on the basis of (4.44). If there are no true charges on the interlayer the relation

$$\oiint \boldsymbol{E}(\boldsymbol{r}) \cdot \mathbf{d}\boldsymbol{f} = 4\pi k_e Q_{\mathrm{pol}}$$

yields for a flat Gaussian box around this surface

$$E_{a,\mathrm{n}} - E_{i,\mathrm{n}} = 4\pi k_e \sigma_{\mathrm{pol}}\,,$$

where σ_{pol} represents either a simple or a dipole layer of charges. With the material equation (4.48) and the statement concerning the normal component of $\boldsymbol{D}$ one finds on the other hand

$$\varepsilon_a\; E_{a,\mathrm{n}} - \varepsilon_i\; E_{i,\mathrm{n}} = 0\,.$$

Combination of the relations allows to write

$$\sigma_{\rm pol} = \frac{1}{4\pi k_e}\left(\frac{\varepsilon_i - \varepsilon_a}{\varepsilon_a}\right) E_{i,\rm n} = \frac{1}{4\pi k_e}\left(\frac{\varepsilon_i - \varepsilon_a}{\varepsilon_i}\right) E_{a,\rm n}\,.$$

A similar equation can be obtained for the polarisation field. The equation

$$\boldsymbol{P} = \frac{1}{4\pi k_d}\boldsymbol{D} - \frac{1}{4\pi k_e}\boldsymbol{E}$$

corresponds to

$$P_{a,\rm n} - P_{i,\rm n} = -\sigma_{\rm pol}\,.$$

Here, one has the additional equation

$$\boldsymbol{D} = \frac{\varepsilon}{(\varepsilon - 1)} 4\pi k_d\, \boldsymbol{P},$$

so that one finds

$$\frac{\varepsilon_a}{\varepsilon_a - 1}\,P_{a,\rm n} - \frac{\varepsilon_i}{\varepsilon_i - 1}\,P_{i,\rm n} = 0\,.$$

The discussion of the tangential components starts with the statement $\operatorname{rot}\boldsymbol{E} = \boldsymbol{0}$, from which follows (see Chap. 2.1)

$$E_{a,\rm t} - E_{i,\rm t} = 0\,.$$

The tangential components of the true electric field are continuous. This implies the relation

$$\frac{D_{a,\rm t}}{\varepsilon_a} - \frac{D_{i,\rm t}}{\varepsilon_i} = 0$$

for the components of the $\boldsymbol{D}$-field. Finally one obtains for the tangential components of the $\boldsymbol{P}$-field with

$$\boldsymbol{E} = \frac{4\pi k_e}{(\varepsilon - 1)}\,\boldsymbol{P}$$

the relation

$$\frac{1}{(\varepsilon_a - 1)}\,P_{a,\rm t} - \frac{1}{(\varepsilon_i - 1)}\,P_{i,\rm t} = 0\,.$$

The various discontinuities of the different components of the fields can be summarised in the following ratios

$$\left.\begin{aligned} D_{a,\mathrm{n}} &= D_{i,\mathrm{n}} \\ \frac{D_{a,\mathrm{t}}}{\varepsilon_a} &= \frac{D_{i,\mathrm{t}}}{\varepsilon_i} \end{aligned}\right. \quad \longrightarrow \quad \frac{D_{a,\mathrm{n}}}{D_{a,\mathrm{t}}} = \frac{\varepsilon_i}{\varepsilon_a}\frac{D_{i,\mathrm{n}}}{D_{i,\mathrm{t}}}$$

$$\left.\begin{aligned} \varepsilon_a E_{a,\mathrm{n}} &= \varepsilon_i E_{i,\mathrm{n}} \\ E_{a,\mathrm{t}} &= E_{i,\mathrm{t}} \end{aligned}\right. \quad \longrightarrow \quad \frac{E_{a,\mathrm{n}}}{E_{a,\mathrm{t}}} = \frac{\varepsilon_i}{\varepsilon_a}\frac{E_{i,\mathrm{n}}}{E_{i,1}}$$

$$\left.\begin{aligned} \frac{\varepsilon_a}{(\varepsilon_a - 1)} P_{a,\mathrm{n}} &= \frac{\varepsilon_i}{(\varepsilon_i - 1)} P_{i,\mathrm{n}} \\ \frac{1}{(\varepsilon_a - 1)} P_{a,\mathrm{t}} &= \frac{1}{(\varepsilon_i - 1)} P_{i,\mathrm{t}} \end{aligned}\right. \quad \longrightarrow \quad \frac{P_{a,\mathrm{n}}}{P_{a,\mathrm{t}}} = \frac{\varepsilon_i}{\varepsilon_a}\frac{P_{i,\mathrm{n}}}{P_{i,\mathrm{t}}} .$$

The relative discontinuity of the three fields is the same, even if their individual components behave differently. One can write the normal and the tangential component in the form

$$F_{k,\mathrm{n}} = F_k \cos\theta_k \qquad F_{k,\mathrm{t}} = F_k \sin\theta_k \quad (k = i, a)\,,$$

where θ_k is the angle between the field vector and the direction of the normal (Fig. 4.25). For all three fields one then finds

$$\frac{\tan\theta_i}{\tan\theta_a} = \frac{\varepsilon_i}{\varepsilon_a}\,. \tag{4.50}$$

These relations are known as the **laws of diffraction** of the field lines.

The formulation of the realistic potential problem with a distribution of dielectric materials is even a bit more complicated, as the polarisation charges can not be given explicitly. They adapt themselves through the response of the material to an external situation. For this reason the macroscopic electrostatics is founded on the following pragmatic basic equations:

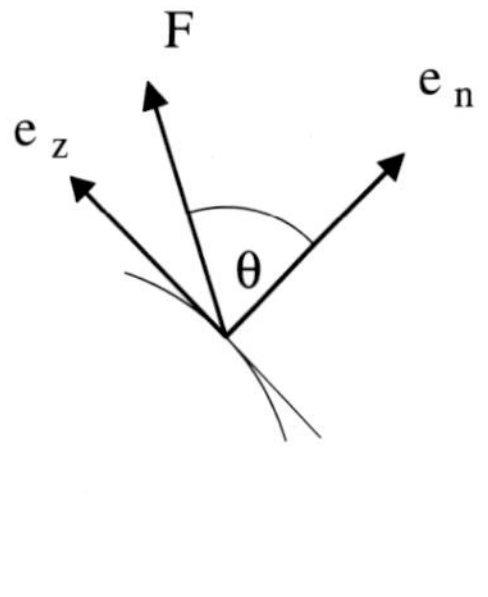

Fig. 4.25 The law of diffraction of field lines

Field equations:

$$\nabla \cdot \boldsymbol{D} = 4\pi k_d \rho_{\mathrm{tr}}(\boldsymbol{r}) \qquad \nabla \times \boldsymbol{E} = \boldsymbol{0} \tag{4.51}$$

Material equation:

$$\boldsymbol{D} = \frac{k_d}{k_e}\, \hat{\varepsilon}\, \boldsymbol{E}\,. \tag{4.52}$$

Additional input specifications are: Besides the distribution of the true charges one has to state boundary conditions for the measurable $\boldsymbol{E}$-field. The irrotational behaviour of the electric field (4.51) allows the use of a description in terms of a potential

$$\boldsymbol{E}(\boldsymbol{r}) = -\nabla V(\boldsymbol{r})\,,$$

so that one is faced with a modified Poisson-/Laplace problem, if one uses the simplest material equation

$$\Delta V(\boldsymbol{r}) = -\frac{4\pi k_e}{\varepsilon}\rho_{\mathrm{tr}}(\boldsymbol{r})\,.$$

If one has found the solution of the differential equation (for a given set of boundary conditions), one can calculate all other quantities of interest, as e.g.

$$\boldsymbol{D}(\boldsymbol{r}) = -\varepsilon\, \frac{k_d}{k_e}\, \nabla V(\boldsymbol{r})$$

$$\boldsymbol{P}(\boldsymbol{r}) = -\frac{1}{4\pi k_e}\,(\varepsilon - 1)\, \nabla V(\boldsymbol{r})$$

$$\sigma_{\mathrm{pol}} = P_{i,\mathrm{n}} - P_{a,\mathrm{n}}\,.$$

The polarisation field only enters in the game, if one relies on a microscopic theory of the polarisation in order to avoid the use of the material equation.

The presence of dielectric materials requires a modification of the discussion of the energy situation in Chap. 2.5. The Poisson equation has the form

$$\rho(\boldsymbol{r}) = -\frac{\varepsilon}{4\pi k_e}\Delta V(\boldsymbol{r})\,,$$

if one uses the simplest material equation. The modification of the energy density

$$W = \frac{1}{2}\iiint \rho(\boldsymbol{r})V(\boldsymbol{r})\,\mathrm{d}V$$

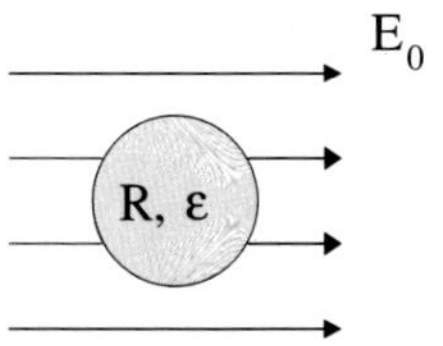

Fig. 4.26 Dielectric sphere in a uniform electric field

is then simply

$$w(\boldsymbol{r}) = \frac{\varepsilon}{8\pi k_e} |\boldsymbol{E}(\boldsymbol{r})|^2 = \frac{1}{8\pi k_d} (\boldsymbol{E}(\boldsymbol{r}) \cdot \boldsymbol{D}(\boldsymbol{r})) . \tag{4.53}$$

The calculation of the two fields and the polarisation in the presence of dielectric materials can be illustrated with the (classic) example: A sphere of dielectric material (radius R, dielectricity constant ε) is placed in a uniform electric field (Fig. 4.26). The problem is: Calculate the modification of the field which is caused by the polarisation of the sphere.

The following steps are necessary in order to solve this problem:

(i) This is a problem without true charges, if one accepts the statement that the external field can be described with the aid of boundary conditions (plates with an infinite separation). The Laplace equation is valid in the interior and the exterior of the sphere

$$\Delta V_{\mathrm{i,a}}(\boldsymbol{r}) = 0 .$$

(ii) This is a problem with azimuthal symmetry, which should be handled with the simpler multipole expansion.
(iii) The technical aspects are not different from those used for the problem of a metal sphere in a uniform field (see p. 158), but the result is quite different.

The solution can than be given in the form. In the interior of the sphere one can write

$$V_{\mathrm{i}}(\boldsymbol{r}) = \sum_{l=0}^{\infty} A_l r^l P_l(x) \qquad (x = \cos\theta) .$$

Terms with $1/(r^{l+1})$ are not possible, as $V_{\mathrm{i}}(\boldsymbol{r})$ would be singular for $r \to 0$. The solution in the exterior

$$V_{\mathrm{a}}(\boldsymbol{r}) = \sum_{l=0}^{\infty} \frac{B_l}{r^{l+1}} P_l(x) - E_0 r P_1(x)$$

satisfies the boundary condition

$$\boldsymbol{E}_a(\boldsymbol{r}) = -\nabla V_a(\boldsymbol{r}) \xrightarrow{r\to\infty} \boldsymbol{E}_0 .$$

The condition of a discontinuity at the surface of the sphere allows the determination of the expansion coefficients

$$1.\ E_{\mathrm{a,n}} = \varepsilon\, E_{\mathrm{i,n}} \longrightarrow \quad -\left.\frac{\partial V_{\mathrm{a}}}{\partial r}\right|_R = -\varepsilon \left.\frac{\partial V_{\mathrm{i}}}{\partial r}\right|_R$$

$$2.\ E_{\mathrm{a,t}} = E_{\mathrm{i,t}} \quad \longrightarrow -\left.\frac{1}{R}\frac{\partial V_{\mathrm{a}}}{\partial \theta}\right|_R = -\left.\frac{1}{R}\frac{\partial V_{\mathrm{i}}}{\partial \theta}\right|_R .$$

The first of these conditions gives

$$\sum_l \left[E_0\,\delta_{l,1} + \frac{(l+1)}{R^{l+2}}\,B_l\right] P_l(x) = -\varepsilon \sum_l l\, A_l\, R^{l-1}\, P_l(x)$$

or with comparison of the coefficients

$$\begin{aligned} &l=1 && \varepsilon A_1 + \frac{E_0}{k} + \frac{2B_1}{R^3} = 0 \\ &l \neq 1 && \varepsilon l A_l + \frac{(l+1)B_l}{R^{2l+1}} = 0 \qquad (l = 0, 2, \ldots) . \end{aligned}$$

The second condition leads to

$$\sum_l \left[A_l R^l + E_0 R \delta_{l,1} - \frac{B_l}{R^{l+1}}\right] \frac{\mathrm{d}P_l(x)}{\mathrm{d}x} = 0 .$$

The derivatives of the Legendre polynomial are also linearly independent functions, so that the relations

$$\begin{aligned} &l=1 && A_1 + E_0 - \frac{B_1}{R^3} = 0 \\ &l \neq 1 && A_l = \frac{B_l}{R^{2l+1}} \qquad (l = 0, 2, \ldots) \end{aligned}$$

follow.

For $l \neq 1$ one can write directly

$$\frac{1}{R^{2l+1}}\,(\varepsilon l + (l+1))\, B_l = 0 .$$

This leads to

$$B_l = A_l = 0 \qquad \text{for } l \neq 1 \, .$$

For $l = 1$ one has to solve the simple system of equations

$$\varepsilon A_1 + \frac{2}{R^3} B_1 = -E_0$$

$$A_1 - \frac{1}{R^3} B_1 = -E_0 \, .$$

The solution is

$$A_1 = -\frac{3}{(\varepsilon + 2)} E_0 \qquad\qquad B_1 = \frac{(\varepsilon - 1)}{(\varepsilon + 2)} R^3 E_0 \, ,$$

so that one obtains for the potential

$$V_{\mathrm{i}}(\boldsymbol{r}) = -\frac{3}{(\varepsilon + 2)} E_0 r \cos\theta = -\frac{3}{(\varepsilon + 2)} E_0 z$$

$$V_{\mathrm{a}}(\boldsymbol{r}) = -E_0 z + \frac{(\varepsilon - 1)}{(\varepsilon + 2)} E_0 \frac{R^3}{r^3} z \, .$$

In the limit $\varepsilon = 1$, that is, one replaces the dielectric sphere by a spherical vacuum, the result is as expected

$$V_{\mathrm{i}}(\boldsymbol{r}) = -E_0 \, z \qquad\qquad V_{\mathrm{a}}(\boldsymbol{r}) = -E_0 \, z \, .$$

In the limit $\varepsilon \to \infty$ (a material with an infinite ability to polarise) follows

$$V_{\mathrm{i}}(\boldsymbol{r}) = 0 \qquad V_{\mathrm{a}}(\boldsymbol{r}) = -E_0 z + E_0 \frac{R^3}{r^3} z \, .$$

This is exactly the result for a metal sphere in a uniform field. The limit $\varepsilon \to \infty$ recovers the case of a metallic sphere. The charges can move freely. Such a material can be fully polarised.

The three fields $\boldsymbol{E}$, $\boldsymbol{D}$ and $\boldsymbol{P}$, as well as the distribution of the polarisation charges can be found from the potential.

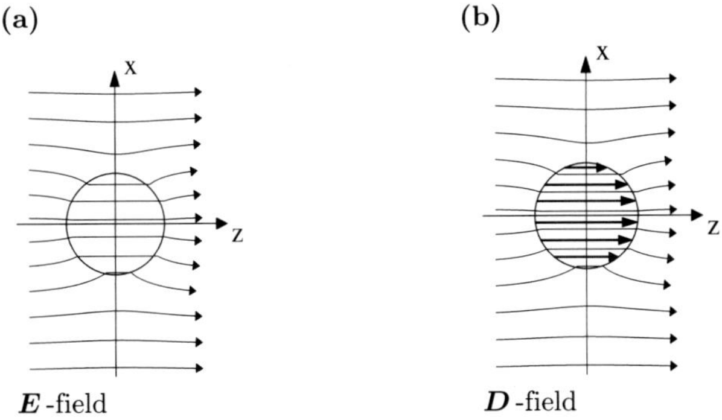

Fig. 4.27 Polarisation of a dielectric sphere

The Cartesian components of the electric field $\boldsymbol{E}$ calculated with $\boldsymbol{E} = -\nabla V$, are

$$\boldsymbol{E}_{\mathrm{i}} = \left(0,\ 0,\ \frac{3}{(\varepsilon+2)} E_0\right)$$

$$\boldsymbol{E}_{\mathrm{a}} = (0,\ 0,\ E_0) + \left(\frac{3xz}{r^5},\ \frac{3yz}{r^5},\ \frac{3z^2 - r^2}{r^5}\right)\left(\frac{\varepsilon - 1}{\varepsilon + 2}\right) E_0 R^3 .$$

The field in the interior of the sphere is constant and points in the z-direction (Fig. 4.27a). It is weaker by a factor $3/(\varepsilon + 2)$, which is smaller than 1 for $\varepsilon > 1$ than the $\boldsymbol{E}_0$-field. The exterior field is a superposition of the constant field $\boldsymbol{E}_0$ and a dipole field. The dipole field decreases rapidly with growing distance from the sphere. On the surface of the sphere there is

$$|E_{\mathrm{a}}|_R > |E_{\mathrm{i}}|_R \ .$$

This can be interpreted as: One 'part of the field lines' ends and starts at the polarisation charges. Another part runs through the sphere and satisfies the law of diffraction (continuous tangential component, discontinuous normal component). They are not perpendicular to the surface of the sphere (except in the limit $\varepsilon \to \infty$).

The $\boldsymbol{D}$-field (not a measurable quantity) has the form

$$\boldsymbol{D}_{\mathrm{i}}(\boldsymbol{r}) = \varepsilon \,\frac{k_d}{k_e}\, \boldsymbol{E}_{\mathrm{i}}(\boldsymbol{r}) = \frac{3\varepsilon k_d}{k_e(\varepsilon + 2)}\, \boldsymbol{E}_0$$

$$\boldsymbol{D}_{\mathrm{a}}(\boldsymbol{r}) = \frac{k_d}{k_e}\, \boldsymbol{E}_{\mathrm{a}}(\boldsymbol{r}) .$$

The $\boldsymbol{E}$-field and the $\boldsymbol{D}$-field agree in the exterior, if the CGS-system is employed. In the SI-system the two fields differ by a factor ε_0. The $\boldsymbol{D}$-field in the interior is

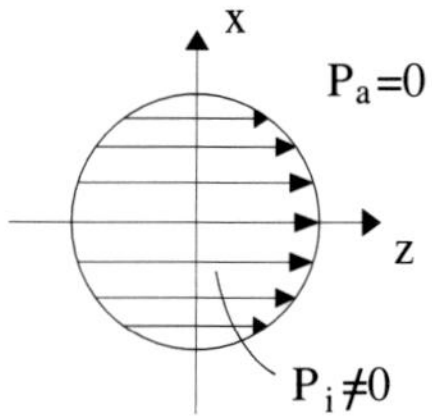

Fig. 4.28 Polarisation of a dielectric sphere: ***P***-field

stronger by a factor ε as the ***E***-field. This expresses the fact, that the ***D***-field lines do not end or begin on the polarisation charges. All field lines of that kind enter and leave the sphere (Fig. 4.27b). The kink in the field lines shown in the figure is due to the discontinuity of the tangential component, the normal component is continuous.

The result for the polarisation ***P*** is

$$\boldsymbol{P}_{\rm i} = \frac{1}{4\pi k_d}\boldsymbol{D}_{\rm i} - \frac{1}{4\pi k_e}\boldsymbol{E}_{\rm i} = \frac{3}{4\pi k_e}\left(\frac{\varepsilon-1}{\varepsilon+2}\right)\boldsymbol{E}_0$$

$$\boldsymbol{P}_{\rm a} = \frac{1}{4\pi k_d}\boldsymbol{D}_{\rm a} - \frac{1}{4\pi k_e}\boldsymbol{E}_{\rm a} = \boldsymbol{0}\,.$$

The polarisation field exists only in the dielectric material. The field itself is uniform. The field lines begin and end on the polarisation charges (note the sign in Fig. 4.28). The distribution of the polarisation charges can be calculated with the condition for the discontinuity

$$\left(P_{\rm a,n} - P_{\rm i,n}\right) = -\sigma_{\rm pol}$$

with the result

$$\sigma_{\rm pol} = \frac{3}{4\pi k_e}E_0\left(\frac{\varepsilon-1}{\varepsilon+2}\right)\cos\theta\,.$$

The polarisation charges are distributed on the metal sphere according to the cosine law (maximal density on the poles no charge at the equator, Fig. 4.15a). As the relation

$$0 \le \frac{\varepsilon-1}{\varepsilon+2} \le 1$$

still holds, one obtains

$$\sigma_{\rm pol}(\text{vacuum}) = 0 \le \sigma_{\rm pol}(\text{dielectric}) \le \sigma_{\rm pol}(\text{metal}) = \frac{3}{4\pi k_e}E_0\cos\theta\,.$$

The density is strongest for the metal sphere.

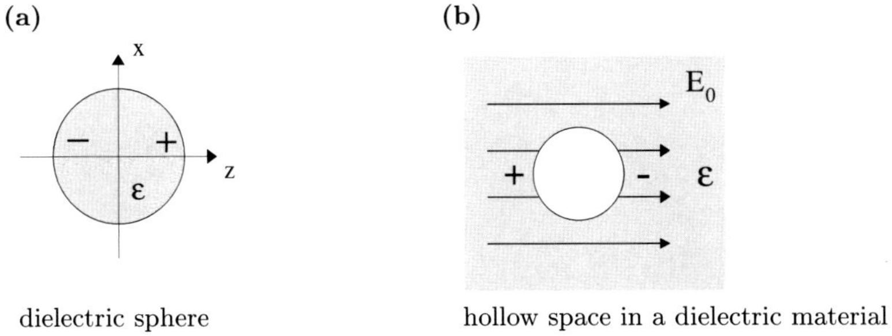

Fig. 4.29 Comparison of polarisation charges: sphere and spherical hollow space

The results presented can be taken over directly for a similar problem: In a dielectric material, which has been placed in a uniform field exists a spherical hollow space (Fig. 4.29b). The only difference in the discussion of this problem is the connection of the fields on the surface

$E_{\rm a,n} = \varepsilon E_{\rm i,n}$	$\varepsilon E_{\rm a,n} = E_{\rm i,n}.$
dielectric sphere	hollow space in dielectric material

One obtains the results for this variant by the replacement of ε by $1/\varepsilon$.

In order to illustrate the difference between a spherical sphere (Fig. 4.29a) and a spherical hollow space (Fig. 4.29b) in a dielectric one can look at the distribution of the polarisation charges

$$\sigma_{\text{pol, hs}} = -\frac{3}{4\pi k_e} E_0 \left(\frac{\varepsilon - 1}{2\varepsilon + 1}\right) \cos\theta .$$

The positive charges are found on the rear part of the spherical shell.

For $\varepsilon \geq 1$ one finds

$$\left(\frac{\varepsilon - 1}{\varepsilon + 2}\right) \geq \left(\frac{\varepsilon - 1}{2\varepsilon + 1}\right) .$$

The charge density on the sphere is (for a given material) larger than the charge density on the walls of the hollow space.

4.6 The Technique of Complex Potentials

The method of complex potentials is an elegant method for the treatment of problems, which are invariant under translations in one coordinate direction, as they can be projected onto a plane without essential loss of the structure of the solution

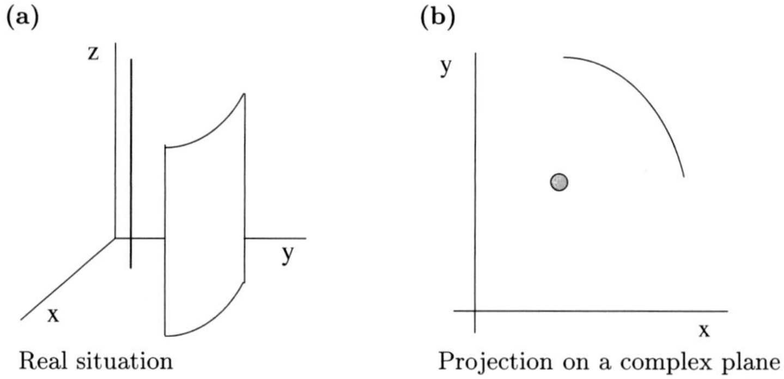

Fig. 4.30 Use of the technique of complex potentials

(Fig. 4.30). The method is based on the application of the theory of complex analytic functions. Such functions are defined in the following fashion:

> A function $w = f(z) = f(x + \mathrm{i}y)$ is called analytic or regular in a domain $\mathcal{G}$ of the complex z-plane, if it is defined and can be differentiated in this plane.

Such functions $w = f(z)$ are a means to generate a conformal mapping of a domain $\mathcal{G}$ of the z-plane into a domain of a w-plane. The word 'conformal' indicates, that a map of curves in the z-plane is true to scale and isogonal in the w-plane. In a more descriptive way one can say, that the images of three points, which are close together in the z-plane form a nearly similar triangle in the w-plane. If such a function is decomposed into a real and an imaginary part

$$w = V(x, y) + \mathrm{i}\, U(x, y)\,,$$

where V and U are real functions of real variables, the differential quotient

$$\frac{\mathrm{d}w}{\mathrm{d}z} = f'(z) = \lim_{\zeta \to z} \frac{f(z) - f(\zeta)}{z - \zeta}$$

can be written in the form

$$= \lim_{\varepsilon_x, \varepsilon_y \to 0} \frac{V(x + \varepsilon_x, y + \varepsilon_y) - V(x, y) + \mathrm{i}\,(U(x + \varepsilon_x, y + \varepsilon_y) - U(x, y))}{\varepsilon_x + \mathrm{i}\varepsilon_y}\,.$$

Differentiability of complex functions implies (as in the case of real functions of two variables), that the value of the derivative is independent of the direction, in which one approaches the position $z = x + \mathrm{i}y$. It is possible e.g. to do this parallel to the real axis ($\varepsilon_y = 0$), or to the imaginary axis ($\varepsilon_x = 0$),

$$\frac{\partial V}{\partial x} + \mathrm{i}\frac{\partial U}{\partial x} = \frac{1}{\mathrm{i}}\left(\frac{\partial V}{\partial y} + \mathrm{i}\frac{\partial U}{\partial y}\right).$$

If one separates the real and imaginary parts, one obtains the real differential equations

$$\frac{\partial V}{\partial x} = \frac{\partial U}{\partial y} \qquad \frac{\partial V}{\partial y} = -\frac{\partial U}{\partial x}\,, \tag{4.54}$$

which are known as the **Cauchy-Riemann differential equations**. For the second derivatives of the functions V and U follows

$$\frac{\partial^2 V}{\partial x^2} = \frac{\partial^2 U}{\partial x \partial y} \qquad \frac{\partial^2 V}{\partial y^2} = -\frac{\partial^2 U}{\partial y \partial x}\,.$$

The order of the differentiation with respect to the two real variables can be exchanged under the assumption, that the mixed second derivatives are continuous. Addition of the two equations gives in this case

$$\Delta V(x, y) = \left(\frac{\partial^2}{\partial x^2} + \frac{\partial^2}{\partial y^2}\right) V(x, y) = 0\,.$$

The function $V(x, y)$ satisfies the (two-dimensional) Laplace equation. In the same way one can verify the equation

$$\Delta U(x, y) = 0\,.$$

The real as well as the imaginary part of an analytic function satisfy a Laplace equation. The two functions can, however, not be chosen arbitrarily, as they are connected by the Cauchy-Riemann conditions (4.54).

These conditions are also responsible for the statement, that the set of curves $V(x, y) = \text{const.}$ and $U(x, y) = \text{const.}$ intersect each other at right angles. The scalar product of the gradients can be written in the form

$$\nabla V(x, y) \cdot \nabla U(x, y) = \frac{\partial V}{\partial x}\frac{\partial U}{\partial x} + \frac{\partial V}{\partial y}\frac{\partial U}{\partial y} = -\frac{\partial V}{\partial x}\frac{\partial V}{\partial y} + \frac{\partial V}{\partial y}\frac{\partial V}{\partial x} = 0\,.$$

The gradients in a point of the complex plane are perpendicular to the curves of the two sets and are orthogonal with respect to each other. The curves $V(x, y) = \text{const.}$ represent the equipotential lines of the problem under discussion (which can be projected on the plane), if the function $V(x, y)$ is identified with the potential. The curves $U(x, y) = \text{const.}$ represent the electric field lines (Fig. 4.31a). The components of the electric field, which are given by the tangents on the field lines, can be obtained through the relation $\boldsymbol{E}(x, y) = -\boldsymbol{\nabla} V(x, y)$ or otherwise with

$$E_x = -\frac{\partial V}{\partial x} = -\frac{\partial U}{\partial y} \qquad E_y = -\frac{\partial V}{\partial y} = \frac{\partial U}{\partial x}\,. \tag{4.55}$$

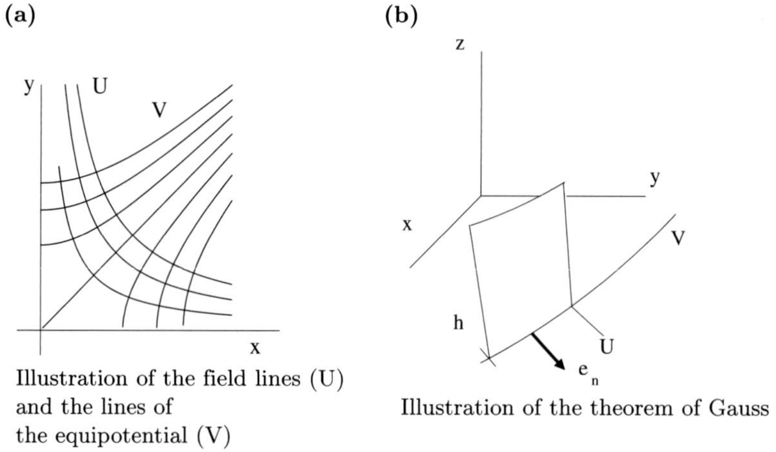
(a)
y
U
V
x
Illustration of the field lines (U) and the lines of the equipotential (V)
(b)
z
y
x
V
h
U
e n
Illustration of the theorem of Gauss

Fig. 4.31 The complex potential method

The following statement can be made, if an equipotential line (even if it is finite) is the intersection of a metal surface with the plane: The flux through a surface over a finite section of an equipotential line corresponds to the charge on this surface

$$\Phi = \iint \boldsymbol{E} \cdot \mathbf{d}\boldsymbol{f} = 4\pi k_e Q \,.$$

The connection with the quantities projected onto the plane can be established by the relations

$$\mathbf{d}\boldsymbol{f} = \boldsymbol{e}_n \,\mathrm{d}s\, h \qquad\qquad Q = \sigma\, l\, h = q_l\, h \,.$$

The following quantities are involved in this relation (see Fig. 4.31b): $\boldsymbol{e}_n$ is the vector, normal to the infinitesimal section of the curve ds representing an equipotential line, h the extension of the surface over the plane, l the length of the sector of the equipotential line, which is limited by the values U_1 and U_2 and q_l is the surface charge σ corresponding to the linear charge along the curve in the plane. If the normal vector is written in the form $\boldsymbol{e}_n = (n_x, n_y)$, the associated tangential vector on the curve would be $\boldsymbol{e}_t = (-n_y, n_x)$ and the statement concerning the flux projected on the plane is

$$q_l = \frac{1}{4\pi k_e} \int_1^2 \mathrm{d}s\, \boldsymbol{E} \cdot \boldsymbol{e}_n = -\frac{1}{4\pi k_e} \int_1^2 \mathrm{d}s\, \boldsymbol{\nabla} V \cdot \boldsymbol{e}_n$$

$$= -\frac{1}{4\pi k_e} \int_1^2 \mathrm{d}s \left(\frac{\partial V}{\partial x} n_x + \frac{\partial V}{\partial y} n_y \right)$$

$$= -\frac{1}{4\pi k_e} \int_1^2 \mathrm{d}s \left(\frac{\partial U}{\partial x} n_x - \frac{\partial U}{\partial x} n_y \right)$$

$$= -\frac{1}{4\pi k_e} \int_1^2 \mathrm{d}s\ \boldsymbol{\nabla} U \cdot \boldsymbol{e}_t = \frac{1}{4\pi k_e} (U_1 - U_2)\ .$$

The difference of the imaginary part of the complex potential determines the flux.

General methods or the calculation of the real and the imaginary complex potential in two space dimensions are the **harmonic analysis** and special techniques of transformation. In the harmonic analysis polar coordinates in the plane lead to the partial differential equation (e.g. for $V(r, \varphi)$)

$$r \frac{\partial}{\partial r} \left(r \frac{\partial V}{\partial r} \right) + \frac{\partial^2 V}{\partial \varphi^2} = 0\,,$$

which, with the standard separation $V(r, \varphi) = R(r)S(\varphi)$, leads to two ordinary differential equations

$$\frac{\mathrm{d}^2 S(\varphi)}{\mathrm{d}\varphi^2} + k^2 S(\varphi) = 0$$

$$r^2 \frac{\mathrm{d}^2 R(r)}{\mathrm{d}r^2} + r \frac{\mathrm{d}R(r)}{\mathrm{d}r} - k^2 R(r) = 0\,.$$

The domain of definition of the first equation is generally restricted to a finite interval, e.g. to $[0, 2\pi]$. Only discrete values of the separation parameter are possible

$$k \longrightarrow k_1, k_2, \ldots, k_n, \ldots\,.$$

The general solution of the two ordinary differential equations are

$$\left.\begin{aligned} S_n(\varphi) &= a_{1n} \cos k_n\varphi + a_{2n} \sin k_n\varphi \\ R_n(r) &= b_{1n} r^{k_n} + b_{2n} r^{-k_n} \end{aligned}\right\} \quad \text{for} \quad k_n > 0$$

and

$$\left.\begin{aligned} S_0(\varphi) &= a_{10} + a_{20}\varphi \\ R_0(r) &= b_{10} + b_{20} \ln r \end{aligned}\right\} \quad \text{for} \quad k_n = 0.$$

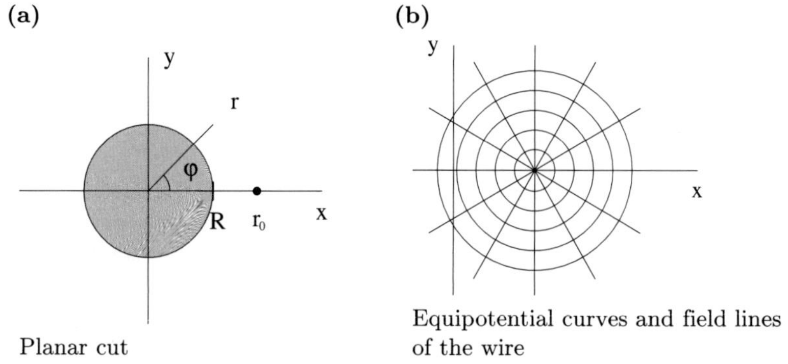

Fig. 4.32 Charged wire parallel to a dielectric cylinder

The general solution of the partial differential equation is therefore the Fourier series

$$V(r,\varphi) = \sum_{n=0}^{\infty} R_n(r) S_n(\varphi)\,. \tag{4.56}$$

The constants have to be determined by boundary and connecting conditions.

A typical problem, which can be solved with the help of the harmonic analysis, is:

> Calculate the electric field of a long thin wire with a uniform, linear charge density λ, which is parallel to and a distance r_0 from a long dielectric cylinder (dielectric constant ε, Radius $R < r_0$, see Fig. 4.32a, which shows a cross section of this arrangement).

The strategy for the solution is similar to the treatment of the mirror charge problem in Chap. 4.2. The first step is the calculation of the potential of the wire along the z-axis without consideration of the cylinder. The second step consists of an ansatz for the potential, which is composed of the potential for the wire and a general solution of the Laplace equation. The coefficients of the ansatz are determined by conditions governed by the connection and the boundary.

The potential of a wire along the z-axis is obtained first in real form (see Detail 4.7.6.1). For points in the x-y plane (the intersecting plane of interest) one finds on using polar coordinates

$$V(r,\varphi) = -2k_e\lambda \ln r\,,$$

if the wire goes through the origin. The potential of a wire, which is offset a distance r_0 along the x-axis, is then

$$V_1(r,\varphi) = -2k_e\lambda \ln|\boldsymbol{r} - \boldsymbol{r}_0| = -2k_e\lambda \ln[r^2 + r_0^2 - 2rr_0\cos\varphi]^{1/2}\,. \tag{4.57}$$

In order to find a complex representation, one has to transcribe the results. In the complex plane one can write for the potential of a straight wire, which is perpendicular to this plane and crosses it in the point z_0,

$$w = -2k_e\lambda \ln(z - z_0)\,.$$

With the form of the complex variables

$$z = r\,\mathrm{e}^{\mathrm{i}\varphi} \qquad \text{and} \qquad z_0 = r_0\,\mathrm{e}^{\mathrm{i}\varphi_0}\,,$$

one finds for the real part according to (4.57),

$$V = \frac{1}{2}\left(w + w^*\right) = -2k_e\lambda\left(\ln(z - z_0)^{1/2} + \ln(z^* - z_0^*)^{1/2}\right)$$

$$= -2k_e\lambda \ln\left[(z - z_0)(z^* - z_0^*)\right]^{1/2}$$

$$= -2k_e\lambda \ln[r^2 + r_0^2 - 2rr_0\cos(\varphi - \varphi_0)]^{1/2}\,.$$

The equipotential lines are circles about the point z_0 (see Fig. 4.32b).

The imaginary part, which represents the field lines, is

$$U = \frac{1}{2\mathrm{i}}\left(w - w^*\right) = -\frac{k_e\lambda}{\mathrm{i}}\ln\left[\frac{z - z_0}{z^* - z_0^*}\right]$$

$$= -2k_e\lambda \arctan\left(\frac{r\sin\varphi - r_0\sin\varphi_0}{r\cos\varphi - r_0\cos\varphi_0}\right),$$

where the relation

$$\frac{1}{2\mathrm{i}}\ln\left(\frac{x + \mathrm{i}y}{x - \mathrm{i}y}\right) = \arctan\frac{y}{x}$$

has been applied. The imaginary part represents a set of straight lines through the point z_0.

In order to write the boundary conditions in the form of (4.56), one has to transform the potential (4.57) into an appropriate Fourier series. One uses for this purpose the relation, which will be prepared in Detail 4.7.6.3

$$\frac{1}{2}\ln(1 - 2x\cos\varphi + x^2) = \ln[1 - 2x\cos\varphi + x^2]^{1/2}$$

$$= -\sum_{n=1}^{\infty}\frac{x^n}{n}\cos n\varphi \qquad x < 1\,. \qquad (4.58)$$

One finds for the domains inside and outside of a circle with the radius r_0 about the origin

$$V_{1i}(r,\varphi) = 2k_e\lambda\left\{\sum_{n=1}^{\infty}\frac{1}{n}\left(\frac{r}{r_0}\right)^n\cos n\varphi - \ln r_0\right\} \qquad r < r_0$$

$$V_{1a}(r,\varphi) = 2k_e\lambda\left\{\sum_{n=1}^{\infty}\frac{1}{n}\left(\frac{r_0}{r}\right)^n\cos n\varphi - \ln r\right\} \qquad r > r_0\,. \tag{4.59}$$

The terms with $\sin n\varphi$ are not present because of the symmetry of the arrangement with respect to the x-axis. The potentials (4.59) yield the correct value for $r = 0$ and correspond to the boundary condition for $r \to \infty$.

The ansatz for the potential with the dielectric cylinder can be undertaken with the general form, which divides a plane perpendicular to the wire into three domains (all circles about the point $z = (r_0, 0)$)

$$V_{ii}(r,\varphi) = V_{1i}(r,\varphi) + V_{2i}(r,\varphi)$$

$$V_{iz}(r,\varphi) = V_{1i}(r,\varphi) + V_{2z}(r,\varphi)$$

$$V_{aa}(r,\varphi) = V_{1a}(r,\varphi) + V_{2a}(r,\varphi)\,.$$

The domain within the cylinder (ii) with radius R, the domain in the intermediate region, that is (iz), with a radius between R and r_0, and outside of a circle with the radius r_0, that is (aa). The functions $V_{2x}(r,\varphi)$, which are the contribution to the potential by the dielectric, are represented by the Fourier series

$$V_{2i}(r,\varphi) = \sum_{n=1}^{\infty} a_n r^n \cos n\varphi + a_0 \qquad r \le R$$

$$V_{2z}(r,\varphi) = \sum_{n=1}^{\infty} b_n r^{-n} \cos n\varphi + b_0 \qquad R \le r \le r_0$$

$$V_{2a}(r,\varphi) = \sum_{n=1}^{\infty} c_n r^{-n} \cos n\varphi + c_0 \ln r \qquad r_0 \le r \quad .$$

This ansatz incorporates the symmetry with respect to $\varphi = 0$ and the nonsingular behaviour of all parts. The coefficients are obtained by demanding that the potential and the electric field (the derivative in the direction of the normal) are determined by the connecting condition on the circle with $r = R$ and the potential alone on a circle with $r = r_0$

$$V_{ii}(R,\varphi) = V_{iz}(R,\varphi) \qquad \left.\frac{\partial V_{iz}(r,\varphi)}{\partial r}\right|_R = \varepsilon \left.\frac{\partial V_{ii}(r,\varphi)}{\partial r}\right|_R$$

$$V_{iz}(r_0,\varphi) = V_{aa}(r_0,\varphi)\,.$$

A direct calculation (Detail 4.7.6.2) gives for $n \geq 1$

$$a_n = 2k_e\lambda \frac{(1-\varepsilon)}{(1+\varepsilon)}\frac{1}{n}\left(\frac{1}{r_0}\right)^n$$

$$b_n = 2k_e\lambda \frac{(1-\varepsilon)}{(1+\varepsilon)}\frac{1}{n}\left(\frac{R^2}{r_0}\right)^n$$

$$c_n = b_n\,,$$

as well as $a_0 = b_0 = c_0 = 0$. The result

$$V_{ii}(r,\varphi) = \frac{4k_e\lambda}{(1+\varepsilon)}\sum_{n=1}^{\infty}\left(\frac{r}{r_0}\right)^n \cos n\varphi - 2k_e\lambda \ln r_0 \qquad r < R$$

$$V_{iz}(r,\varphi) = 2k_e\lambda \left\{\sum_{n=1}^{\infty}\frac{1}{n}\left\{\left(\frac{r}{r_0}\right)^n + \frac{(1-\varepsilon)}{(1+\varepsilon)}\left(\frac{R^2}{rr_0}\right)^n\right\}\cos n\varphi \right.$$

$$\left. - \ln r_0\right\} \qquad r_0 > r > R$$

$$V_{aa}(r,\varphi) = 2k_e\lambda \left\{\sum_{n=1}^{\infty}\frac{1}{n}\left\{\left(\frac{r_0}{r}\right)^n + \frac{(1-\varepsilon)}{(1+\varepsilon)}\left(\frac{R^2}{rr_0}\right)^n\right\}\cos n\varphi - \ln r\right\}$$

$$r > r_0$$

can be interpreted in a fashion similar to the method of mirror charges. One adds and subtracts the term

$$-2k_e\lambda \frac{(1-\varepsilon)}{(1+\varepsilon)}\ln r$$

to the result in the intermediate domain and finds after some rearrangement the result

$$
\begin{aligned}
V_{iz}(r,\varphi) &= -2k_e\lambda\frac{(\varepsilon-1)}{(1+\varepsilon)}\ln r \\
&\quad +2k_e\lambda\frac{(1-\varepsilon)}{(1+\varepsilon)}\left[\sum_{n=1}^{\infty}\frac{1}{n}\left(\frac{R^2}{rr_0}\right)^n\cos n\varphi-\ln r\right] \\
&\quad +2k_e\lambda\left[\sum_{n=1}^{\infty}\frac{1}{n}\left(\frac{r}{r_0}\right)^n\cos n\varphi-\ln r_0\right] \\
&= -2k_e\lambda\frac{(\varepsilon-1)}{(1+\varepsilon)}\ln r \\
&\quad -2k_e\lambda\frac{(1-\varepsilon)}{(1+\varepsilon)}\ln\left[r^2+\left(\frac{R^2}{r_0}\right)^2-\frac{2rR^2}{r_0}\cos\varphi\right]^{1/2} \\
&\quad -2k_e\lambda\ln\left[r^2+r_0^2-2rr_0\cos\varphi\right]^{1/2} .
\end{aligned}
$$

This result corresponds to a potential of three wires parallel to the z-axis, that is

- one wire along the z-axis with the effective charge density

$$\lambda_{\text{eff}} = \frac{(\varepsilon-1)}{(1+\varepsilon)}\lambda ,$$

- one wire, which passes through the mirror point $(x, y) = (R^2/r_0, 0)$ with the effective charge density

$$\lambda_{\text{eff}} = \frac{(1-\varepsilon)}{(1+\varepsilon)}\lambda$$

and

- a wire with the charge density λ through the point $(x, y) = (r_0, 0)$.

A corresponding rearrangement can be performed for the potential in the exterior domain

$$V_{aa}(r,\varphi) = -2k_e\lambda\frac{(\varepsilon-1)}{(1+\varepsilon)}\ln r$$

$$-2k_e\lambda\frac{(1-\varepsilon)}{(1+\varepsilon)}\ln\left[r^2+r_0^2-2rr_0\cos\varphi\right]^{1/2}$$

$$-2k_e\lambda\frac{(1-\varepsilon)}{(1+\varepsilon)}\ln\left[r^2+\left(\frac{R^2}{r_0}\right)^2-\frac{2rR^2}{r_0}\cos\varphi\right]^{1/2}.$$

The potential V_{ii} in the inner domain

$$V_{ii}(r,\varphi) = 2k_e\lambda\frac{(1-\varepsilon)}{(1+\varepsilon)}\ln r_0$$

$$+2k_e\lambda\left\{\frac{2}{(1+\varepsilon)}\ln\left[r^2+r_0^2-2rr_0\cos\varphi\right]^{1/2}\right\}$$

corresponds (up to a constant) to the potential of one single wire through the point $(x, y) = (r_0, 0)$, with the effective charge density

$$\lambda_{\text{eff}} = \frac{2}{(1+\varepsilon)}\lambda\,.$$

The potential of the present problem can be represented by a superposition of potentials with different effective, linear charge densities. The resulting equipotential lines (and the associated field lines) in the exterior regions are found to be a superposition of the potential of three wires with effective charges, which go over into the potential of a single wire with the effective charge $2\lambda/(\varepsilon+1)$ in the interior region. Figure 4.33 shows the resulting situation. One notices, how the modification of the equipotential lines of the wire can be observed in the vicinity of the cylinder.

The simplification of the geometry of a given problem by **conformal mapping** is an alternative method for the solution of two dimensional potential problems. The method can be illustrated with the following example: Consider the following mapping

$$z_1 = z_2^{\gamma} \tag{4.60}$$

of the z_2-plane into the z_1-plane, with γ an arbitrary real number. The form of complex numbers with magnitude and phase

$$r_1 = r_2^{\gamma} \quad \text{and} \quad \varphi_1 = \gamma\varphi_2$$

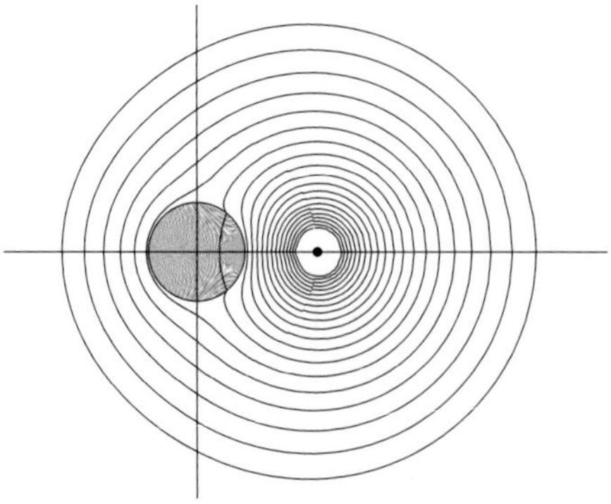

Fig. 4.33 Equipotential lines of a long, straight wire with a dielectric cylinder at the distance r_0

indicate that the positive real axis of the z_2-plane ($\varphi_2 = 0$) is mapped onto the positive real axis of the z_1-plane ($\varphi_1 = 0$). A ray in the upper half-plane of the z_2-plane is mapped onto a ray of the z_1-plane with an angle $\gamma\varphi_2$. The negative real axis of the z_2-plane is mapped onto a ray with the angle $\pi\gamma$. The transformation (4.60) has a branch point for $z_2 = 0$, but is otherwise analytic in the complete z_2-plane.

A transformation as (4.60) can be used to make the transition from a simple geometry in the z_2-plane to a complicated geometry in the z_1-plane. If one has found the solution of a problem with a simpler geometry, one can obtain the solution in the z_1-plane by the map $z_1 = z_2^{\gamma}$.

A direct though simple example is the calculation of the potential in the first quadrant of the complex plane, which is created by two uniformly charged, perpendicular metal plates positioned along the positive real and positive imaginary axes (Fig. 4.34a). The solution of the problem can be started by the generation of a uniform electric field E (e.g. with $E > 0$) in the y_2-direction by inserting a uniform (linear) charge density along the real axis in the z_2-plane. The corresponding

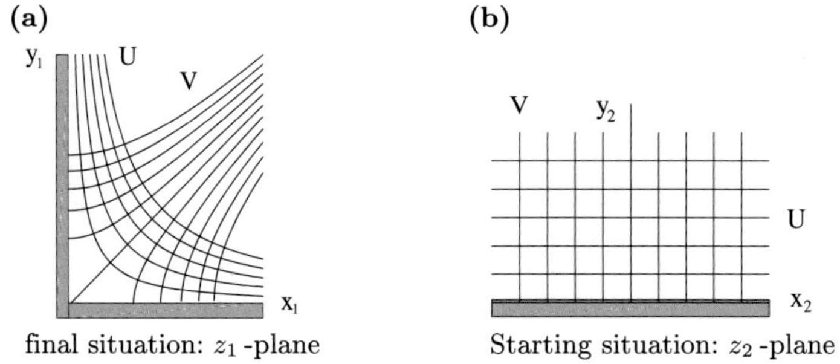

Fig. 4.34 Application of the map $z_1 = c\sqrt{z_2}$

complex potential is

$$w = \mathrm{i}\, E\, z_2 \quad \text{with} \quad V = -E\, y_2 \qquad U = E\, x_2 \,.$$

The equipotential lines are parallel to the real axis, the field lines are parallel to the imaginary axis (Fig. 4.34b). One has

$$E_{x_2} = -\frac{\partial V}{\partial x_2} = -\frac{\partial U}{\partial y_2} = 0 \qquad E_{y_2} = -\frac{\partial V}{\partial y_2} = \frac{\partial U}{\partial x_2} = E \,.$$

The transformation, which changes the simple geometry into the final geometry, is a rotation of the negative real axis of the z_2-plane by $\pi/2$. This is achieved with $\gamma = 1/2$ and

$$z_1 = z_2^{1/2} \quad \text{resp.} \quad z_2 = z_1^2 \,,$$

leading to

$$w = \mathrm{i}\, E z_1^2 \quad \text{with} \quad V = -2E\, x_1 y_1 \qquad U = E\,(x_1^2 - y_1^2) \,.$$

The equipotential lines and the field lines of this problem are elements of two orthogonal sets of hyperbolas (Fig. 4.34a). The components of the electric field in the first quadrant of the z_1-plane are

$$E_{x_1} = 2E\, y_1 \qquad E_{y_1} = 2E\, x_1 \,.$$

As stated, this example is only a special example. The rotation of the negative real axis by an arbitrary angle could have been discussed in the same fashion.

Further generalisations of this method are possible, e.g. one can use different classes of analytic functions

$$z_1 = f(z_2) \quad \text{with the inverse} \quad z_2 = g(z_1) \tag{4.61}$$

or the more exotic **Schwarz-Christoffel transformation**, which maps the interior of a polygon in the z_1-plane onto the upper half-plane of the z_2-plane.

An explicit example for the use of other elementary functions is the mapping of two vertical lines with a separation of $\pm A\pi/2$ from the imaginary axis of the z_1-plane onto the positive and the negative real axis of the z_2-plane (see Fig. 4.35a and b, as well as Detail 4.7.6.3). The maps of the two corner points $z = \pm A\pi/2$ into the z_2-plane is generated with the differential equation

$$\frac{\mathrm{d}z_1}{\mathrm{d}z_2} = \mathrm{i}\, A(z_2 - a)^{-1/2}(z_2 + a)^{-1/2} = \frac{A}{[a^2 - z_2^2]^{1/2}} \,.$$

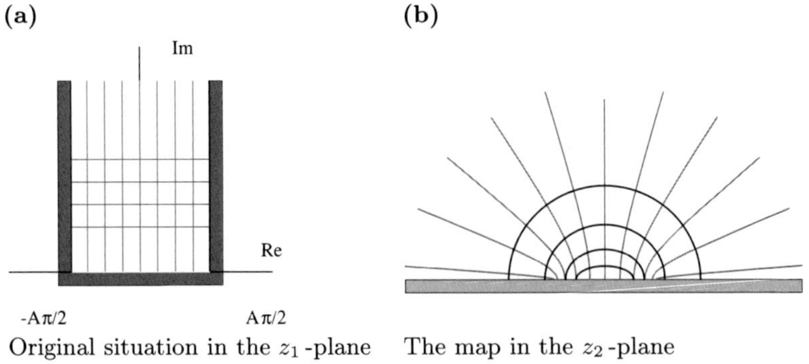

Fig. 4.35 The map $z_2 = a\ \sin(z_1/A)$

The solution is

$$z_1 = A \arcsin\left(\frac{z_2}{a}\right) \qquad z_2 = a \sin\left(\frac{z_1}{A}\right) .$$

The interior of a strip with the width $A\pi$ in the upper half of the of the z_1-plane is mapped onto the upper half of the z_2-plane and vice versa. A constant field in the strip of the z_1-plane corresponds to a complex potential of a linear charge distribution between the points $-a$ and $+a$ of the z_2-plane.

This concludes the discussion of the electrostatic potential problem. While electrostatics (though in a less precise form) was assembled already in the last quarter of the eighteenth century, the counterpart, the quantitative formulation of magnetostatics was only started in the second quarter of the nineteenth century. The word counterpart refers to the fact that electric fields are vortex free fields with sources, while magnetic fields (of stationary currents) are source free fields with vortices. The topic magnetostatics will be taken up in the next chapter.

4.7 Details

4.7.1 Surface Charge of a Grounded Metal Sphere Induced by a Point Charge

The first step of the solution of the problem is: Determine the potential of a point charge q at a separation a from the centre of a grounded metal sphere with radius R with the method of mirror charges

$$V_a(\boldsymbol{r}) = k_e \left\{ \frac{q}{|\boldsymbol{r} - \boldsymbol{a}|} + \frac{q'}{|\boldsymbol{r} - \boldsymbol{a}'|} \right\} \qquad r \geq R . \tag{4.62}$$

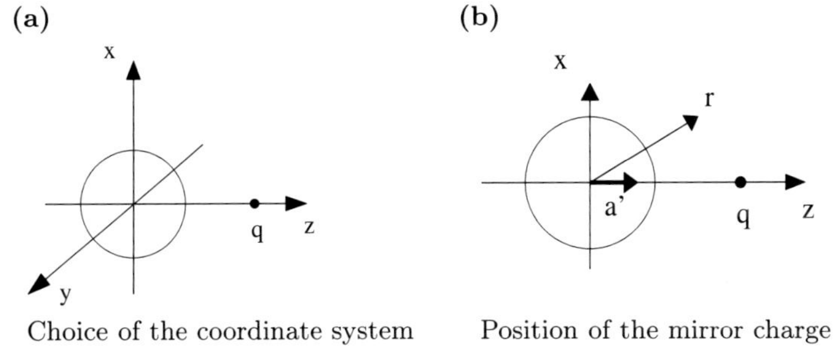

Fig. 4.36 Metal sphere and point charge

The mirror charge and its position are (Fig. 4.36)

$$q' = -q\,\frac{R}{a} \qquad \text{and} \quad \boldsymbol{a}' = (0,\ 0,\ R^2/a)\,,$$

if the position of the charge is $\boldsymbol{a} = (0, 0, a)$. The second task is the calculation of the induced surface charge distribution on the sphere using the normal derivative

$$\sigma(R,\theta) = -\frac{1}{4\pi k_e}\frac{\partial V_a}{\partial n}\bigg|_R = -\frac{1}{4\pi k_e}\frac{\partial V_a}{\partial r}\bigg|_{r=R} \tag{4.63}$$

and the total induced charge on the sphere

$$Q_{\text{ind}} = \oiint_{\text{sp}} \sigma(R,\theta)\,\mathrm{d}f\,. \tag{4.64}$$

4.7.1.1 Charge Distribution

With the choice of the coordinate system indicated above and

$$\boldsymbol{r} = (r\cos\varphi\sin\theta,\ r\sin\varphi\sin\theta,\ r\cos\theta)$$

the magnitude of the difference of the two vectors is (use $x = \cos\theta$)

$$|\boldsymbol{r} - \boldsymbol{a}| == \left[(r\sin\theta)^2 + (r\cos\theta - a)^2\right]^{1/2} = \left[r^2 - 2arx + a^2\right]^{1/2}.$$

The potential in the exterior has the form

$$V_a(\boldsymbol{r}) = k_e\left(\frac{q}{\left[r^2 - 2arx + a^2\right]^{1/2}} + \frac{q'}{\left[r^2 - 2a'rx + a'^2\right]^{1/2}}\right).$$

The calculation of the normal derivative requires the steps

$$\left.\frac{\partial V_a}{\partial n}\right|_{r=R} = k_e \frac{\partial}{\partial r}\left(\frac{q}{\left[r^2 - 2arx + a^2\right]^{1/2}} + \frac{q'}{\left[r^2 - 2a'rx + a'^2\right]^{1/2}}\right)\Bigg|_{r=R}$$

$$= -\frac{k_e}{2}\left(\frac{2(r-ax)q}{\left[r^2 - 2arx + a^2\right]^{3/2}} + \frac{2(r-a'x)q'}{\left[r^2 - 2a'rx + a'^2\right]^{3/2}}\right)\Bigg|_{r=R}$$

$$\text{(insert here} \quad a' = \frac{R^2}{a}, q' = -q\frac{R}{a} \quad \text{and} \quad r = R)$$

$$= -k_e q\left(\frac{(R-ax)}{\left[R^2 - 2aRx + a^2\right]^{3/2}}\right.$$

$$\left. - \frac{\frac{R}{a}\left(R - \frac{R^2}{a}x\right)}{\left[\frac{R^2}{a^2}\left(a^2 - 2aRx + R^2\right)\right]^{3/2}}\right)$$

$$= \frac{qk_e}{\left[R^2 - 2Rax + a^2\right]^{3/2}}\left(-R + ax + \left(R - \frac{R^2}{a}x\right)\frac{a^2}{R^2}\right)$$

$$= \frac{qk_e}{R\left[R^2 - 2Rax + a^2\right]^{3/2}}\left(a^2 - R^2\right).$$

The induced surface charge distribution has a sign opposite to the sign of q

$$\sigma(R, \theta) = -\frac{q}{4\pi R}\frac{\left(a^2 - R^2\right)}{\left[R^2 - 2Rax + a^2\right]^{3/2}}$$

as $a \geq R$.

4.7.1.2 Integration

The surface integral (4.64)

$$Q_{\text{ind}} = \oiint_{\text{sp}} \sigma(R, \theta)\, \mathrm{d}f$$

is

$$= \iint \sigma(R,\,\theta) R^2 \,\mathrm{d}\varphi\; \mathrm{d}\cos\theta\,.$$

The φ-integration can be executed directly

$$= 2\pi\, R^2 \int_{-1}^{1} \sigma(R,\,\theta)\; \mathrm{d}\cos\theta\,.$$

The remaining integral can be evaluated via ($x = \cos\theta$)

$$\begin{aligned}
Q_{\text{ind}} &= -\frac{q\,R}{2}\int_{-1}^{1} \frac{(a^2 - R^2)}{\left[a^2 - 2aRx + R^2\right]^{3/2}}\,\mathrm{d}x \\
&= -\frac{q}{2}\frac{R\,(a^2 - R^2)}{aR}\,\frac{1}{\left[a^2 - 2aRx + R^2\right]^{1/2}}\Bigg|_{-1}^{1} \\
&= -\frac{q}{2}\frac{(a^2 - R^2)}{a}\left(\frac{1}{\left[(a-R)^2\right]^{1/2}} - \frac{1}{\left[(a+R)^2\right]^{1/2}}\right) \\
&= -\frac{q}{2}\frac{(a^2 - R^2)}{a}\left(\frac{1}{(a-R)} - \frac{1}{(a+R)}\right) \qquad a > R(!) \\
&= -\frac{R}{a}\,q = q'\,.
\end{aligned}$$

The total induced charge on the surface of the metal sphere is equal to the mirror charge.

4.7.2 The Potential of Two Point Charges Outside of a Grounded Metal Sphere

This problem can be solved directly by superposition. The two point charges are placed in the position $\boldsymbol{a}_1$ and $\boldsymbol{a}_2$. The potentials $V_1(\boldsymbol{r})$ and $V_2(\boldsymbol{r})$ should be added in order to obtain the total potential $V(\boldsymbol{r})$, which is expected to satisfy the boundary condition $V(\boldsymbol{R}) = 0$ on the surface of the sphere.

The calculation of the individual potentials $V_i(\boldsymbol{r})$ $(i = 1, 2)$ follows the pattern indicated in Chap. 4.2 with the difference that the two point charges should not necessarily be placed on the z-axis. It should e.g. be possible to choose a coordinate system, so that the two vectors $\boldsymbol{a}_1$ and $\boldsymbol{a}_2$ lie in the x - z plane (Fig. 4.37), so that the

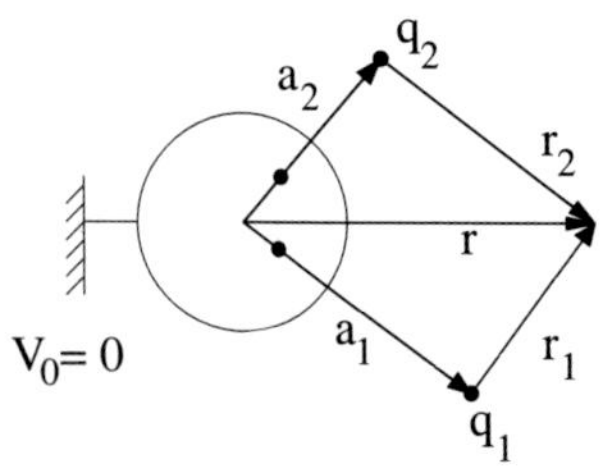

Fig. 4.37 Position of the charges and the mirror charges

discussion is actually not more difficult. It is necessary to employ the general form of the multipole expansion (3.35) in view of the statements

$$\boldsymbol{r} = (r\cos\varphi\sin\theta,\ r\sin\varphi\sin\theta,\ r\cos\theta)$$
$$\boldsymbol{a}_i = (a_i\cos\varphi_i\sin\theta_i,\ a_i\sin\varphi_i\sin\theta_i,\ a_i\cos\theta_i) \qquad i = 1, 2\,.$$

The standard notation for this expansion is

$$\frac{1}{|\boldsymbol{r} - \boldsymbol{r}'|} = \sum_{l=0}^{\infty} \frac{4\pi}{(2l+1)} \frac{r_<^l}{r_>^{l+1}} \sum_{m=-l}^{l} Y_{lm}(\Omega) Y_{lm}^*(\Omega')\,,$$

so that the details of the ansatz

$$V_i(\boldsymbol{r}) = V_{i,\text{mirror}}(\boldsymbol{r}) + \frac{k_e\, q_i}{|\boldsymbol{r} - \boldsymbol{a}_i|} \qquad r \geq R$$

for the mirror charges are:

- As one expects, that the position of the mirror charges will be in the interior of the sphere, one can write for the contribution of the mirror charges (a direct extension of (4.6), Chap. 4.2)

$$V_{i,\text{mirror}}(\boldsymbol{r}) = k_e \sum_{l=0}^{\infty} \frac{4\pi}{(2l+1)} \frac{A_{i,l}}{r^{l+1}} \sum_{m=-l}^{l} Y_{lm}(\Omega) Y_{lm}^*(\Omega_i) \qquad r \geq R\,.$$

- The potential of the point charges on the surface of the sphere are

$$\frac{1}{|\boldsymbol{R} - \boldsymbol{a}_i|} = \sum_{l=0}^{\infty} \frac{4\pi}{(2l+1)} \frac{R^l}{a_i^{l+1}} \sum_{m=-l}^{l} Y_{lm}(\Omega) Y_{lm}^*(\Omega_i)\,.$$

The coefficients $A_{i,l}$ are determined by the specified boundary values

$$V_{\text{sp}}(\boldsymbol{R}) = \sum_{i=1}^{2} V_i(\boldsymbol{R}) = 0\,,$$

which are

$$\sum_{l=0}^{\infty} \frac{4\pi}{(2l+1)} \sum_{m=-l}^{l} \left\{ \left[\frac{A_{1,l}}{R^{l+1}} + \frac{q_1 R^l}{a_1^{l+1}} \right] Y_{lm}^*(\Omega_1) \right.$$
$$\left. + \left[\frac{A_{2,l}}{R^{l+1}} + \frac{q_2 R^l}{a_2^{l+1}} \right] Y_{lm}^*(\Omega_2) \right\} Y_{lm}(\Omega) \stackrel{!}{=} 0\,.$$

The linear independence of the spherical harmonics $Y_{lm}(\Omega)$ guaranties that the expression in curly brackets has to be equal to zero. This statement should be valid for each charge q_i and its position $\boldsymbol{a}_i$. This is only possible, if each of the terms in the square brackets vanish individually. The consequence is

$$A_{i,l} = -q_i \frac{R^{2l+1}}{a_i^{l+1}}\,.$$

The result for the contribution of the mirror charges

$$V_{i,\text{mirr}}(\boldsymbol{r}) = -\frac{k_e q_i R}{a_i} \sum_{l,m} \frac{4\pi}{(2l+1)} \left(\frac{R^2}{a_i} \right)^l \frac{1}{r^{l+1}} Y_{lm}(\Omega) Y_{lm}^*(\Omega_i)$$

can be resumed with the definition of these charges

$$q_i' = -q_i \,\frac{R}{a_i}$$

and their position (with $a_i' \leq R$!)

$$\boldsymbol{a_i}' = \frac{R^2}{a_i^2} \boldsymbol{a}_i$$

as

$$V_{i,\text{mirr}}(\boldsymbol{r}) = \frac{k_e q_i'}{|\boldsymbol{r} - \boldsymbol{a}_i'|}\,.$$

The individual potentials are therefore

- The point charge q_1 at the position $\boldsymbol{a}_1$ produces a mirror charge q_1' at the position $\boldsymbol{a}_1'$ with the potential

$$V_1(\boldsymbol{r}) = k_e \left\{ \frac{q_1}{|\boldsymbol{r} - \boldsymbol{a}_1|} + \frac{q_1'}{|\boldsymbol{r} - \boldsymbol{a}_1'|} \right\} \qquad r \geq R\,. \tag{4.65}$$

- The second charge q_2 at the position $\boldsymbol{a}_2$ produces the potential

$$V_2(\boldsymbol{r}) = k_e \left\{ \frac{q_2}{|\boldsymbol{r} - \boldsymbol{a}_2|} + \frac{q_2'}{|\boldsymbol{r} - \boldsymbol{a}_2'|} \right\} \qquad r \geq R\,. \tag{4.66}$$

The total potential, which is produced by the addition of two real and two induced point charges satisfies the boundary condition

$$V(R) = V_1(R) + V_2(R) = 0\,.$$

As each of the individual potentials also satisfies the condition $V_i(\boldsymbol{R}) = 0$, this means, that the coefficients $A_{i,l}$ could have been determined by a *set* of individual conditions. It could happen, that the conditions

$$V_1(\boldsymbol{R}) = -V_2(\boldsymbol{R}) = F(\boldsymbol{R})$$

give different individual potentials. Their sum is nonetheless $V(\boldsymbol{R}) = 0$. This is guaranteed by the uniqueness of the solution of a Dirichlet problem.

4.7.3 Comments on the Green's Function of Stationary Potential Problems

This topic has already been discussed in Chap. 4.3, but it is appropriate to present some additional remarks and a more detailed discussion of this Dirichlet problem with spherical symmetry.

4.7.3.1 Potential Problems and Green's Functions

The potential problem of electrostatics is to find a solution of the Poisson equation, a linear, inhomogeneous partial differential equation, for a given set of boundary conditions

$$\Delta V(\boldsymbol{r}) = -4\pi\, k_e\, \rho(\boldsymbol{r}).$$

The solution of this differential equation can be given in the form

$$V(\boldsymbol{r}) = V_{\text{hom}}(\boldsymbol{r}) + V_{\text{part}}(\boldsymbol{r}) = V_{\text{hom}}(\boldsymbol{r}) + \iiint \mathrm{d}V' \; G(\boldsymbol{r}, \boldsymbol{r}')\rho(\boldsymbol{r}') \,.$$

The function V_{hom} is a general solution of the Laplace equation

$$\Delta V_{\text{hom}}(\boldsymbol{r}) = 0 \,.$$

The Green's function $G(\boldsymbol{r}, \boldsymbol{r}')$ satisfies the differential equation

$$\Delta_{\boldsymbol{r}} G(\boldsymbol{r}, \boldsymbol{r}') = -4\pi \, k_e \delta(\boldsymbol{r} - \boldsymbol{r}') \tag{4.67}$$

and, depending on the specification, one of the following boundary conditions:

- In the case of simple boundary conditions, in which the asymptotic behaviour of the potential is required to be

$$V(\boldsymbol{r}) \xrightarrow{r \to \infty} 0$$

  one has to use a Green's function with the property

$$G(\boldsymbol{r}', \boldsymbol{r}) \xrightarrow{r' \to \infty} 0 \,.$$

- In the case of Dirichlet boundary conditions the potential is supposed to have specified values on the surfaces O of an arbitrary domain B

$$V(\boldsymbol{R}) = f(\boldsymbol{R}) \qquad\qquad \boldsymbol{R} \in \mathrm{O}(B) \,.$$

  In this case one has to demand the boundary conditions (Chap. 4.3, (4.29))

$$G(\boldsymbol{r}', \boldsymbol{r}) = 0 \qquad \text{for} \quad \boldsymbol{r}' = \boldsymbol{R} \in \mathrm{O}(B) \,.$$

- If the normal derivative of the potential $\partial V/\partial n$ is supposed to have specified values on the surfaces of a domain B, one is dealing with a von Neumann boundary condition. The simplest choice of the boundary condition is in this case

$$\frac{\partial G(\boldsymbol{r}', \boldsymbol{r})}{\partial n'} = \text{const.} \qquad \text{for} \quad \boldsymbol{r}' = \boldsymbol{R} \in \mathrm{O}(B) \,,$$

  where the constant is determined by the surface of the domain B

$$\text{const.} = -\frac{4\pi k_e}{F_{\mathrm{O}}(B)} \,.$$

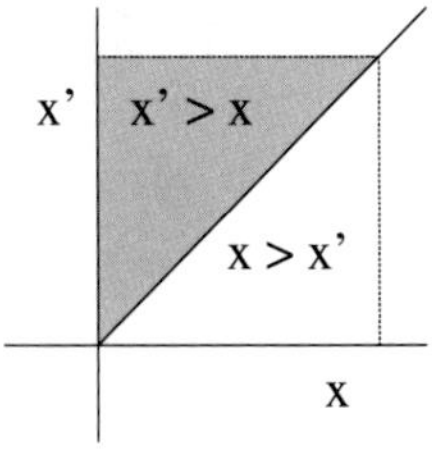

Fig. 4.38 Basic domain of a Green's function

The Green's function is defined in a six-dimensional domain. In order to visualise the situation, it is convenient to use the two-dimensional equivalent $G(x,\, x')$ with the differential equation

$$\frac{\mathrm{d}^2}{\mathrm{d}x^2} G(x,\, x') = -4\pi\, k_e\, \delta(x - x')\,.$$

The singularity, which is caused by the δ-function, divides the domain of definition of this function of two variables into two areas (Fig. 4.38). The Green's function satisfies the Laplace equation in the domain with $x' > x$ (domain 1) as well as in the domain with $x > x'$ (domain 2). The function G can be constructed in each of these domains by (infinitely many) particular solutions of the homogeneous differential equation. The solutions G_{I} and G_{II} in these two domains have to be connected along the straight line $x' = x$ in a certain fashion (details below) using the fact, that the Green's function is a symmetric function

$$G(\boldsymbol{r}, \boldsymbol{r}') = G(\boldsymbol{r}', \boldsymbol{r})\,.$$

4.7.3.2 A Dirichlet Problem with Spherical Symmetry

The following problem has been posed:

> Calculate the potential in a domain between two spherical surfaces (radius $R_1 = R$ and $R_2 \to \infty$) with an arbitrary distribution of point and space charges (Fig. 4.39) using the Green's function.

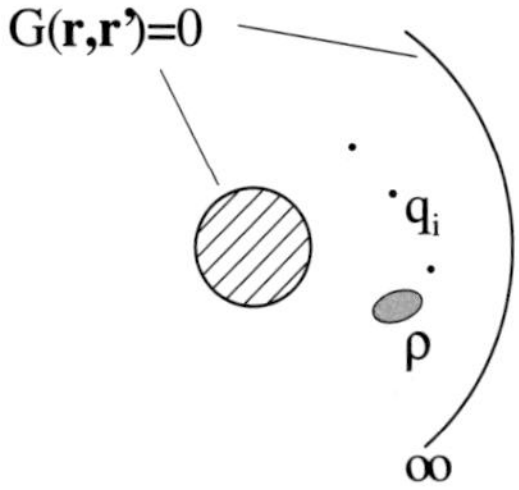

Fig. 4.39 The input for the Dirichlet problem

The solution of this problem can be obtained with an ansatz in the form of a multipole expansion

$$G(\boldsymbol{r},\, \boldsymbol{r}') = \sum_{l',m'} g_{l'}(r, r') Y_{l',m'}(\Omega) Y^*_{l',m'}(\Omega')\,. \tag{4.68}$$

The task is the determination of the expansion coefficients $g_{l'}(r, r')$. If this ansatz is inserted into the differential equation (4.67) for the Green's function, one finds after some steps a differential equation for the radial part. The first step results in

$$\sum_{l',m'} \left\{ \frac{1}{r^2} \frac{\mathrm{d}}{\mathrm{d}r} \left(r^2 \frac{\mathrm{d}}{\mathrm{d}r} g_{l'}(r, r') \right) - \frac{l'(l'+1)}{r^2} g_{l'}(r, r') \right\} Y_{l',m'}(\Omega) Y^*_{l',m'}(\Omega')$$

$$= -\frac{4\pi\, k_e}{r^2} \delta(r - r')\, \delta(\varphi - \varphi')\, \delta(\cos\theta - \cos\theta')\,, \tag{4.69}$$

where the angular part of the Laplace operator

$$\Delta_{\theta,\varphi} = \frac{1}{r^2 \sin\theta} \left\{ \frac{\mathrm{d}}{\mathrm{d}\theta} \left(\sin\theta \frac{\mathrm{d}}{\mathrm{d}\theta} \right) \right\} + \frac{1}{r^2 \sin^2\theta} \frac{\mathrm{d}^2}{\mathrm{d}\varphi^2}$$

has already been applied to the spherical harmonics

$$\Delta_{\theta,\varphi} Y_{l',m'}(\Omega) = \frac{l'(l'+1)}{r^2} Y_{l',m'}(\Omega)\,.$$

In order to sort the sum, one can multiply Eq. (4.69) with $Y^*_{l,m}(\Omega) Y_{l,m}(\Omega')$ and integrate over both sets of angular coordinates. The orthogonality relation of the spherical harmonics leads on the left hand side to a factor $\delta_{ll'}\delta_{mm'}$, on the right hand side one finds for the angular part

$$\iint \mathrm{d}\Omega \iint \mathrm{d}\Omega' \;\; Y^*_{l,m}(\Omega) \delta(\varphi - \varphi')\, \delta(\cos\theta - \cos\theta') Y_{l,m}(\Omega')$$

$$= \iint \mathrm{d}\Omega\, Y^*_{l,m}(\Omega) Y_{l,m}(\Omega) = 1\,.$$

The differential equation for the radial parts is therefore

$$\frac{\mathrm{d}}{\mathrm{d}r} \left(r^2 \frac{\mathrm{d}}{\mathrm{d}r} g_l(r, r') \right) - l(l+1) g_l(r, r') = -4\pi\, k_e\, \delta(r - r') \tag{4.70}$$

$$(l = 0, 1, 2, \ldots)\,.$$

The boundary value problem, which is addressed with this ordinary differential equation, is called **Sturm's boundary value problem**. The solution is obtained by a standard procedure, which consists of the following steps:

The first step involves finding the solutions of the differential equation in the domains (1) with $r < r'$ and (2) with $r' < r$, in which the differential equation is homogeneous. It is similar to the radial part of the Laplace problem, with the difference that there is an additional variable r'. The dependence of the solution on this variable is not determined by the differential equation. On the other hand one can write down the solution of the radial equation directly, if one regards r' as a parameter (Chap. 3.2, (3.14))

$$g_l(r, r'_{\text{fixed}}) = a_l r^l + b_l \frac{1}{r^{l+1}} \,.$$

As r' is not really a parameter, it is necessary to interpret the integration constants as functions of r'. This suggests to write for the solution in the domains (1) and (2) (which are supposed to be different)

$$g_{1l}(r, r') = a_{1l}(r') r^l + b_{1l}(r') \frac{1}{r^{l+1}} \qquad r < r'$$

$$g_{2l}(r, r') = a_{2l}(r') r^l + b_{2l}(r') \frac{1}{r^{l+1}} \qquad r' < r \,.$$

The conditions, which are still available

- the boundary conditions
- the symmetry
- the connection of the two solutions at points with $r = r'$

are then used to find the coefficient functions.

The boundary conditions are

$$g_l(R_1, r') = 0 \qquad \lim_{R_2 \to \infty} g_l(R_2, r') = 0 \,.$$

For the inner spherical shell one has $r = R_1 < r'$. This implies, that the first boundary condition can be used to determine the function g_{1l}.

The condition on the outer spherical shell concerns the function g_{2l}, as r' is $r' < r = R_2 \to \infty$. This argument establishes a relation between the functions. The statement $g_{1l}(R_1, r') = 0$ implies

$$b_{1l}(r') = -a_{1l}(r') R_1^{2l+1} \,.$$

The second condition can only be satisfied, if the factor of r^l, the function $a_{2l}(r')$, vanishes. The intermediate result due to the application of the boundary conditions is therefore (use from now on $R_1 = R$)

$$g_{1l}(r, r') = a_{1l}(r')\left[r^l - \frac{R^{2l+1}}{r^{l+1}}\right] \qquad r < r'$$

$$g_{2l}(r, r') = b_{2l}(r')\frac{1}{r^{l+1}} \qquad r' < r\,.$$

The ansatz (4.68) for the Green's function $G(\boldsymbol{r}, \boldsymbol{r}')$ is symmetric in the angular variables, as one has

$$\sum_m Y_{lm}(\Omega)Y^*_{lm}(\Omega') = \sum_m Y^*_{lm}(\Omega)Y_{lm}(\Omega') = \frac{(2l+1)}{4\pi}P_l(\cos\alpha)\,.$$

The symmetry condition demands therefore for the radial part

$$g_{1l}(r, r') = g_{2l}(r', r)\,.$$

This relation for the radial parts connects the functions in the two domains and leads thus to the statement

$$a_{1l}(r')\left[r^l - \frac{R^{2l+1}}{r^{l+1}}\right] = b_{2l}(r)\frac{1}{(r')^{l+1}}\,,$$

which amounts to a dependence of the remaining coefficient function on the variable r' up to a constant factor c_l

$$a_{1l}(r') = c_l\frac{1}{(r')^{l+1}} \qquad b_{2l}(r) = c_l\left[r^l - \frac{R^{2l+1}}{r^{l+1}}\right].$$

The resulting functions

$$g_{1l}(r, r') = c_l\left[\frac{r^l}{(r')^{l+1}} - \frac{R^{2l+1}}{(rr')^{l+1}}\right] \qquad r < r'$$

$$g_{2l}(r, r') = c_l\left[\frac{r'^l}{r^{l+1}} - \frac{R^{2l+1}}{(rr')^{l+1}}\right] \qquad r' < r$$

are continuous along the dividing line of the two domains, that is

$$g_{1l}(r, r') = g_{2l}(r', r) \qquad \text{for} \quad r' = r\,.$$

The differential equation for the radial parts is of second order and singular at the positions with $r = r'$. For the final connection of the solutions a statement on the first derivatives of the functions g_{il} is required for the points on the dividing line. This relation can be obtained from the differential equation (4.71) for the functions g_l. With

$$\frac{\mathrm{d}}{\mathrm{d}r}\left(r^2\frac{\mathrm{d}g}{\mathrm{d}r}\right) = r\frac{\mathrm{d}^2}{\mathrm{d}r^2}(rg)$$

this equation is

$$\frac{\mathrm{d}^2}{\mathrm{d}r^2}\left(rg_l(r,r')\right) - \frac{l(l+1)}{r}g_l(r,r') = -4\pi\, k_e\,\frac{\delta(r-r')}{r}\,.$$

Integration over the critical point between the limits $r' - \varepsilon$ and $r' + \varepsilon$, gives

$$\frac{\mathrm{d}}{\mathrm{d}r}\;\left(rg_l(r,r')\right)_{r=r'+\varepsilon} - \frac{\mathrm{d}}{\mathrm{d}r}\left(rg_l(r,r')\right)_{r=r'-\varepsilon}$$

$$-l(l+1)\int_{r'-\varepsilon}^{r'+\varepsilon}\frac{\mathrm{d}r}{r}g_l(r,r') = -\frac{4\pi\, k_e}{r'}\,.$$

In the first term (with $r > r'$) one has to insert g_{2l}, in the second term (with $r < r'$) g_{1l}. For the limiting value with $\varepsilon \to 0$, there is no contribution to the remaining integral due to the continuity of the integrand. This argument leads to

$$\lim_{\varepsilon\to 0}\left\{\frac{\mathrm{d}}{\mathrm{d}r}\left(rg_l(r,r')\right)_{r=r'+\varepsilon} - \frac{\mathrm{d}}{\mathrm{d}r}\left(rg_l(r,r')\right)_{r=r'-\varepsilon}\right\} = -\frac{4\pi\, k_e}{r'}\,.$$

In words: The first derivative of the radial parts of the Green's function is discontinuous at the critical positions. If all results for the radial parts, obtained so far, are inserted into this condition, one finds

$$\left\{\left[-\frac{l}{r'} + \frac{lR^{2l+1}}{(r')^{2l+2}}\right] - \left[\frac{l+1}{r'} + \frac{lR^{2l+1}}{(r')^{2l+2}}\right]\right\} c_l = -\frac{4\pi\, k_e}{r'}$$

or after sorting

$$c_l = \frac{4\pi\, k_e}{2l+1}\,.$$

The final result can, as a result of the symmetry, be expressed either by g_{1l} or by g_{2l}. For $r < r'$ the result is (α is the angle between the two directions in space, which is

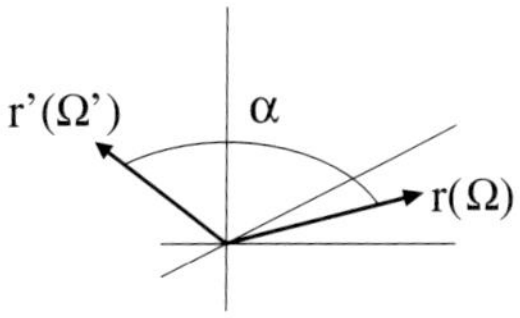

Fig. 4.40 Geometry of the angles in space

given by Ω and Ω', Fig. 4.40)

$$G(\boldsymbol{r},\,\boldsymbol{r}') = \sum_{lm} \frac{4\pi\, k_e}{(2l+1)} \left[\frac{r^l}{(r')^{l+1}} - \frac{R^{2l+1}}{(rr')^{l+1}} \right] Y_{lm}(\Omega) Y^*_{lm}(\Omega')$$

$$= k_e \sum_l \left[\frac{r^l}{(r')^{l+1}} - \frac{R^{2l+1}}{(rr')^{l+1}} \right] P_l(\cos\alpha)\,. \tag{4.71}$$

For $r > r'$ one obtains the corresponding result

$$G(\boldsymbol{r},\,\boldsymbol{r}') = k_e \sum_l \left[\frac{r'^l}{(r)^{l+1}} - \frac{R^{2l+1}}{(rr')^{l+1}} \right] P_l(\cos\alpha)\,. \tag{4.72}$$

The first term in these expansions represents the inverse of the separation, e.g. for $r < r'$

$$\sum_l \frac{r^l}{(r')^{l+1}} P_l(\cos\alpha) = \frac{1}{|\boldsymbol{r} - \boldsymbol{r}'|}\,.$$

The second term can be discussed with the same expansion

$$\frac{R}{rr'} \sum_l \left(\frac{R^2}{rr'} \right)^l P_l(\cos\alpha) = \frac{R}{rr'} \left[\frac{1}{1 - 2\frac{R^2}{rr'}\cos\alpha + \left(\frac{R^2}{rr'}\right)^2} \right]^{1/2}$$

$$= \frac{R}{r'} \left[\frac{1}{r^2 - 2r\left(\frac{R^2}{r'^2} r'\right)\cos\alpha + \left(\frac{R^2}{r'^2} r'\right)^2} \right]^{1/2}$$

$$= \frac{R}{r'} \frac{1}{\left| \boldsymbol{r} - \frac{R^2}{r'^2}\boldsymbol{r}' \right|}\,.$$

The Green's function approach gives the same result as the application of the method of mirror images.

4.7.4 Evaluation of the General Formula for a Potential in the Case of Spherical Symmetry

The application of the general formula for the solution of a Dirichlet problem ((4.32) in Chap. 4.3) with spherical symmetry requires the calculation of the Green's function on lateral surfaces. If the outer sphere has an infinite radius, one finds

$$G(\boldsymbol{r}', \boldsymbol{r}) = \frac{k_e}{|\boldsymbol{r} - \boldsymbol{r}'|} - \frac{k_e\, R}{r'} \frac{1}{\left|\boldsymbol{r} - \dfrac{R^2}{r'^2}\boldsymbol{r}'\right|} . \tag{4.73}$$

This shows, that one has to deal only with the inner sphere. The simplest way to obtain the derivative (pay heed to the sign of the normal derivative)

$$\frac{\partial G(\boldsymbol{r}', \boldsymbol{r})}{\partial n'} = - \left.\frac{\partial G(\boldsymbol{r}', \boldsymbol{r})}{\partial r'}\right|_{r'=R}$$

uses the initial expansion

$$G(\boldsymbol{r}', \boldsymbol{r}) = k_e \sum_l \left[\frac{(r')^l}{r^{l+1}} - \frac{R^{2l+1}}{(rr')^{l+1}}\right] P_l(\cos\alpha) . \tag{4.74}$$

The derivative of this function (applying an interchange of differentiation and summation) is

$$\begin{aligned}\left.\frac{\partial G(\boldsymbol{r}', \boldsymbol{r})}{\partial r'}\right|_{r'=R} &= k_e \sum_l \left[\frac{l(r')^{l-1}}{r^{l+1}} + \frac{(l+1)R^{2l+1}}{r^{l+1}(r')^{l+2}}\right]_{r'=R} P_l(\cos\alpha) \\ &= k_e \sum_l \left[\frac{(2l+1)R^{l-1}}{r^{l+1}}\right] P_l(\cos\alpha) .\end{aligned} \tag{4.75}$$

If one starts with the expansion of the separation of two points (Chap. 3.2, (3.20))

$$\sum_l \frac{(r')^l}{r^{l+1}} P_l(\cos\alpha) = \frac{1}{\left[r^2 - 2rr'\cos\alpha + r'^2\right]^{1/2}} ,$$

one obtains by differentiation with respect to r' the relation

$$\sum_l \frac{l(r')^{l-1}}{r^{l+1}} P_l(\cos\alpha) = \frac{r\cos\alpha - r'}{\left[r^2 - 2rr'\cos\alpha + r'^2\right]^{3/2}} ,$$

which corresponds to the first term in (4.75). The second term can be treated with the expansion of the separation

$$\sum_l \frac{R^{l-1}}{r^{l+1}} P_l(\cos\alpha) = \frac{1}{R}\frac{1}{\left[r^2 - 2rR\cos\alpha + R^2\right]^2}$$

$$= \frac{\dfrac{r^2}{R} - 2r\cos\alpha + R}{\left[r^2 - 2rR\cos\alpha + R^2\right]^{3/2}} .$$

The relation (4.75) yields therefore (compare Detail 4.7.1)

$$\left.\frac{\partial G(\boldsymbol{r}', \boldsymbol{r})}{\partial r'}\right|_{r'=R} = k_e \left\{ \frac{2(r\cos\alpha - R) + (\frac{r^2}{R} - 2r\cos\alpha + R)}{\left[r^2 - 2rR\cos\alpha + R^2\right]^{3/2}} \right\}$$

$$= \frac{k_e\,(r^2 - R^2)}{R\left[r^2 - 2rR\cos\alpha + R^2\right]^{3/2}} .$$

Consider as an example the problem: Calculate the potential, which is defined by specification of the potential on two concentric spherical surfaces (radius R and ∞) with

$$V(\boldsymbol{R_1} = \boldsymbol{R}) = V_0 \qquad V(\boldsymbol{R_2} \to \infty) = 0 .$$

In the domain B between these surfaces there are no point or space charges

$$\rho(\boldsymbol{r}) = 0 \qquad \text{in } B .$$

The potential is found, if this input is inserted into the formula (4.35) in Chap. 4.3 for the solution, adapted to the particular symmetry

$$V(\boldsymbol{r}) = k_e \iiint \frac{\rho(\boldsymbol{r}')}{|\boldsymbol{r} - \boldsymbol{r}'|}\,\mathrm{d}V' - k_e \iiint \frac{R\rho(\boldsymbol{r}')}{r'|\boldsymbol{r} - \dfrac{R^2}{r'^2}\boldsymbol{r}'|}\,\mathrm{d}V' \tag{4.76}$$

$$+\frac{1}{4\pi}\iint V(R, \theta', \varphi')\,\frac{R(r^2 - R^2)}{\left[r^2 + R^2 - 2rR\cos\alpha(\theta', \varphi')\right]^{3/2}}\,\mathrm{d}\Omega'$$

and the necessary integrations are performed. The problem has azimuthal symmetry, so that $\cos\alpha(\theta', \varphi') = \cos\theta'$. For the case of a charge density $\rho(\boldsymbol{r}) = 0$, only the surface term needs to be considered. One finds

$$
\begin{aligned}
V(\boldsymbol{r}) &= \frac{1}{4\pi}\iint V(R, \theta', \varphi')\,\frac{R(r^2 - R^2)}{\left[r^2 + R^2 - 2rR\cos\alpha(\theta', \varphi')\right]^{3/2}}\,\mathrm{d}\Omega' \\
&= \frac{V_0}{4\pi}R(r^2 - R^2)\,2\pi\int_{-1}^{1}\frac{1}{\left[r^2 + R^2 - 2rR\cos\theta'\right]^{3/2}}\,\mathrm{d}\cos\theta' \\
&= \frac{V_0}{2rR}R(r^2 - R^2)\left[\frac{1}{\left[r^2 + R^2 - 2rR\cos\theta'\right]^{1/2}}\right]_{-1}^{1} \\
&= \frac{V_0}{2r}(r^2 - R^2)\left[\frac{1}{r - R} - \frac{1}{r + R}\right] \qquad (r > R) \\
&= \frac{V_0 R}{r} = k_e\frac{Q}{r}\,.
\end{aligned}
$$

The last step is a consequence of the statement: The surface charge density of a sphere with the radius R and the potential value V_0 is constant

$$
\sigma = -\frac{1}{4\pi k_e}\left.\frac{\partial V(\boldsymbol{r})}{\partial r}\right|_{r=R},
$$

in the present case

$$
\sigma = \frac{V_0}{4\pi k_e R}\,.
$$

Integration over the surface, or more simply multiplication by $4\pi R^2\sigma$, shows, that the total charge of the surface is

$$
Q = \frac{V_0 R}{k_e}\,.
$$

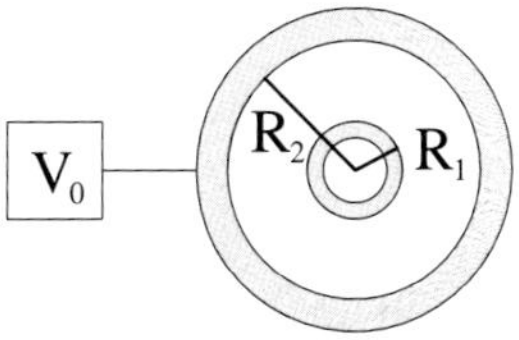

Fig. 4.41 A cut through a spherical capacitor

4.7.5 The Capacity of a Spherical and a Cylindrical Capacitor

The capacity of an arrangement of metal surfaces equals the ratio of charge and voltage (C=Q /V). The potential in the space between the surfaces and therefore the voltage can be calculated readily for arrangements with a simple geometry.

4.7.5.1 The Spherical Capacitor, Simple Calculation

A spherical capacitor (compare Chap. 1.4) consists of two concentric metal spheres (inner sphere with outer radius R_1, outer sphere with inner radius R_2). The inner sphere carries the charge $Q = 4\pi\,\sigma_1\,R_1^2$, the outer sphere is kept at a constant potential V_0 (Fig. 4.41). The calculation of the potential between the two surfaces, a Dirichlet problem, can be obtained by using the results of the electrical field of the capacitor of Chap. 1.4

$$\boldsymbol{E}(\boldsymbol{r}) = k_e \frac{Q}{r^2}\boldsymbol{e}_r \quad \text{for} \quad R_1 \le r < R_2$$

and determining the potential by line integration. The integration from r to R_2 in the radial direction gives

$$V(R_2) - V(r) = -\int_r^{R_2} \boldsymbol{E}\cdot \mathbf{d}\boldsymbol{r}' = -k_e Q \int_r^{R_2} \frac{\mathrm{d}r'}{r'^2}$$

$$= \frac{k_e Q}{r'}\bigg|_r^{R_2} = \frac{k_e Q}{R_2} - \frac{k_e Q}{r}\,.$$

As the potential of the outer sphere is $V(R_2) = V_0$, the potential in the space between the two spherical shells is

$$V(r) = \frac{k_e Q}{r} - \frac{k_e Q}{R_2} + V_0\,.$$

An alternative way to this result can be found by adaption of the formula of Chap. 4.3 for a Dirichlet problem with spherical symmetry (with different boundary conditions). It will be treated in Chap. 4.7.5.3.

4.7.5.2 The Spherical Capacitor, Alternative Calculation of the Potential

One uses the general formula (4.35) for the solution, which has been found in Chap. 4.3

$$V(\boldsymbol{r}) = k_e \iiint \frac{\rho(\boldsymbol{r}')}{|\boldsymbol{r}-\boldsymbol{r}'|}\,\mathrm{d}V' - k_e \iiint \frac{R\rho(\boldsymbol{r}')}{r'|\boldsymbol{r} - \frac{R^2}{r'^2}\boldsymbol{r}'|}\,\mathrm{d}V' \tag{4.77}$$

$$+\frac{1}{4\pi}\iint V(R,\theta',\varphi')\frac{R(r^2-R^2)}{\left[r^2+R^2-2rR\cos\alpha(\theta',\varphi')\right]^{3/2}}\,\mathrm{d}\Omega' .$$

This formula is appropriate for a situation, in which the potential on the surface of a sphere with radius R and the charge density in a domain with $R \le r \le \infty$ are given.

In order to apply this formula for the calculation of the potential of a spherical capacitor with the radii $R_1 < R_2$, it is necessary to identify the radius R with the outer radius R_2 of the spherical capacitor and to adjust the electric field for $r \le R_2$, so that the charge Q is distributed on a sphere with the radius $R_1 < R$. This can be done by placing a point charge with $\rho(\boldsymbol{r}') = Q\delta(\boldsymbol{r}')$ at the origin and choosing Q as $Q = 4\pi R_1^2\sigma$. It is possible in this way to simulate the field of the capacitor. One can now choose the value V_0 on the spherical surface with R_2, but has to recall that the direction of a normal vector, which usually points to the direction of the centre of the concentric spherical surfaces, has to be reverted, so that the region with $r > R_2$ is now the outer region of the problem (Fig. 4.42). With these replacements

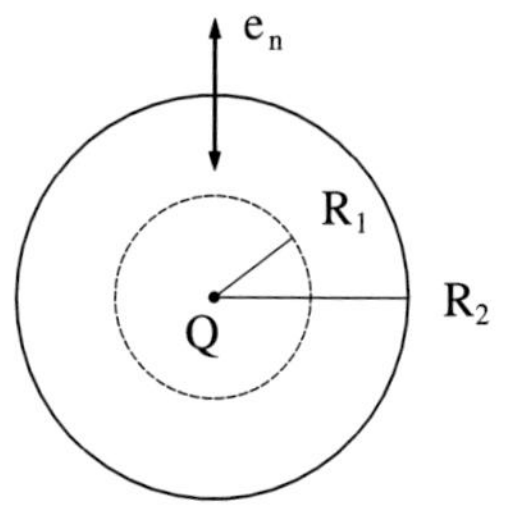

Fig. 4.42 The modification of the Dirichlet problem in Chap. 4.3 for the discussion of the spherical capacitor ($R = R_2$, Eq. (4.77))

the equation (4.77) is changed to

$$V(\boldsymbol{r}) = k_e \iiint \frac{Q\,\delta(\boldsymbol{r}')}{|\boldsymbol{r}-\boldsymbol{r}'|}\,\mathrm{d}V'$$

$$-k_e \iiint \frac{R_2 Q\,\delta(\boldsymbol{r}')}{\left[r^2 r'^2 - 2rr'R_2^2\cos\alpha + R_2^4\right]^{1/2}}\,\mathrm{d}V'$$

$$-\frac{1}{4\pi}\iint V(R,\theta',\varphi')\,\frac{R(r^2-R^2)}{\left[r^2+R_2^2-2rR_2\cos\theta'\right]^{3/2}}\,\mathrm{d}\Omega' .$$

Evaluation yields the result of part 4.7.5.1

$$V(\boldsymbol{r}) = k_e\frac{Q}{r} - k_e\frac{Q}{R_2} - \frac{V_0}{2}R_2(r^2-R_2^2)\left[\frac{1}{\left[r^2+R_2^2-2rR_2\cos\theta'\right]^{1/2}}\right]_{-1}^{1}$$

$$= k_e\frac{Q}{r} - k_e\frac{Q}{R_2} - \frac{V_0}{2}\frac{(r^2-R_2^2)}{r}\left[\frac{1}{R_2-r}-\frac{1}{R_2+r}\right] \qquad (r < R_2)$$

$$= k_e\frac{Q}{r} - k_e\frac{Q}{R_2} + V_0 .$$

4.7.5.3 The Cylindrical Capacitor

The calculation of the capacity of any capacitor arrangement requires the calculation of the voltage between the two metallic surfaces of the capacitor. Application of the Gauss' theorem plus simple line integration is sufficient for a simple cylinder geometry.

The cylindrical capacitor is composed of a metal cylinder with the radius R_1, the length L and the charge Q, which is surrounded by a hollow metal cylinder with an inner radius R_2 (and length L) (Fig. 4.43a).

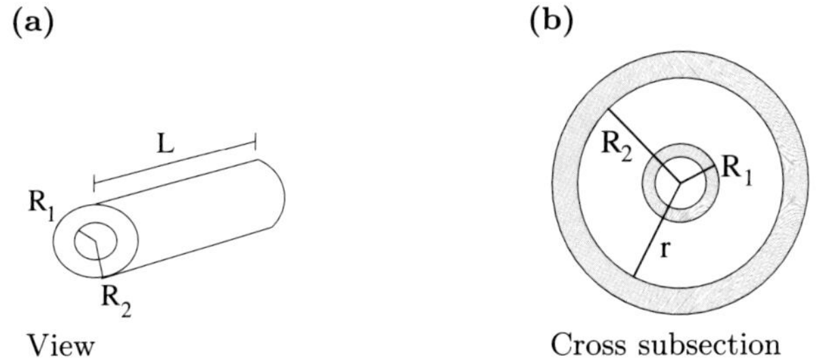

Fig. 4.43 Cylindrical capacitor

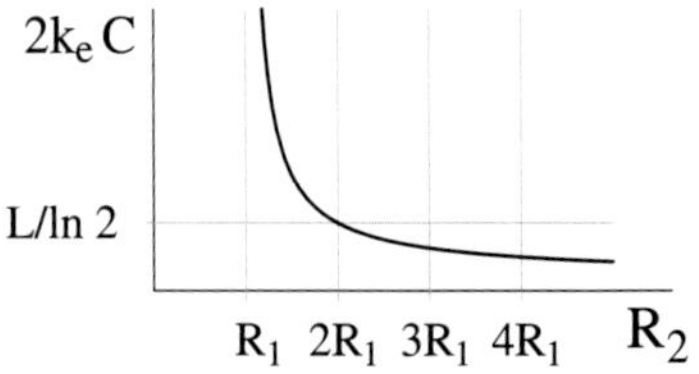

Fig. 4.44 The capacity ($2C$) of a cylindrical capacitor as a function of the outer radius for a fixed inner radius

The Gaussian surface is chosen to be the surface of a cylinder in the intermediate space. Then the field is $\boldsymbol{E} = E\boldsymbol{e}_n$, with the normal $\boldsymbol{n}$ of the surface. If edge effects are neglected one finds for the field between the two cylinders with the Gauss' theorem due to the symmetry

$$4\pi\, k_e Q = \oiint_{\text{cyl}} \boldsymbol{E}\cdot\mathbf{d}\boldsymbol{f} = E(r) \oiint_{\text{cyl}} \boldsymbol{e}_n\, \mathbf{d}\boldsymbol{f} = E(r)\, 2\pi\, r L$$

$$\Longleftrightarrow \qquad E(r) = \frac{2k_e Q}{Lr}\,.$$

The corresponding electric potential between the cylinders is obtained by integration in the radial direction (Fig. 4.43b)

$$U(R_1, R_2) = \frac{2k_e Q}{L}\int_{R_1}^{R_2} \frac{\mathrm{d}r}{r} = \frac{2k_e Q}{L}\ln\frac{R_2}{R_1}\,.$$

The capacity of this arrangement is

$$C = \frac{Q}{U} = \frac{1}{2k_e}\frac{L}{\ln(R_2/R_1)}\,.$$

For a fixed value of the potential U and the parameter R_1 and L the capacity of cylindrical capacitors decreases with increasing R_2 as shown in Fig. 4.44.

4.7.6 Potential of a Straight Wire with a Dielectric Cylinder

4.7.6.1 Potential of a Straight Wire Parallel to the z-Axis Through the Point $(x, y) = (r_0, 0)$

One first calculates the electric field

$$\boldsymbol{E}(\boldsymbol{r}) = E(r)\boldsymbol{e}_r$$

of a long, straight wire (Fig. 4.45a) along the z-axis using the Gauss' theorem. For a linear charge density λ one obtains for a Gaussian cylinder with the length h and

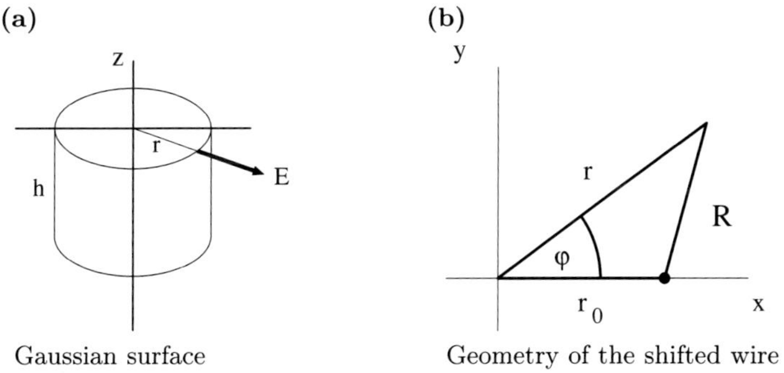

Fig. 4.45 Potential of a long straight wire

a radius r around the z-axis for the present symmetry

$$\iint \boldsymbol{E} \cdot \mathbf{d}\boldsymbol{f} = E(r) r\, h \int \mathrm{d}\varphi = 4\pi k_e \lambda\, h$$

and thus

$$E(r) = \frac{2k_e\lambda}{r}\,.$$

The corresponding electric potential is

$$V(r) = -\int^r E(r')\mathrm{d}r' = -2k_e\lambda \ln r\,.$$

For a wire, which runs parallel to the z-axis and cuts the x-axis in a point $(r_0, 0)$, the distance of a field point R from the point of crossing the x-y plane using the theorem of cosines (Fig. 4.45b)

$$R = [r^2 + r_0^2 - 2rr_0\cos\varphi]^{1/2}\,.$$

The quantity r is the distance of the field point from the origin of the coordinate system and φ the azimuthal angle. In order to get the desired Fourier series for the potential

$$V(r,\varphi) = -2k_e\lambda \ln R(r,\varphi)$$

one needs an expansion of the function

$$\ln(1 - 2x\cos\varphi + x^2) = -2\sum_{n=1}^{\infty} \frac{x^n}{n}\cos n\varphi\,. \tag{4.78}$$

The first terms of this series can be obtained with the formula of Taylor. The general proof of the correctness of the expansion is a bit more tedious. It can be gained with

- One verifies first the related expansion

$$\frac{(1 - x^2)}{(1 - 2x\cos\varphi + x^2)} = 1 + 2\sum_{n=1}^{\infty} x^n \cos n\varphi\,, \tag{4.79}$$

  by multiplication of the equation with

$$1 - x^2 = 1 - 2x\cos\varphi + x^2 + 2\sum_{n=1}^{\infty} x^n \cos n\varphi$$
$$-4\sum_{n=1}^{\infty} x^{n+1}\cos n\varphi\cos\varphi + 2\sum_{n=1}^{\infty} x^{n+2}\cos n\varphi$$

  and separate the term with the product of the cosine functions into two contributions

$$2\cos n\varphi\cos\varphi = \cos(n+1)\varphi + \cos(n-1)\varphi$$

  On the right hand side (RHS) one finds after sorting

$$RHS = 1 - x^2 + \left[2\sum_{n=1}^{\infty} x^n \cos n\varphi - 2x\cos\varphi - 2\sum_{m=2}^{\infty} x^m \cos m\varphi\right]$$
$$+\left[2\sum_{m=2}^{\infty} x^{m+1}\cos(m-1)\varphi + 2x^2\right.$$
$$\left. -2\sum_{m=1}^{\infty} x^{m+1}\cos(m-1)\varphi\right].$$

  The last term in the first bracket is equal to a contribution with $\cos(n+1)\varphi$ (after renaming). The first term in the second bracket is equal to the last term of the starting equation (after renaming). One sees that the terms in the brackets cancel each other, so that the right and then left sides are equal. One can prove, that the series (4.79) converge uniformly for $|x| < 1$ and can therefore be integrated term by term.
- One then rearranges the relation (4.79) by shifting the 1 to the left side and multiplies the resulting equation by $1/2x$. The resulting relation can be integrated

in a direct manner

$$\frac{(\cos\varphi - x)}{(1 - 2x\cos\varphi + x^2)} = \sum_{n=1}^{\infty} x^{n-1}\cos n\varphi\,,$$

so that one obtains, after another sorting, the desired series expansion

$$\int^{x} \frac{(\cos\varphi - x')\mathrm{d}x'}{(1 - 2x'\cos\varphi + x'^2)} = -\frac{1}{2}\ln(1 - 2x\cos\varphi + x^2)$$

$$= \sum_{n=1}^{\infty} \frac{x^n}{n}\cos n\varphi\,.$$

It is necessary to bring the function for R into the form

$$R = r_0\left[1 - 2\frac{r}{r_0}\cos\varphi + \left(\frac{r}{r_0}\right)^2\right]^{1/2} \qquad r < r_0$$

respectively

$$R = r\left[1 - 2\frac{r_0}{r}\cos\varphi + \left(\frac{r_0}{r}\right)^2\right]^{1/2} \qquad r_0 < r$$

in order to obtain the expansion of the logarithmic potential

$$R = r_0\left[1 - 2\frac{r}{r_0}\cos\varphi + \left(\frac{r}{r_0}\right)^2\right]^{1/2} \qquad r < r_0$$

and finds with

$$\ln r_0\left[1 - 2\frac{r}{r_0}\cos\varphi + \left(\frac{r}{r_0}\right)^2\right]^{1/2} = \ln r_0$$

$$+\frac{1}{2}\left[1 - 2\frac{r}{r_0}\cos\varphi + \left(\frac{r}{r_0}\right)^2\right]$$

and (4.78) for $r < r_0$ the expansions

$$V_i(r,\varphi) = 2k_e\lambda\left\{\sum_{n=1}^{\infty}\frac{1}{n}\left(\frac{r}{r_0}\right)^n \cos n\varphi - \ln r_0\right\} \qquad r < r_0$$

resp. (4.80)

$$V_a(r,\varphi) = 2k_e\lambda\left\{\sum_{n=1}^{\infty}\frac{1}{n}\left(\frac{r_0}{r}\right)^n \cos n\varphi - \ln r\right\} \qquad r > r_0\,.$$

4.7.6.2 Wire and Dielectric Cylinder: Applying the Conditions for the Connections of the Three Areas in Space

$$V_{ii} = 2k_e\lambda\left\{\sum_{n=1}^{\infty}\frac{1}{n}\left(\frac{r}{r_0}\right)^n \cos n\varphi - \ln r_0\right\} + a_0 + \sum_{n=1}^{\infty} a_n r^n \cos n\varphi$$

$$V_{iz} = 2k_e\lambda\left\{\sum_{n=1}^{\infty}\frac{1}{n}\left(\frac{r}{r_0}\right)^n \cos n\varphi - \ln r_0\right\} + b_0 + \sum_{n=1}^{\infty} b_n r^{-n} \cos n\varphi$$

$$V_{aa} = 2k_e\lambda\left\{\sum_{n=1}^{\infty}\frac{1}{n}\left(\frac{r_0}{r}\right)^n \cos n\varphi - \ln r\right\} + c_0 \ln r + \sum_{n=1}^{\infty} c_n r^{-n} \cos n\varphi\,.$$

The conditions for the connection on the surface of the dielectric cylinder with radius R are the continuity of the potential and the continuity of the normal derivative of the dielectric shift, with the representation in terms of the potential

$$\varepsilon \left.\frac{\partial V_{ii}(r,\varphi)}{\partial r}\right|_R = \left.\frac{\partial V_{iz}(r,\varphi)}{\partial r}\right|_R .$$

These conditions lead, due to the orthogonality of the cosine functions, to the system of equations ($n \geq 1$)

$$a_n R^n - b_n \frac{1}{R^n} = 0$$

$$\varepsilon\, n a_n R^{n-1} + n b_n \frac{1}{R^{n+1}} = 2k_e\,\lambda(1-\varepsilon)\frac{R^{n-1}}{r_0^n}\,,$$

with the solution

$$a_n = 2k_e\,\lambda\,\frac{(1-\varepsilon)}{n(1+\varepsilon)}\left(\frac{1}{r_0}\right)^n \qquad b_n = 2k_e\,\lambda\,\frac{(1-\varepsilon)}{n(1+\varepsilon)}\left(\frac{R^2}{r_0}\right)^n .$$

The constants a_0 and b_0 are not determined, but can be set equal to zero as a consequence of the gauge freedom of the potential.

For the connection of the solutions at $r = r_0$ one can only use the condition

$$V_{iz}(r_0, \varphi) = V_{aa}(r_0, \varphi) .$$

The normal component of the dielectric shift can not be used, as it is not continuous due to the presence of a true charge on the border. The only condition gives

$$c_0 = 0 \qquad c_n = b_n .$$

4.7.6.3 The Mapping $z_2 = a\sin(z_1/A)$

The conformal mapping

$$z_2 = a\sin\left(\frac{z_1}{A}\right)$$

with the reverse

$$z_1 = A\arcsin\left(\frac{z_2}{a}\right)$$

is analysed at this point in some detail. The simplest approach is to specify curves in the z_1-plane and to look at the maps in the z_2-plane. The combination of the individual maps should be able to provide an impression of the total mapping.

If one uses

$$z_1 = x + \mathrm{i}y ,$$

one finds

$$z_2 = a\sin\left(\frac{x}{A} + \mathrm{i}\frac{y}{A}\right) .$$

Application of the addition theorem gives

$$z_2 = a\left[\sin\left(\frac{x}{A}\right)\cos\left(\mathrm{i}\frac{y}{A}\right) + \cos\left(\frac{x}{A}\right)\sin\left(\mathrm{i}\frac{y}{A}\right)\right] .$$

For the conversion of the trigonometric functions with an imaginary argument one uses

$$\cos(\mathrm{i}t) = \frac{1}{2}\left(\mathrm{e}^{\mathrm{i}(\mathrm{i}t)} + \mathrm{e}^{-\mathrm{i}(\mathrm{i}t)}\right)$$

$$= \frac{1}{2}\left(\mathrm{e}^{-t} + \mathrm{e}^{t}\right) = \cosh t$$

$$\sin(\mathrm{i}t) = \frac{1}{2\mathrm{i}}\left(\mathrm{e}^{\mathrm{i}(\mathrm{i}t)} - \mathrm{e}^{-\mathrm{i}(\mathrm{i}t)}\right)$$

$$= \frac{1}{2\mathrm{i}}\left(\mathrm{e}^{-t} - \mathrm{e}^{t}\right)$$

$$= \mathrm{i}\frac{1}{2}\left(\mathrm{e}^{t} - \mathrm{e}^{-t}\right) = \mathrm{i}\sinh t\,.$$

The result in the z_2-plane is therefore

$$z_2 = a\left[\sin\left(\frac{x}{A}\right)\cosh\left(\frac{y}{A}\right) + \mathrm{i}\cos\left(\frac{x}{A}\right)\sinh\left(\frac{y}{A}\right)\right].$$

In the following figures, the curves in the z_1-plane (black) and the corresponding maps in the z_2-plane (grey) are drawn on one sheet.

In particular, the following curves of the z_1-plane and their maps are considered:

- The two lines in the upper half of the $z_1$plane at a distance $\pm A\pi/2$ parallel to the imaginary axis are mapped onto two parts of the real axis of the z_2-plane, which extends from the point $\pm a$ to infinitely distant points (Fig. 4.46).

$$z_1 = A\frac{\pi}{2} + \mathrm{i}y \qquad z_2 = a\cosh\frac{y}{A} \qquad z_2 \geq a$$

$$z_1 = -A\frac{\pi}{2} + \mathrm{i}y \qquad z_2 = -a\cosh\frac{y}{A} \qquad z_2 \leq -a\,.$$

- The imaginary axis of the z_1-plane is mapped to the imaginary axis of the z_2-plane (with a rescaling)

$$z_1 = \mathrm{i}y \qquad z_2 = \mathrm{i}a\sinh\frac{y}{A}\,.$$

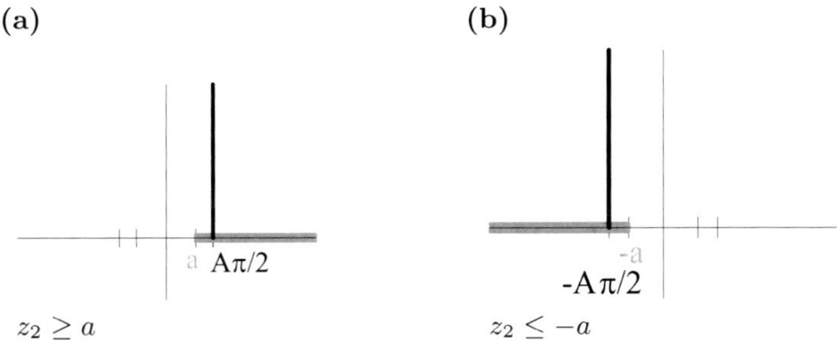

Fig. 4.46 Mapping of lines parallel to the imaginary axis of the z_1-plane

- The interval $[-A\pi/2,\ A\pi/2]$ of the z_1-plane is mapped to the interval $[-a,\ a]$ of the z_2-plane (Fig. 4.47)

$$z_1 = x \qquad -\frac{\pi}{2}A \leq x \leq \frac{\pi}{2}A$$

$$z_2 = a \sin\frac{x}{A} \qquad -a \leq z_2 \leq a\,.$$

- Parallels to the imaginary axis of the z_1-plane, whose real part lies in the interval $[-A\pi/2,\ A\pi/2]$ (Fig. 4.48), correspond to rays that begin on the real z_2-axis in the area $-a \leq \mathrm{Re}(z_2) \leq a$. After an initial curvature, the mapped curves turn into straight lines with a gradient of

$$\tan\alpha = \frac{\cos\left(\frac{c}{A}\right)}{\sin\left(\frac{c}{A}\right)}$$

$$z_1 = c + \mathrm{i}y \qquad -\frac{\pi}{2}A \leq c \leq \frac{\pi}{2}A$$

$$z_2 = a\left[\sin\left(\frac{c}{A}\right)\cosh\left(\frac{y}{A}\right) + \mathrm{i}\cos\left(\frac{c}{A}\right)\sinh\left(\frac{y}{A}\right)\right].$$

- Straight line segments in the interval $[-A\pi/2,\ A\pi/2]$ parallel to the real axis of the z_1-plane merge into ellipse-like curves in the z_2-plane, whereby the images of two straight line segments with $y = \pm const.$ complement each other to form

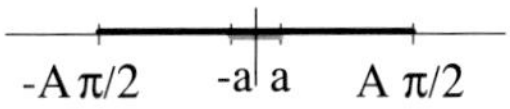

Fig. 4.47 Mapping of $[-A\pi/2,\ A\pi/2]$ of the z_1-plane

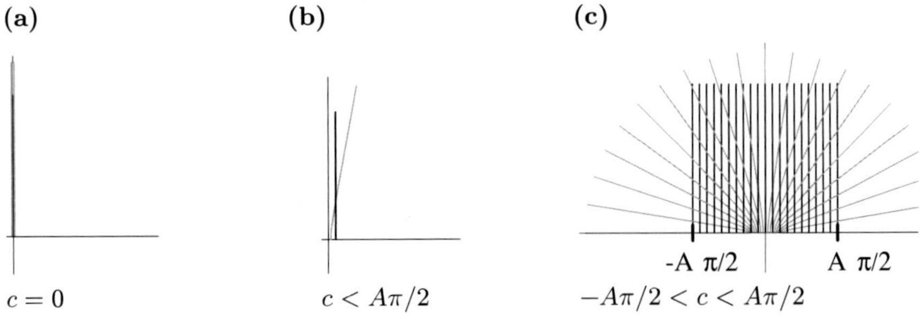

Fig. 4.48 Mapping of parallels to the imaginary axis of the z_1-plane with real part in $[-A\pi/2,\ A\pi/2]$

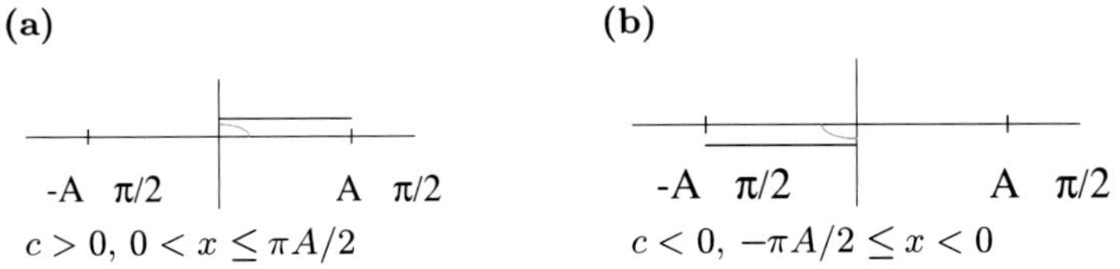

Fig. 4.49 Parts of the mapping of straight line segments in the interval $[-A\pi/2,\ A\pi/2]$ parallel to the real axis of the z_1-plane

an ellipse. If the distance from the real axis of the z_1-plane is large, the ellipses turn into circles (Figs. 4.49 and 4.50)

$$z_1 = x + \mathrm{i}c \qquad -\frac{\pi}{2}A \le x \le \frac{\pi}{2}A$$

$$z_2 = a\left[\sin\left(\frac{x}{A}\right)\cosh\left(\frac{c}{A}\right) + \mathrm{i}\cos\left(\frac{x}{A}\right)\sinh\left(\frac{c}{A}\right)\right].$$

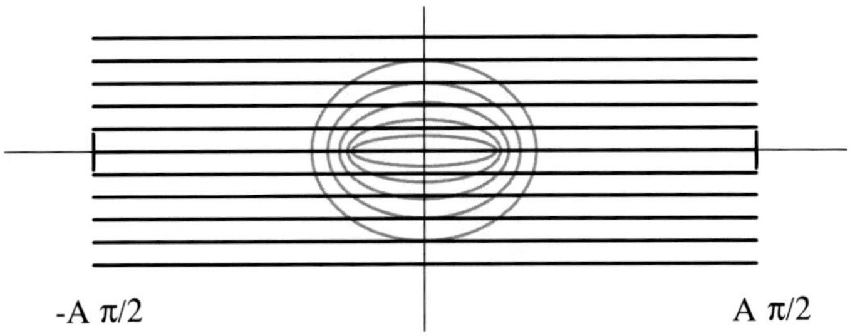

Fig. 4.50 Mapping of straight line segments in the interval $[-A\pi/2,\ A\pi/2]$ parallel to the real axis of the z_1-plane

Magnetostatics

5

The word ‘magnetic’ is said to be associated with the town of Magnesia in Asia Minor. In this region ‘stones’ have been found in ancient times that did attract chips of iron. The name of the mineral with the chemical formula Fe_2O_3 is magnetite or lodestone In the long period between the ancient and the modern era only one additional feature was added to our knowledge of magnetism. The fact, that the earth has a magnetic field was discovered around 1600 and has been used since that time for navigation. The first systematic investigation of magnetism is due to Hans Christian Ørstedt. He found in 1820 that a magnetic needle was not only aligned by the magnetic stones, but also by electric currents in conductors: Electric currents produce magnetic fields (Fig. 5.1a,b).

This chapter starts (Chap. 5.1) with a brief excursion on electric currents. The basic equation of magnetostatics, Ampère’s law, in integral and in differential form is then assembled (Chap. 5.2). A general solution can only be given for the case of relatively simple boundary conditions. Magnetostatics is more intricate in comparison with electrostatic due to the vortex structure of the magnetic field. An approximation, the law of Biot-Savart, is a practical alternative in many cases. The second basic law of magnetostatics, which expresses the fact that the magnetic field is free of sources, allows a representation of the field in terms of a vector potential (Chap. 5.3). It is also necessary, to extend the discussion of magnetic fields in a vacuum to the consideration of fields in spaces filled with dia- and paramagnetic materials (Chap. 5.4)—in analogy to the situation for electric fields. The more complicated material equations for ferromagnetic materials can actually not be handled fully without recourse to the microscopic aspects of this property. The last section (Chap. 5.5) deals with remarks on magnetic forces between charges and conductors.

© The Author(s), under exclusive license to Springer-Verlag GmbH, DE, part of Springer Nature 2024

R. M. Dreizler, C. S. Lüdde, *Electrostatics and Magnetostatics*,
https://doi.org/10.1007/978-3-662-69933-1_5

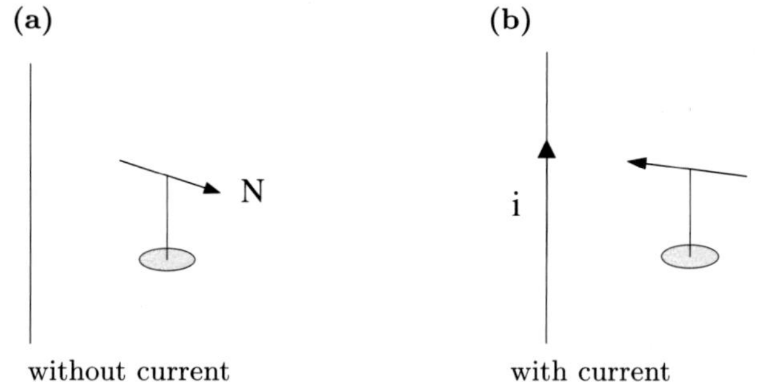

Fig. 5.1 Illustration of the experiment by Ørstedt: Alignment of a magnetic needle

5.1 The Electric Current

The free electrons in a conductor (e.g. a metal wire) are in thermal motion. This motion is statistical. In the cross section of the wire, one would find, that as many free electrons move to the left as to the right through the cross section of the wire per unit time. One can produce an electric field in a piece of wire, if one applies a potential difference between the ends e.g. with a battery (Fig. 5.2). This field produces another type of motion. The electric field is, due to the relation $\boldsymbol{E} = -\nabla V$, directed from the higher value of the potential to the lower value. A force $\boldsymbol{F}_{\mathrm{e}} = -e_0 \boldsymbol{E}$ acts on the electrons. The electrons move under the influence of this field (on the average) from the lower value of the potential to the higher one. An electric current is flowing

$$i = \left| \frac{\mathrm{d}q}{\mathrm{d}t} \right| . \tag{5.1}$$

Concerning the direction of the current one finds: Electrons move from points with a lower potential (V_1) to points with a higher potential (V_2)—for the situation shown in Fig. 5.2 from the right side to the left side. As one initially thought, that positive charges were moving in the wire, the direction of current is (for historic reason) defined to take place exactly in the opposite direction.

▶ The flow of a current in a given direction corresponds to the motion of the free electrons in the opposite direction.

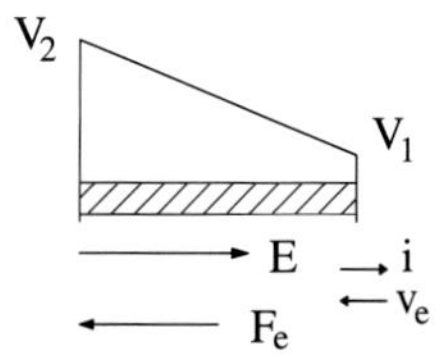

Fig. 5.2 The direction of an electric current

Some details concerning the velocity of the electrons in the conductor might be useful. The thermal motion is quite fast. This statement follows from a relation between the mean velocity of the electrons and the temperature, established in the kinetic theory of gases (consult a suitable text on statistical physics). The mean kinetic energy is proportional to the temperature T, which is usually given in the Kelvin scale

$$\frac{m}{2}\bar{v}^2 = \frac{3}{2}kT\,.$$

The constant k with the value $k = 1.38\,10^{-16}$ erg Kelvin^{-1} is the Boltzmann constant. As the mass of an electron is $m_\mathrm{e} = 9.11\,10^{28}$ g, one obtains a mean thermal velocity $\bar{v}$ of

$$\bar{v}_\mathrm{therm} \approx 10^7 \text{ cm/s}$$

at a room temperature of $T = 300°$ Kelvin.

The motion under the influence of a potential difference is by contrast relatively slow. The following argument leads to a formula for an estimate of this velocity. The free charge in a volume element of the conductor (length l, cross section F, number n of electrons per cm^3) is

$$q = -n\,e_0\,l\,F\,.$$

Every free electron passes the cross section of the wire in the time $t = l/v_e$, so that the current is

$$i = \frac{q}{t} = -n\,e_0\,F\,v_e \qquad \longrightarrow \qquad v_e = -\frac{i}{n\,e_0\,F}\,. \tag{5.2}$$

For a copper wire with a cross section of 0.4 cm^2 and an electron density of $n \approx 10^{21}$ cm^{-3} the velocity is

$$|\bar{v}_\mathrm{therm}| \approx 10^{-1} \text{ cm/s}$$

at a current strength of 10 Ampère.

A realistic description of the motion of the electrons in the wire should therefore sound like this: The electrons execute a fast thermal motion. No direction of the motion stands out because of the collision with the metal ions. The electric field produces a rather slow motion of the free electrons towards the positive clamp, which is superimposed on the statistical motion.

For the formal discussion of the electric current one uses a vector, the **current density** $\boldsymbol{j}$, with the definition: If the flow of the current is uniform across the complete cross section of the conductor with a cross section A, the vector is defined by the statements (Fig. 5.3)

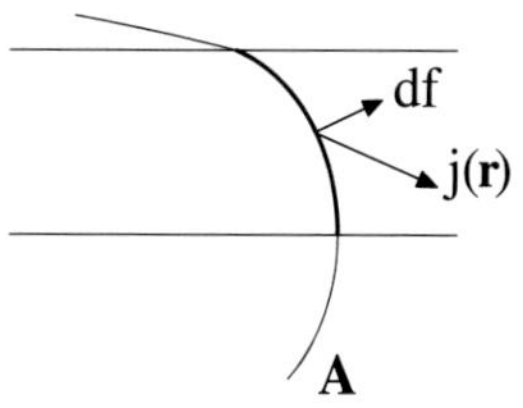

Fig. 5.3 Definition of the current density

▶ The vector points in the direction of the flow of the current caused by the potential difference, its magnitude is $|j| = |i|/A$.

The general relation between current and current density is

$$i = \iint_A \boldsymbol{j}(\boldsymbol{r}) \cdot \mathbf{d}\boldsymbol{f} \tag{5.3}$$

as the current density can vary across the cross section of the conductor A. The surface integral is evaluated over any surface A, which covers the cross section. The following points should be kept in mind:

(a) The magnitude of the current is a scalar quantity (a direction of the current can be expressed by $\pm$, if necessary), the current density is a vector quantity.
(b) Magnitude and density of the current are associated with the motion of the free electrons under the influence of a potential difference. They are quantities, which have been averaged with respect to microscopic aspects.

The charge density and the current density are connected by the equation of continuity. This equation is e.g. obtained with the following argument: Consider an arbitrary volume V in a conductor. First, calculate the total charge in this volume at a time t_1

$$Q(t_1) = \iiint_V \rho(\boldsymbol{r}, t_1)\, \mathrm{d}V \,.$$

At a later time t_2 the charge in the volume is

$$Q(t_2) = \iiint_V \rho(\boldsymbol{r}, t_2)\, \mathrm{d}V \,.$$

The change of the charge in time in the volume is given by the net flow through the surface of the volume, which is defined by the surface integral of the current density

$$\left.\frac{\mathrm{d}Q}{\mathrm{d}t}\right|_{t_1} = \lim_{t_2 \to t_1} \left[\frac{Q(t_2) - Q(t_1)}{t_2 - t_1}\right] = -\oint\!\!\!\oint_{\mathrm{O(V)}} \boldsymbol{j}(\boldsymbol{r}, t_1) \cdot \mathbf{d}\boldsymbol{f} \,. \tag{5.4}$$

The minus sign is a consequence of the standard definition of the direction of the current. The fact, that electrons entering the volume, is equivalent to the fact that positive charges are leaving the volume. This expresses charge conservation in a classical theory: Charges can move, but can not be created or destroyed.

Equation (5.4) can be rewritten using

$$\frac{\mathrm{d}Q}{\mathrm{d}t} = \frac{\mathrm{d}}{\mathrm{d}t} \iiint_{\mathrm{V}} \rho(\boldsymbol{r}, t)\, \mathrm{d}V = \iiint_{\mathrm{V}} \frac{\partial \rho(\boldsymbol{r}, t)}{\partial t}\, \mathrm{d}V\,.$$

Application of the theorem of Gauss to the integral of the charge density in (5.4) gives

$$\iiint_{\mathrm{V}} \left\{ \frac{\partial \rho(\boldsymbol{r}, t)}{\partial t} + \nabla \cdot \boldsymbol{j}(\boldsymbol{r}, t) \right\} \mathrm{d}V = 0$$

or as the volume V can be chosen freely

$$\frac{\partial \rho(\boldsymbol{r}, t)}{\partial t} + \operatorname{div} \boldsymbol{j}(\boldsymbol{r}, t) = 0 \qquad \frac{\partial \rho(\boldsymbol{r}, t)}{\partial t} + \nabla \cdot \boldsymbol{j}(\boldsymbol{r}, t) = 0\,. \tag{5.5}$$

This is the equation of continuity. It expresses charge conservation in a differential form: A change of the charge density with time is only possible, if there is a flow of charges. An alternative formulation is: a gain or loss of the charges can be interpreted as sources or sinks of the current density.

One can think of two situations, for which the charge density does not change with time: the time derivative of the charge density vanishes

$$\frac{\partial}{\partial t} \rho(\boldsymbol{r}, t) = 0\,,$$

if

- there is no current density $\boldsymbol{j} = \boldsymbol{0}$. This is the case of a stationary charge distribution, which has been the subject of the first four chapters.
- One has $\boldsymbol{j} \neq \boldsymbol{0}$, and $\nabla \cdot \boldsymbol{j} = 0$. This is the case of a stationary current distribution. Charges flow continuously into and out of a volume of space, so that the charge density in the volume does not change. That is the situation, which is addressed by magnetostatics.

Two practical consequences of the equation of continuity are the rule of Kirchhoff and the law of Ohm:

(i) The **rule of Kirchhoff** states: A sum of currents entering a branch point in an electric circuit (Fig. 5.4a) equals the sum of currents leaving the branch point

$$\sum_n i_n = 0\,. \tag{5.6}$$

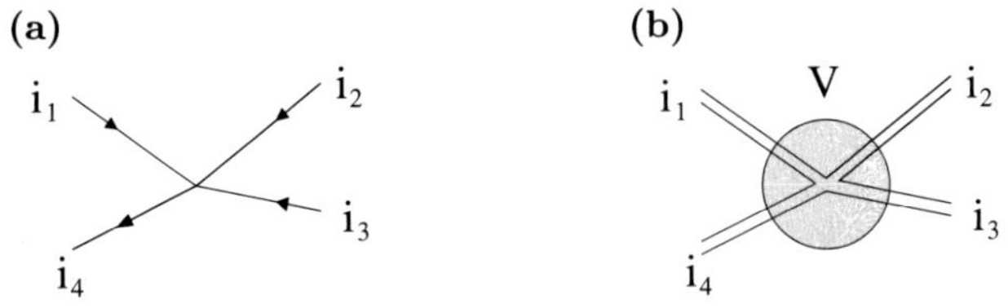

Fig. 5.4 The rule of Kirchhoff

This rule is a direct consequence of the equation of continuity for stationary currents. It follows from the statement

$$0 = \iiint_V \nabla \cdot j(\boldsymbol{r}, t)\mathrm{d}V = \sum_n \iint_{O(V)} j_n \cdot \mathrm{d}\,f = \sum_n i_n$$

which is valid for any volume around the branch point (Fig. 5.4b) by application of the divergence theorem

Only the cross sections of the current carrying conductors contribute to this relation.

(ii) The practical form of the law of Ohm is:
The potential difference across a resistor is equal to the value of the resistance R multiplied by the strength of the current flowing through the resistor

$$U = R \cdot i\,.$$

The differential form

$$\boldsymbol{j}(\boldsymbol{r}) = \sigma(\boldsymbol{r})\boldsymbol{E}(\boldsymbol{r})$$

states that the current density is proportional to the electric field. If this is the case, the material of the resistor is referred to as an ohmic conductor.

The material constant σ is the **specific electric conductivity**. Its inverse $\rho(\boldsymbol{r}) = 1/\sigma(\boldsymbol{r})$ is called the specific electric resistance. All quantities involved can in principle vary with the position in the conductor.

The connection between the two forms can be established with the following argument. Look at a line integral of the differential form between two points $\boldsymbol{r}_1$ and $\boldsymbol{r}_2$ in a conductor (Fig. 5.5). Instead of actually evaluating the integral

$$\int_1^2 \mathrm{d}\boldsymbol{r} \cdot \boldsymbol{j}(\boldsymbol{r})\, \rho(\boldsymbol{r}) = \int_1^2 \mathrm{d}\boldsymbol{r} \cdot \boldsymbol{E}(\boldsymbol{r})\,,$$

argue as follows: The integral on the right hand side yields the potential difference between the two points, as the value the integral does not depend on the path

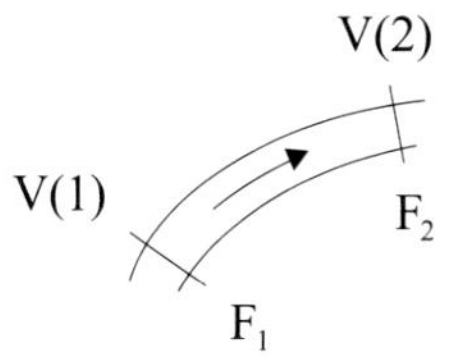

Fig. 5.5 Ohm's law

between the points.

$$\int_1^2 \mathrm{d}\boldsymbol{r} \cdot \boldsymbol{E}(\boldsymbol{r}) = -\int_1^2 \mathrm{d}\boldsymbol{r} \cdot \mathrm{grad}(V(\boldsymbol{r}) = V_1(\boldsymbol{r}) - V_2(\boldsymbol{r}) = U_{12}(\boldsymbol{r})\,.$$

The left hand side can be simplified with the assumption

1. Choose the path, so that $\mathrm{d}\boldsymbol{r}$ points in the direction of the current density $\boldsymbol{j}(\boldsymbol{r})$, so that one has

$$\mathrm{d}\boldsymbol{r} \cdot \boldsymbol{j}(\boldsymbol{r}) = dr\ j(r)$$

2. If, in addition, the current density is constant along the path of integration $j(r) \rightarrow j$ and if it is expressed by the current divided by the cross section of the wire $j = iA$, one arrives at the law of Ohm

$$U_{12} = R_{12}\, i\,.$$

The electric resistance R_{12} is given by

$$R_{12} = \int_1^2 \mathrm{d}r\ \frac{\rho(r)}{A}\,,$$

with a reminder, that it can depend on the length of the path and the cross section of the wire.

According to the arguments above, the motion of the free electron can be described like this: The electrons in a volume of the size (cross section by a length L) move on the average in the direction perpendicular the cross section, but quite slowly. If the resolution of the viewer is considerably improved, he/she will notice the fast statistical motion of the electrons colliding with the lattice ions, with other electrons and the walls of the wire. This scenario can be investigated in some detail with models of the statistical motion.

The basic equations of magnetostatics can be discussed after these introductory remarks on electric currents. The remaining material of this chapter is a summary of the experiments, which have been started by H.C. Ørstedt and have been continued and interpreted by the French physicists J.B. Biot and F. Savart. The results of

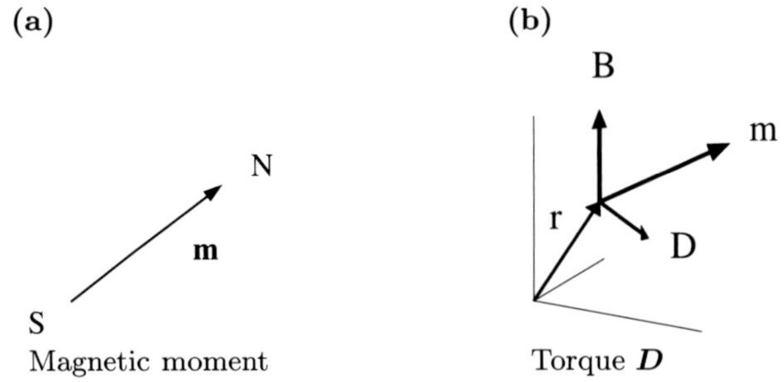

Fig. 5.6 Operative definition of a magnetic field

these endeavours are the basis of the quantitative foundation of the theory by A.M. Ampère.

5.2 Stationary Magnetic Fields

The first step is a look at the simplest magnetic fields, the stationary magnetic fields.

5.2.1 Experimental Basis

The discussion of electric fields was started with the introduction of point charges. A magnetic analogue, a magnetic monopole does not seem to exist.[1] Natural permanent magnets as well as simple electromagnets have a dipole structure. Magnetic dipoles are characterised by a magnetic moment $\boldsymbol{m}$, which is directed from the South pole to the North pole (Fig. 5.6a). If a magnetic dipole is placed in a magnetic field (the standard nomenclature is $\boldsymbol{B}(\boldsymbol{r})$, a precise classification of magnetic fields is given on p. 280), it is aligned by a torque due to the field (Fig. 5.6b)

$$\boldsymbol{D}(\boldsymbol{r}) = \boldsymbol{m} \times \boldsymbol{B}(\boldsymbol{r}) .$$

The direction towards the north pole of a magnet is the local direction of the magnetic field. The operative definition of the strength (magnitude) of a magnetic field is expressed by the relation

$$|\boldsymbol{B}| = \lim_{m \to 0} \frac{|\boldsymbol{D}|}{|m| \sin \varphi} .$$

[1] Quite a number of experiments were undertaken with the hope to observe magnetic monopoles.

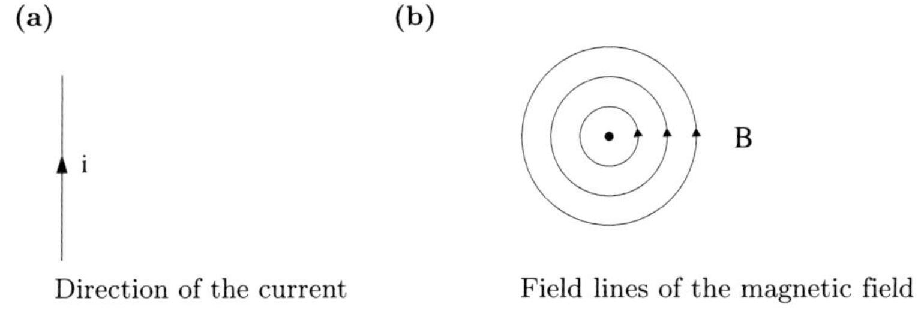

Fig. 5.7 Basic experiment of magnetostatics

One should measure the torque for the angle of deflection and divide by the moment of a calibrated dipole. The basic experiment of magnetostatics, the measurement of the magnetic field of a long, straight conductor, in which a stationary current flows, can be carried out in this fashion (Fig. 5.7a). The result of this experiment can be stated as follows:

1. The magnetic field lines are concentric circles around the conductor (Fig. 5.7b). The direction of the field corresponds to the right hand rule: the thumb points in the direction of the current, the fingers of the curved right hand indicate the direction of the field.
2. The strength of the field depends on the strength of the current i, which produces the field, and the distance r from the centre of the long, straight conductor in the form

$$B(r) = 2k_m \frac{i}{r} \,. \tag{5.7}$$

The choice of the constant k_m fixes the unit for the measurement of the magnetic field. In the SI-system the constant

$$k_{m,\mathrm{SI}} = \frac{\mu_0}{4\pi} = 10^{-7}\,\frac{\mathrm{kg\,m}}{\mathrm{C}^2}$$

is used. The quantity μ_0 is the (magnetic) **permeability of the vacuum**. It is an inverse counter part of the dielectric constant in electrostatics. The unit of the strength of the magnetic field is thus

$$[B]_{\mathrm{SI}} = \frac{\mathrm{kg}}{\mathrm{C}^2}\frac{\mathrm{m\,C}}{\mathrm{s}}\frac{1}{\mathrm{m}} = \frac{\mathrm{kg}}{\mathrm{C\,s}} = \mathrm{Tesla}\,.$$

The magnetic dipole moment is measured in this system in the units

$$[m]_{\mathrm{SI}} = \left[\frac{D}{B}\right]_{\mathrm{SI}} = \frac{\mathrm{C\,m}^2}{\mathrm{s}} = [i]_{\mathrm{SI}} \cdot \mathrm{m}^2\,.$$

In the extended CGS- or Gauss system one uses

$$k_{m,\mathrm{CGS}} = \frac{1}{c} \quad (\text{where } c \approx 2.998 \cdot 10^{10}\,\mathrm{cm/s})$$

and in consequence

$$[B]_{\mathrm{CGS}} = \frac{\mathrm{s}}{\mathrm{cm}} \frac{\mathrm{cm}^{3/2}\,\mathrm{g}^{1/2}}{\mathrm{s}^2} \frac{1}{\mathrm{cm}} = \frac{\mathrm{g}^{1/2}}{\mathrm{cm}^{1/2}\,\mathrm{s}} .$$

In the CGS-system the same unit is used for the electric and the magnetic field strength

$$[B] = [E] ,$$

but in order to distinguish the two phenomena different names are used for the two quantities

$$[E] = \frac{\mathrm{statvolt}}{\mathrm{cm}} \qquad\qquad [B] = \mathrm{Gauss} .$$

The unit of a magnetic dipole in the CGS-system is

$$[m]_{\mathrm{CGS}} = \left[\frac{D}{B}\right] = \frac{\mathrm{g} \cdot \mathrm{cm}^2}{\mathrm{s}^2} \frac{\mathrm{cm}^{1/2}\,\mathrm{s}}{\mathrm{g}^{1/2}} = \frac{\mathrm{g}^{1/2}\mathrm{cm}^{3/2}}{\mathrm{s}} \cdot \mathrm{cm} = [q]_{\mathrm{CGS}} \cdot \mathrm{cm} .$$

The unit is the same as for the electric dipole in the CGS-system

$$[m] = [p] ,$$

however it is different from the unit of the magnetic dipole in the SI-system. An alternative unit of the CGS-system, that is used extensively, is

$$[m]_{\mathrm{CGS}} = \frac{\mathrm{erg}}{\mathrm{Gauss}} .$$

In the two systems of unit the result of the Ørstedt experiment in (1820) is

$$B_{\mathrm{CGS}}(r) = \frac{2i}{cr} \quad \text{and} \quad B_{\mathrm{SI}}(r) = \frac{\mu_0}{2\pi} \frac{i}{r} .$$

The result (5.7) can also be written in the form

$$B(r)\, 2\pi\, r = 4\pi\, k_m\, i ,$$

as the field has the same value for each point on a concentric circle. It is also possible to use

$$\oint_{\mathrm{C}} \boldsymbol{B}(\boldsymbol{r}) \cdot \mathbf{d}\boldsymbol{s} = 4\pi i\, k_m\,. \tag{5.8}$$

This is a special form of a relation, which is referred to as the **law of Ampère**. It is possible to verify by experiment, that it is generally valid. It holds for bent conductors, for closed loops of conductors etc. It says in short, that every conductor, in which a current flows, is surrounded by a magnetic field.

This summary of these experiments can serve as a starting point for the development of magnetostatics.

5.2.2 The Law of Ampère

The first question, that should be answered, is the question of the differential form of the law of Ampère. This question can be answered by rewriting the line integral with the theorem of Stokes

$$\oint_{\mathrm{K}} \boldsymbol{B}(\boldsymbol{r}) \cdot \mathbf{d}\boldsymbol{s} = \oint_{\mathrm{O(K)}} (\nabla \times \boldsymbol{B}(\boldsymbol{r})) \cdot \mathbf{d}\boldsymbol{f}\,.$$

The surface integral has to evaluated for an arbitrary, open and oriented surface, whose boundary is the oriented curve K. The orientation of the curve and the surface are related by the right hand rule (Fig. 5.8). The integral over the current density, taken over the surface of the cross section of the density, represents the current

$$i = \iint \boldsymbol{j}(\boldsymbol{r}) \cdot \mathbf{d}\boldsymbol{f} = \iint_{\mathrm{O(K)}} \boldsymbol{j}(\boldsymbol{r}) \cdot \mathbf{d}\boldsymbol{f}\,.$$

As the current density vanishes in the space outside the conductor, the integral over the cross section can be replaced by an integral over $O(K)$

$$\iint_{\mathrm{O(K)}} \Big[\nabla \times \boldsymbol{B}(\boldsymbol{r}) - 4\pi k_m\, \boldsymbol{j}(\boldsymbol{r})\Big] \cdot \mathbf{d}\boldsymbol{f} = 0\,.$$

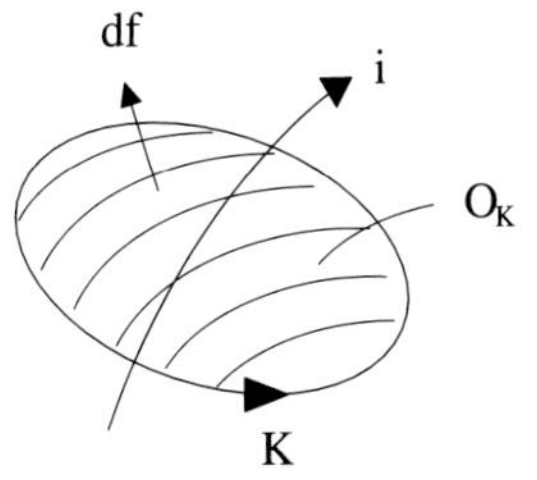

Fig. 5.8 Application of the theorem of Stokes

As every surface $O(K)$ with the boundary K gives the same result, one can also use

$$\operatorname{rot} \boldsymbol{B}(\boldsymbol{r}) = 4\pi k_m \, \boldsymbol{j}(\boldsymbol{r}) \qquad \nabla \times \boldsymbol{B}(\boldsymbol{r}) = 4\pi k_m \, \boldsymbol{j}(\boldsymbol{r}) \, . \tag{5.9}$$

This equation, the **differential equation of Ampère**, states, that the $\boldsymbol{B}$-field is a vortex field. The current density determines the structure of the vortex. The law of Ampère in the form

$$\frac{\partial}{\partial y} B_{\mathrm{z}}(\boldsymbol{r}) - \frac{\partial}{\partial z} B_{\mathrm{y}}(\boldsymbol{r}) = 4\pi k_m \, j_{\mathrm{x}}(\boldsymbol{r})$$

$$\frac{\partial}{\partial z} B_{\mathrm{x}}(\boldsymbol{r}) - \frac{\partial}{\partial x} B_{\mathrm{z}}(\boldsymbol{r}) = 4\pi k_m \, j_{\mathrm{y}}(\boldsymbol{r})$$

$$\frac{\partial}{\partial x} B_{\mathrm{y}}(\boldsymbol{r}) - \frac{\partial}{\partial y} B_{\mathrm{x}}(\boldsymbol{r}) = 4\pi k_m \, j_{\mathrm{z}}(\boldsymbol{r})$$

is a set of differential equations for the components of the field. For the solution of this system of equations one can state: For a current distribution $\boldsymbol{j}(\boldsymbol{r})$, which is restricted to a finite domain one can use the boundary condition

$$\boldsymbol{B}(\boldsymbol{r}) \xrightarrow{r \to \infty} \boldsymbol{0} \qquad \boldsymbol{j}(\boldsymbol{r}) \xrightarrow{r \to \infty} \boldsymbol{0} \, ,$$

which yields a solution of the differential equations above of the form

$$\boldsymbol{B}(\boldsymbol{r}) = k_m \nabla \times \iiint \frac{\boldsymbol{j}(\boldsymbol{r}')}{|\boldsymbol{r} - \boldsymbol{r}'|} \, \mathrm{d} V' \, . \tag{5.10}$$

The proof of this statement can be accomplished by insertion into the differential equations and use of some standard relations of vector analysis.[2] The first step of the proof is to calculate the rotation of (5.10)

$$\nabla \times \boldsymbol{B}(\boldsymbol{r}) = k_m \iiint \nabla \times \left[\nabla \times \frac{\boldsymbol{j}(\boldsymbol{r}')}{|\boldsymbol{r} - \boldsymbol{r}'|} \right] \mathrm{d} V' \, .$$

The interchange of differentiation and integration presents no problem. For the resolution of the product of the two operators involved one uses

$$\nabla \times [\nabla \times \boldsymbol{G}(\boldsymbol{r})] = \nabla \left(\nabla \cdot \boldsymbol{G}(\boldsymbol{r}) \right) - \Delta \boldsymbol{G}(\boldsymbol{r}) \, .$$

[2] A collection of relevant formulae of vector analysis is found in Appendix C.1.

The differential operators act only on the coordinates $\boldsymbol{r}$ in the function $G(\boldsymbol{r})$, so that one obtains

$$\nabla \times \boldsymbol{B}(\boldsymbol{r}) = -k_m \iiint \boldsymbol{j}(\boldsymbol{r}')\Delta\left(\frac{1}{|\boldsymbol{r}-\boldsymbol{r}'|}\right) \mathrm{d}V'$$

$$+k_m \iiint \nabla\left(\boldsymbol{j}(\boldsymbol{r}')\cdot\nabla\left(\frac{1}{|\boldsymbol{r}-\boldsymbol{r}'|}\right)\right) \mathrm{d}V' .$$

In the first term one uses

$$\Delta\left(\frac{1}{|\boldsymbol{r}-\boldsymbol{r}'|}\right) = -4\pi\,\delta(\boldsymbol{r}-\boldsymbol{r}')$$

and obtains

$$T_1(\boldsymbol{r}) = 4\pi k_m\, \boldsymbol{j}(\boldsymbol{r}) .$$

In order to discuss the second term one employs

$$\nabla\left(\frac{1}{|\boldsymbol{r}-\boldsymbol{r}'|}\right) = -\nabla'\left(\frac{1}{|\boldsymbol{r}-\boldsymbol{r}'|}\right)$$

and finds

$$T_2(\boldsymbol{r}) = k_m \nabla\left[\iiint \boldsymbol{j}(\boldsymbol{r}')\cdot\nabla'\left(\frac{1}{|\boldsymbol{r}-\boldsymbol{r}'|}\right) \mathrm{d}V'\right] .$$

The integrand can be treated with the extended product rule

$$\nabla'\cdot\left(\varphi(\boldsymbol{r}')\boldsymbol{j}(\boldsymbol{r}')\right) = \boldsymbol{j}(\boldsymbol{r}')\cdot\nabla'\varphi(\boldsymbol{r}') + \varphi(\boldsymbol{r}')\nabla'\cdot\boldsymbol{j}(\boldsymbol{r}')$$

$$= \boldsymbol{j}(\boldsymbol{r}')\cdot\nabla'\varphi(\boldsymbol{r}') .$$

The term with the factor div $\boldsymbol{j}$ on the right hand side vanishes for a stationary current distribution. There remains

$$T_2(\boldsymbol{r}) = k_m \nabla\left[\iiint \left(\nabla'\cdot\frac{\boldsymbol{j}(\boldsymbol{r}')}{|\boldsymbol{r}-\boldsymbol{r}'|}\right) \mathrm{d}V'\right]$$

or after the application of the divergence theorem

$$T_2(\boldsymbol{r}) = k_m \nabla\left[\oint\!\!\!\oint \mathrm{d}\boldsymbol{f}'\cdot\left(\frac{\boldsymbol{j}(\boldsymbol{r}')}{|\boldsymbol{r}-\boldsymbol{r}'|}\right)\right] .$$

The integrand of the surface integral vanishes on an infinite sphere for a current distribution in a finite section of space, so that

$$T_2(\boldsymbol{r}) = 0 .$$

This completes the proof of the general formula (with simple boundary conditions). First remarks concerning this formula are:

- The gradient of the expression for the separation of two points is

$$\nabla \frac{1}{|\boldsymbol{r}-\boldsymbol{r}'|} = -\frac{(\boldsymbol{r}-\boldsymbol{r}')}{|\boldsymbol{r}-\boldsymbol{r}'|^3} .$$

  As this differential operator acts only on this function, one can also write

$$\boldsymbol{B}(\boldsymbol{r}) = k_m \iiint \boldsymbol{j}(\boldsymbol{r}') \times \frac{(\boldsymbol{r}-\boldsymbol{r}')}{|\boldsymbol{r}-\boldsymbol{r}'|^3} \, \mathrm{d}V' . \tag{5.11}$$

  This formula is similar to the corresponding expression in electrostatics

$$\boldsymbol{E}(\boldsymbol{r}) = k_e \iiint \rho(\boldsymbol{r}') \frac{(\boldsymbol{r}-\boldsymbol{r}')}{|\boldsymbol{r}-\boldsymbol{r}'|^3} \, \mathrm{d}V' .$$

  The difference is due to the fact, that a vector quantity generates a vortex field in the magnetic case in contrast to a scalar quantity in the electric case.
- If one calculates the divergence of the solution (5.10)

$$\nabla \cdot \boldsymbol{B}(\boldsymbol{r}) = k_m \nabla \cdot \left[\nabla \times \iiint \frac{\boldsymbol{j}(\boldsymbol{r}')}{|\boldsymbol{r}-\boldsymbol{r}'|} \, \mathrm{d}V' \right] ,$$

  one may write for the product of two operators

$$\nabla \cdot (\nabla \times \boldsymbol{G}(\boldsymbol{r})) = \operatorname{div} \operatorname{rot} \boldsymbol{G}(\boldsymbol{r}) = 0 .$$

  The action of the operator div rot on a vector function, which is two times continuously differentiable, is zero. Therefore one finds

$$\operatorname{div} \boldsymbol{B}(\boldsymbol{r}) = \nabla \cdot \boldsymbol{B}(\boldsymbol{r}) = 0 . \tag{5.12}$$

  There exist no sources of the magnetic field or expressed differently: Magnetic monopoles do not exist. A corresponding expression in integral form can be obtained with the divergence theorem

$$\iiint_{\mathrm{V}} \nabla \cdot \boldsymbol{B}(\boldsymbol{r}) \, \mathrm{d}V = \oiint_{\mathrm{O(V)}} \boldsymbol{B}(\boldsymbol{r}) \cdot \mathbf{d}\boldsymbol{f} = 0 . \tag{5.13}$$

In analogy to the electric case one refers to the surface integral as the magnetic flow

$$\phi = \oiint_{O(V)} \boldsymbol{B}(\boldsymbol{r}) \cdot \mathbf{d}\boldsymbol{f} \,. \tag{5.14}$$

This means, that the magnetic flow through a closed surface vanishes, or expressed more pictorially: The same number of field lines enter into an arbitrary, closed surface as they leave it.

- The formula (5.11) for the calculation of the magnetic field of a stationary distribution of currents is not very practical. The presence of the vector product in the integrand leads to the fact, that the evaluation can become quite tedious even for simple situations. A practical but approximate method will be derived in Chap. 5.2.1.
- It is instructive, to compare the basic equations of electrostatics and magnetostatics in vacuum. These equations correspond (as will be shown in Dreizler and Lüdde, 2024, Electrodynamics and Special Theory of Relativity (Springer Berlin Heidelberg), Chap. 1.2) to the stationary (time independent) limit of Maxwell's equations

$$\begin{aligned} &\text{Electrostatics:} \quad &&\text{div}\, \boldsymbol{E}(\boldsymbol{r}) = 4\pi k_e\, \rho(\boldsymbol{r}) \quad &&\text{rot}\, \boldsymbol{E}(\boldsymbol{r}) = 0 \\ &\text{Magnetostatics:} \quad &&\text{div}\, \boldsymbol{B}(\boldsymbol{r}) = 0 \quad &&\text{rot}\, \boldsymbol{B}(\boldsymbol{r}) = 4\pi k_m\, \boldsymbol{j}(\boldsymbol{r}). \end{aligned} \tag{5.15}$$

  These equations state:

1. The sources of the $\boldsymbol{E}$-field are the charges. The field is free of vortices. The electric field lines start and end on the charges.
2. The $\boldsymbol{B}$-field has no sources. It has vortices, which are determined by the current density. The magnetic field lines are closed curves around the current distributions of s.

5.2.3 The Formula of Biot-Savart

The practical variant of Eq. (5.11) assumes, that the cross section of the current carrying conductor is sufficiently small, so that one can neglect the variation of the function of the distance between two points (Fig. 5.9a). With this assumption one writes

$$\mathrm{d}V' = \mathrm{d}f'\mathrm{d}s' \quad \text{and} \quad \boldsymbol{j}(\boldsymbol{r}') = j(\boldsymbol{r}')\boldsymbol{e}_{s'} \,.$$

Fig. 5.9 Transition to the formula of Biot-Savart

The current flows only in the direction of the linear extension of the conductor (Fig. 5.9b). Equation (5.11) can then be written

$$\boldsymbol{B}(\boldsymbol{r}) = k_m \iiint \boldsymbol{e}_{s'} \times \frac{(\boldsymbol{r}-\boldsymbol{r}')}{|\boldsymbol{r}-\boldsymbol{r}'|^3} \left(j(\boldsymbol{r}')\mathrm{d}f'\right) \mathrm{d}s' .$$

If the variation of the function for the separation of two points is neglected in the integration over the cross section of the conductor, one can use the abbreviation

$$i = \iint j(\boldsymbol{r}')\, \mathrm{d}\, f' .$$

With $\mathbf{ds}' = \mathrm{d}s'\boldsymbol{e}_{s'}$, one finally obtains

$$\boldsymbol{B}(\boldsymbol{r}) = i\, k_m \int_{\mathrm{K(L)}} \mathbf{ds}' \times \frac{(\boldsymbol{r}-\boldsymbol{r}')}{|\boldsymbol{r}-\boldsymbol{r}'|^3} . \qquad (5.16)$$

The triple integral is reduced to a line integral over the geometry of the conductor. This formula is the **law of Biot-Savart**. It is usually quoted in the form: A line element $\mathbf{ds}'$, at the position $\boldsymbol{r}'$ contributes the infinitesimal value

$$\mathrm{d}\boldsymbol{B}(\boldsymbol{r}) = i\, k_m\, \mathbf{ds}' \times \frac{(\boldsymbol{r}-\boldsymbol{r}')}{|\boldsymbol{r}-\boldsymbol{r}'|^3} \qquad (5.17)$$

to the $\boldsymbol{B}$-field in the point $\boldsymbol{r}$ (Fig. 5.10). The sum of all contributions $\mathrm{d}\boldsymbol{B}(\boldsymbol{r})$ in the point $\boldsymbol{r}$ constitutes the integral form (5.16) of the Biot-Savart law. The following, simple examples illustrate the manner, in which it is applied.

If one uses this formula for the calculation of the magnetic field of a thin, straight conductor with the current i, one should come back to the result (5.7) of the Ørsted-Ampère experiment. The calculation requires the steps: The infinitesimal segment $\mathbf{ds}' = (0, 0, \mathrm{d}z')$ of the conductor along the z-axis (Fig. 5.11a) can, due to the symmetry, be chosen in such a way, that the vector $\boldsymbol{r}$ is the shortest distance between

Fig. 5.10 Illustration of Eq. (5.17)

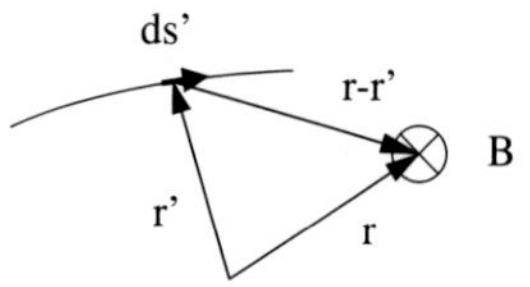

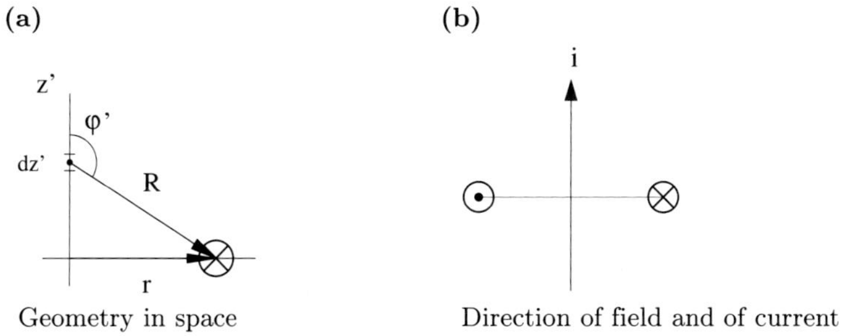

Fig. 5.11 Evaluation of the formula of Biot-Savart for a straight current carrying conductor

the field point and the conductor. The position of the origin of the coordinate system is $\boldsymbol{r}' = (0,\ 0,\ z')$, so that the difference $\boldsymbol{R} = \boldsymbol{r} - \boldsymbol{r}'$ has the length $R = [r^2 + z'^2]^{1/2}$. The contributions $\mathrm{d}\boldsymbol{B}(\boldsymbol{r})$ of all the elements of the conductor point in the same direction. Addition of these scalar contributions to the field yields

$$\mathrm{d}B(\boldsymbol{r}) = i\, k_m\, \frac{\mathrm{d}z'\ \sin\varphi'}{R^2}\,,$$

where φ' is the angle between the z-direction and the difference vector $\boldsymbol{R}$. The sine of the angle φ' is

$$\sin\varphi' = \sin(\pi - \varphi') = \frac{r}{R}\,.$$

Integration over all elements of the conductor leads indeed to the result

$$\boldsymbol{B}(\boldsymbol{r}) = i\, k_m\, r \int_{-\infty}^{\infty} \frac{\mathrm{d}z'}{(z'^2 + r^2)^{3/2}} = \frac{2i\, k_m}{r}\,.$$

The direction of the field is (Fig. 5.11b) for all points of the right/left half plane into/out of the page.

The calculation of the magnetic field of a current ring (strength of the current i, radius a) for points on the axis of the ring proceeds in the same fashion as for the electric equivalent. The geometry (Fig. 5.12a, b) is: The vector $\mathrm{d}\boldsymbol{B}$ is perpendicular to the vector $\mathrm{d}\boldsymbol{s}'$, which points e.g. out of the page and the distance between the element of the ring and a point on the axis of the ring is $\boldsymbol{R}$. The infinitesimal vector of the field is decomposed into components along the axis and perpendicular to the axis

$$\mathrm{d}\boldsymbol{B} = (\mathrm{d}B_A,\ \mathrm{d}B_S)\,.$$

The perpendicular components add up to the value zero, if one integrates over the complete ring. The field in the direction of the axis is calculated by realising, that

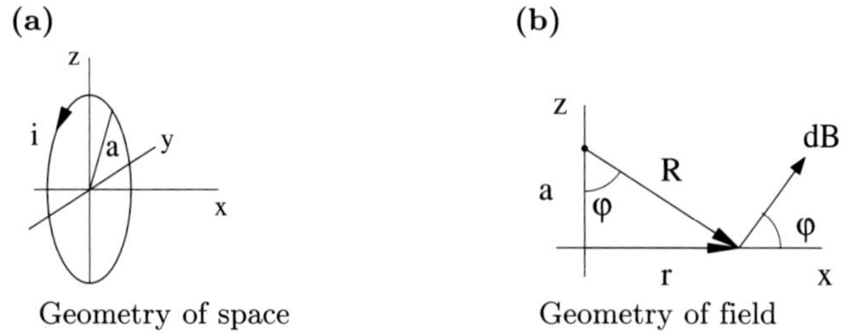

Fig. 5.12 Magnetic field of a circular ring

$\mathbf{d}\boldsymbol{s}'$ is perpendicular to $\boldsymbol{R}$, so that the infinitesimal contribution to the magnetic field in the point $\boldsymbol{r}$ (measured from the centre of the ring) is $\mathrm{d}B = (i k_m \mathrm{d}s')/R^2$. As the component along the axis (with $\cos\varphi = a/R$) is

$$\mathrm{d}B_A = \mathrm{d}B \cos\varphi = i\, k_m \frac{a \mathrm{d}s'}{R^3} .$$

As a and R are constants for a chosen point on the axis, the remaining integration over the ring contributes

$$\int_{\text{ring}} \mathrm{d}s' = 2\pi a$$

and the final result for the ring is

$$B(r) = B_A(r) = 2\pi i\, k_m \frac{a^2}{(a^2 + r^2)^{3/2}} . \tag{5.18}$$

The direction of the field is, for the flow of the current indicated in Fig. 5.12, in the direction of the axis.

If this result is compared with the result for the field of an electric dipole (see Chap. 2.5) one finds, that the introduction of the magnetic moment of a current ring is a useful concept. As a consequence of the dimensional considerations in Chap. 5.2.1 one can write for the magnetic moment of the ring in the two systems

$$m_{SI} = i\pi a^2 \quad \text{resp.} \quad m_{CGS} = i k_m \pi a^2$$

or generally

$$m = i k_f \pi a^2 .$$

The constant k_f is defined as

$$k_{f,SI} \qquad k_{f,CGS} \equiv k_{m,CGS} = \frac{1}{c}\,.$$

Using this definition one finds for the magnetic moment

$$m = i\,k_f \pi a^2 = i\,k_f\,F_{\text{ring}}$$

and for the magnetic field on the axis of the ring

$$B_A(r) = \frac{2mk_m}{k_f(a^2 + r^2)^{3/2}}\,.$$

The far field ($r \gg a$) is

$$B_A(r) = \frac{2mk_m}{k_f r^3}\,. \tag{5.19}$$

A **solenoid** is a coil in the form of a cylinder with a constant number of windings per unit of length (Fig. 5.13). The result for the field on the axis of a current ring (5.18) can be used to find the magnetic field on the axis of the solenoid. A section of the coil with the length $\mathrm{d}z$ is viewed from a point of the axis P under an angle θ (see Fig. 5.14a). The geometry in the figure (same notation as in the case of the ring) shows for the slightly coloured and the small triangle the relation

$$\sin\theta = \frac{a}{R} \qquad \text{resp.} \qquad \sin\theta = \frac{R\mathrm{d}\theta}{\mathrm{d}z}\,.$$

If the number of windings per unit of length is n, then the number of windings in a section $\mathrm{d}z$ is $dn = n\mathrm{d}z$ and thus the total current in the section $\mathrm{d}i = i\mathrm{d}n$, if i is the current in one loop. If the section of the coil is identified with a thin current ring and current $\mathrm{d}i$ of (5.18) is used for the field on the axis one obtains

$$\mathrm{d}B = \frac{2\pi k_m a^2}{R^3} in \frac{R^2}{a} \mathrm{d}\theta = 2\pi i k_m n \sin\theta \mathrm{d}\theta\,.$$

Fig. 5.13 The solenoid

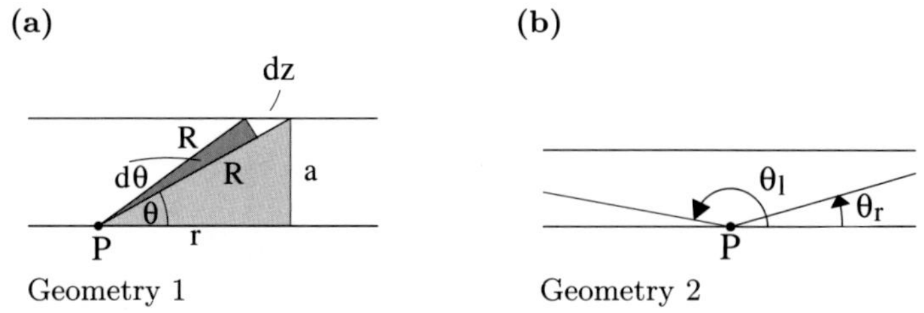

Fig. 5.14 Calculation of the field of a solenoid

Integration over the complete solenoid, which is characterised from the point P by the bounding angles θ_l and θ_r (see Fig. 5.14b) results in the total field

$$B^P = \int_{\theta_r}^{\theta_l} \mathrm{d}B = 2\pi i k_m n(\cos\theta_r - \cos\theta_l)\,.$$

The direction of the field is given by the right hand rule (the thumb marks the direction of the field).

Two limiting situations are of interest:

- If the solenoid is long and the point P in the interior, then the angles are $\theta_r = 0$ and $\theta_l = \pi$. The result is

$$B^{\text{middle}} = 4\pi i k_m n\,. \tag{5.20}$$

- For a point at the end of a long solenoid one should use $\theta_r = 0$ and $\theta_l = \pi/2$. The result is in this case

$$B^{\text{end}} = 2\pi i k_m n\,.$$

The formula (5.20) is applied widely with the assumption, that the field inside the solenoid is uniform. A variant is the replacement of the number of windings per unit length by the total number of windings N divided by the length of the coil.

5.3 The Magnetic Vector Potential

The discussion of electric fields, as e.g. the calculation of fields, could be simplified considerably with the introduction of a scalar function. The relation

$$\text{rot}\,\boldsymbol{B}(\boldsymbol{r}) = \nabla \times \boldsymbol{B}(\boldsymbol{r}) \neq \boldsymbol{0}$$

does not allow to introduce such a function in the case of magnetic fields. The equation

$$\operatorname{div} \boldsymbol{B}(\boldsymbol{r}) = \nabla \cdot \boldsymbol{B}(\boldsymbol{r}) = 0$$

does, however, permit the introduction of a vectorial supplementary function

$$\boldsymbol{B}(\boldsymbol{r}) = \operatorname{rot} \boldsymbol{A}(\boldsymbol{r}) = \nabla \times \boldsymbol{A}(\boldsymbol{r}) \,, \tag{5.21}$$

as the equation

$$\operatorname{div}\operatorname{rot} \boldsymbol{A}(\boldsymbol{r}) = \nabla \cdot (\nabla \times \boldsymbol{A}(\boldsymbol{r})) = 0$$

is valid for all twice continuously differentiable functions.[3] This auxiliary function $\boldsymbol{A}(\boldsymbol{r})$ is the **magnetic vector potential**. It might be a bit astonishing to replace a vector function ($\boldsymbol{B}$) by another vector function ($\boldsymbol{A}$). That such a step is, nonetheless, useful, is shown by the following argument: The solution of the differential equation of Ampère for the case of a simple boundary condition (5.11)

$$\boldsymbol{B}(\boldsymbol{r}) = k_m \iiint \frac{\boldsymbol{j}(\boldsymbol{r}') \times (\boldsymbol{r} - \boldsymbol{r}')}{|\boldsymbol{r} - \boldsymbol{r}'|^3} \, \mathrm{d}V'$$

can be written in the form (5.10)

$$\boldsymbol{B}(\boldsymbol{r}) = k_m \nabla \times \iiint \frac{\boldsymbol{j}(\boldsymbol{r}')}{|\boldsymbol{r} - \boldsymbol{r}'|} \mathrm{d}V' \,.$$

Introduction of the vector potential

$$\boldsymbol{B}(\boldsymbol{r}) = \nabla \times \boldsymbol{A}(\boldsymbol{r})$$

leads to

$$\boldsymbol{A}(\boldsymbol{r}) = k_m \iiint \frac{\boldsymbol{j}(\boldsymbol{r}')}{|\boldsymbol{r} - \boldsymbol{r}'|} \mathrm{d}V' \,. \tag{5.22}$$

The vector potential goes to zero in the asymptotic region, if the current distribution is restricted to a finite region around the origin of the coordinate system

$$\boldsymbol{A}(\boldsymbol{r}) \xrightarrow{r \to \infty} \boldsymbol{0} \,.$$

[3] The proof can be obtained by writing the left hand side in detail.

The calculation of the three integrals for $\boldsymbol{A}$ in (5.22) followed by the construction of the rotation should be simpler than the calculation of the three integrals for $\boldsymbol{B}$ in (5.11). In the formulae for $\boldsymbol{A}$ features only the function for the distance and not a vector product with the difference $(\boldsymbol{r} - \boldsymbol{r}')$. It might also be possible to develop alternative methods for the treatment of problems in terms of the vector potential.

The vector potential is not unique. A situation like this happens also in electrostatics. A function $V(\boldsymbol{r})$ and a function $V'(\boldsymbol{r}) = V(\boldsymbol{r}) + \text{const.}$ lead to the same electric field

$$\boldsymbol{E}(\boldsymbol{r}) = -\nabla V(\boldsymbol{r}) = -\nabla V'(\boldsymbol{r}).$$

As statements concerning experiments depend only on the fields (or on potential differences) there exist no measurable consequences of the lack of uniqueness.

The following statement is valid in magnetostatics: The vector function $\boldsymbol{A}(\boldsymbol{r})$ as well as the vector function $\boldsymbol{A}'(\boldsymbol{r}) = \boldsymbol{A}(\boldsymbol{r}) + \nabla\psi(\boldsymbol{r})$ lead to the same magnetic field, as one has

$$\boldsymbol{B}(\boldsymbol{r}) = \nabla \times \boldsymbol{A}'(\boldsymbol{r}) = \nabla \times \boldsymbol{A}(\boldsymbol{r}) + \nabla \times (\nabla\psi(\boldsymbol{r})) = \nabla \times \boldsymbol{A}(\boldsymbol{r}),$$

if the scalar function $\psi(\boldsymbol{r})$ is twice continuously differentiable.[4] A transformation of auxiliary functions, which do not change the physical content of a theory, are called **gauge transformations**. Thus the gauge transformation $V'(\boldsymbol{r}) = V(\boldsymbol{r}) + \text{const.}$ is the gauge transformation of electrostatics and

$$\boldsymbol{A}'(\boldsymbol{r}) = \boldsymbol{A}(\boldsymbol{r}) + \text{grad}\,\psi(\boldsymbol{r})$$

the gauge transformation of magnetostatics.[5]

The possibility to choose the function $\psi(\boldsymbol{r})$ without changing the physics, that is the magnetic field, plays a role for the discussion of the differential equation for the vector potential. If one inserts the suggestion $\boldsymbol{B} = \nabla \times \boldsymbol{A}$ into the differential equation of Ampère (5.9)

$$\nabla \times \boldsymbol{B}(\boldsymbol{r}) = 4\pi k_m \boldsymbol{j}(\boldsymbol{r}),$$

one obtains the equation

$$\nabla \times (\nabla \times \boldsymbol{A}(\boldsymbol{r})) = 4\pi k_m \boldsymbol{j}(\boldsymbol{r}).$$

[4] The property rot grad $\psi(\boldsymbol{r}) = 0$ always holds for such a function.

[5] Gauge transformations play a special role in quantum field theories. Such theories, which are known under the name of gauge theories, have become a central topic of elementary particle physics.

The vector product with the nabla operator in the combination $\nabla \times (\nabla\times)$ can be decomposed using another formula of vector analysis, however only in terms of the Cartesian components of $\boldsymbol{A}$ (see Detail 5.6.1.1)

$$\nabla \times (\nabla \times \boldsymbol{A}(\boldsymbol{r})) = \nabla\,(\nabla \cdot \boldsymbol{A}(\boldsymbol{r})) - \Delta \boldsymbol{A}(\boldsymbol{r})$$

or in the form

$$\text{rot rot}\,\boldsymbol{A}(\boldsymbol{r}) = \text{grad}(\text{div}\,\boldsymbol{A}(\boldsymbol{r})) - \text{div grad}\,\boldsymbol{A}(\boldsymbol{r})\,.$$

The differential equation is particularly simple, if one chooses, thanks to the gauge freedom

$$\nabla \cdot \boldsymbol{A}(\boldsymbol{r}) = \text{div}\,\boldsymbol{A}(\boldsymbol{r}) = 0\,. \tag{5.23}$$

The differential equation for the vector potential is then

$$\Delta \boldsymbol{A}(\boldsymbol{r}) = -4\pi k_m\, \boldsymbol{j}(\boldsymbol{r})\,, \tag{5.24}$$

which looks definitely like the vector analogon of the Poisson-/Laplace equation. It can only be applied in the Cartesian decomposition

$$\Delta A_{\text{x}}(\boldsymbol{r}) = -4\pi k_m\, j_{\text{x}}(\boldsymbol{r}),\ \ldots$$

The question, which still needs to be answered, is: Can one be sure, that the condition $\nabla \cdot \boldsymbol{A}(\boldsymbol{r}) = 0$ is satisfied in all situations? The answer to the question is: Assume, that a solution of the original differential equation with the operator rot rot has been found, which satisfies

$$\nabla \cdot \boldsymbol{A}(\boldsymbol{r}) = f(\boldsymbol{r}) \neq 0\,.$$

One can then consider the physically absolutely equivalent vector potential

$$\boldsymbol{A}'(\boldsymbol{r}) = \boldsymbol{A}(\boldsymbol{r}) + \text{grad}\,\psi(\boldsymbol{r})\,,$$

for which the relation

$$\nabla \cdot \boldsymbol{A}'(\boldsymbol{r}) = f(\boldsymbol{r}) + \Delta\psi(\boldsymbol{r})$$

is valid. If one chooses now $\psi(\boldsymbol{r})$ (and this is always possible) so that ψ is a solution of a Poisson equation

$$\Delta\psi(\boldsymbol{r}) = -f(\boldsymbol{r})\,,$$

then follows

$$\nabla \cdot \boldsymbol{A}'(\boldsymbol{r}) = 0\,.$$

This means, that it is possible to replace the vector potential, which does not satisfy the condition $\nabla \cdot \boldsymbol{A}(\boldsymbol{r}) = 0$, with the aid of a gauge transformation, which satisfies the condition.

In particular it is possible to demonstrate that the solution (5.22)

$$\boldsymbol{A}(\boldsymbol{r}) = k_m \iiint \frac{\boldsymbol{j}(\boldsymbol{r}')}{|\boldsymbol{r} - \boldsymbol{r}'|}\,\mathrm{d}V'$$

satisfies the condition (see Detail 5.6.1.2)

$$\nabla \cdot \boldsymbol{A}(\boldsymbol{r}) = 0\,.$$

The condition (5.23)

$$\operatorname{div} \boldsymbol{A}(\boldsymbol{r}) = \nabla \cdot \boldsymbol{A}(\boldsymbol{r}) = 0$$

is referred to as the **Coulomb gauge**. This gauge is an appropriate choice of a gauge in the case of magnetostatics. A different gauge will e.g. be more convenient, if one considers the full set of equations (Maxwell's equations) of electrodynamics.

The solution, quoted for the vector potential with simple boundary condition can also be given in the Biot-Savart form, which can, as stated above, be used, if current carrying conductors are sufficiently thin. For such conductors one can use the replacement

$$\boldsymbol{j}(\boldsymbol{r}')\,\mathrm{d}\,V' \longrightarrow i\,\mathbf{d}\boldsymbol{s}'$$

(as before) and finds

$$\boldsymbol{A}(\boldsymbol{r}) = i\,k_m \int_{\text{conductor}} \frac{\mathbf{d}\boldsymbol{s}'}{|\boldsymbol{r} - \boldsymbol{r}'|}\,. \tag{5.25}$$

An example for the use of this formula is the calculation of the vector potential of a circular ring radius a with the current i (Fig. 5.15) in an arbitrary point of space. The ring is e.g. placed in the x- y plane. It is sufficient to consider only one point of the field in the x- z plane because of the cylindrical symmetry with respect to z-axis. The symmetry also suggests to work with cylindrical or with spherical coordinates. For the present problem this means that one has to work with the Cartesian decomposition of $\boldsymbol{A}$ respectively $\mathbf{d}\boldsymbol{s}$. If spherical coordinates are used, the

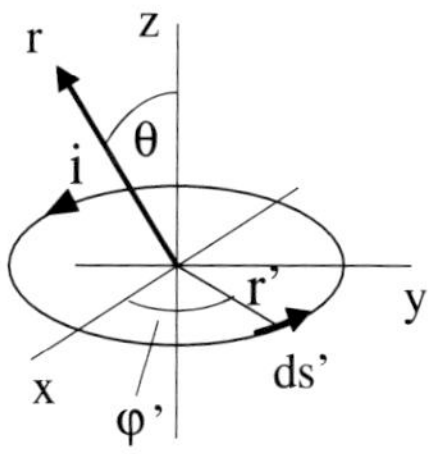

Fig. 5.15 Magnetic field of a current ring: geometry

field point $\boldsymbol{r}$, the point on the ring $\boldsymbol{r}'$ and an element on the ring $\boldsymbol{r}'$ are characterised by

$$\boldsymbol{r} = (r\sin\theta,\ 0,\ r\cos\theta)$$

$$\boldsymbol{r}' = \left(a\cos\varphi',\ a\sin\varphi',\ 0\right)$$

$$\mathrm{d}\,\boldsymbol{s}' = \left(-a\sin\varphi'\mathrm{d}\varphi',\ a\cos\varphi'\mathrm{d}\varphi',\ 0\right)\,.$$

The distance of the field point from the origin r and the azimuthal angle θ are given by

$$r = [x^2 + z^2]^{1/2} \quad \text{resp.} \quad \sin\theta = \frac{x}{r}\,.$$

The vector $\mathbf{d}\boldsymbol{s}'$ has the length $a\mathrm{d}\varphi'$ and is perpendicular to $\boldsymbol{r}'$. The distance between the points and r' is

$$|\boldsymbol{r} - \boldsymbol{r}'| = \left[r^2 + a^2 - 2ra\sin\theta\cos\varphi'\right]^{1/2}\,.$$

The three Cartesian components of the vector product are then

$$A_{\mathrm{x}}(r,\theta) = -i\,k_m a\int_0^{2\pi}\frac{\sin\varphi'\mathrm{d}\varphi'}{\left[r^2 + a^2 - 2ra\sin\theta\cos\varphi'\right]^{1/2}}$$

$$A_{\mathrm{y}}(r,\theta) = i\,k_m a\int_0^{2\pi}\frac{\cos\varphi'\mathrm{d}\varphi'}{\left[r^2 + a^2 - 2ra\sin\theta\cos\varphi'\right]^{1/2}} \qquad (5.26)$$

$$A_{\mathrm{z}}(r,\theta) = 0\,.$$

The integrals, that have to be evaluated are

$$I = \int_0^{2\pi} f(\varphi')\mathrm{d}\varphi' = \int_0^{2\pi} f(2\pi - \varphi')\mathrm{d}\varphi' ,$$

a result, that arose from the substitution

$$\varphi' = 2\pi - \alpha \qquad \mathrm{d}\varphi' = -\mathrm{d}\alpha$$

with

$$I = -\int_{2\pi}^{0} f(2\pi - \alpha)\mathrm{d}\alpha = \int_0^{2\pi} f(2\pi - \alpha)\mathrm{d}\alpha .$$

Because of $\sin(2\pi - \varphi') = -\sin\varphi'$ and $\cos(2\pi - \varphi') = \cos\varphi'$ one obtains only one component

$$A_{\mathrm{x}} = 0 \qquad A_{\mathrm{y}} \neq 0 .$$

There exists a component of the vector potential in the y-direction. It corresponds, in view of the symmetry, to a φ-component in case of an arbitrary choice of the field point

$$A_{\varphi}(r,\theta) \equiv A_{\mathrm{y}}(r,\theta) .$$

The calculation, that follows is basically the same as in the discussion of the electric potential of a uniformly charged ring in Chap. 2.4, but definitely more tedious. The explicit evaluation of the integral in (5.26) leads to a combination of elliptic integrals, rather than one complete elliptic integral. This is due to the additional factor $\cos\varphi'$ in the integrand. The result is (see Detail 5.6.2.1)

$$A_{\varphi}(r,\theta) = i\,k_m a \, \frac{4i\,a\,k_m}{\left[r^2 + a^2 - 2ra\sin\theta\right]^{1/2}}$$

$$\left\{\left(1 - \frac{2}{\kappa^2}\right)K(\kappa) + \frac{2}{\kappa^2}\,E(\kappa)\right\} .$$

The functions $K(\kappa)$ and $E(\kappa)$ are complete elliptic integrals of the first and second kind (see Vol. 1, Math. Chap. 4.3.4) and the parameter κ^2 is defined as

$$\kappa^2 = \frac{-4ar\sin\theta}{\left(r^2 + a^2 - 2ra\sin\theta\right)} .$$

The associated magnetic field for a field point in the x-z plane can be calculated in Cartesian decomposition or in a decomposition in spherical coordinates (the calculation is discussed in Detail 5.6.2.2). The use of the spherical decomposition is less tedious, as only the derivatives in the expression

$$\begin{aligned}\boldsymbol{B}(r,\theta) &= (B_r,\ B_\theta,\ B\varphi)\\ &= \left(\frac{1}{r\sin\theta}\frac{\partial}{\partial\theta}\left(\sin\theta A_\varphi(r,\theta)\right),\ -\frac{1}{r}\frac{\partial}{\partial r}\left(rA_\varphi(r,\theta)\right),\ 0\right)\end{aligned}$$

have to be calculated using the chain rule. If Cartesian coordinates are used, it is necessary to calculate first the vector potential, appealing to the cylinder symmetry, in an arbitrary point of space. This involves the quantities

$$\boldsymbol{A}(r,\theta,\varphi) = (A_x,\ A_y,\ A_z) = (-A(r,\theta)\ \sin\varphi\,,\ A(r,\theta)\ \cos\varphi\,,\ 0)\,,$$

where the distance is now given by

$$r = [x^2 + y^2 + z^2]^{1/2}$$

and the angles by

$$\theta = \arctan\frac{[x^2+y^2]^{1/2}}{z} \qquad \varphi = \arctan\frac{y}{x}\,.$$

The magnetic field is obtained from the derivatives of the magnetic field as

$$\boldsymbol{B}(x,y,z) = (B_x,\ B_y,\ B_z) = \left(-\frac{\partial A_y}{\partial z},\ \frac{\partial A_x}{\partial z},\ \frac{\partial A_y}{\partial x} - \frac{\partial A_x}{\partial y}\right).$$

As the final expressions for the magnetic field is not very transparent, the discussion of a number of approximations might be more useful (see Detail 5.6.2.3). For small values of the parameter κ^2, the usual approximation of the elliptic integrals involved is

$$\begin{aligned}K(\kappa) &= \frac{\pi}{2}\left(1+\frac{\kappa^2}{4}+\frac{9\kappa^4}{64}+\ldots\right)\\ E(\kappa) &= \frac{\pi}{2}\left(1-\frac{\kappa^2}{4}-\frac{3\kappa^4}{64}+\ldots\right).\end{aligned}$$

These expressions are adequate for

- points about the axis of the ring with $\sin\theta \ll 1$,
- points around the coordinate origin, which are characterised by $r \ll a$ and
- points with a sufficient distance from ring with $r \gg a$

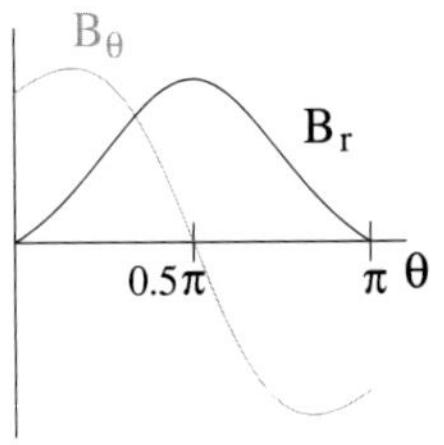

Fig. 5.16 The components of B_θ and B_r in order κ^2 for $r = 8$ and $a = 1$ (arbitrary units for the distances and the field components) as functions of θ

if these expressions are used to order κ^2

$$A(r, \theta) \approx \frac{i\pi a^2 r\, k_m \sin\theta}{[r^2 + a^2 - 2ar \sin\theta]^{3/2}} \,.$$

The associated spherical components of the magnetic field are

$$B_\theta = -\frac{1}{r}\frac{\partial(rA(r,\theta))}{\partial r} = -i\pi a^2 k_m \sin\theta \,\frac{(2a^2 - r^2 - ar\sin\theta)}{[r^2 + a^2 - 2ar\sin\theta]^{5/2}}$$

$$B_r = \frac{1}{r\sin\theta}\frac{\partial \sin\theta\, A(r,\theta)}{\partial\theta} = i\pi a^2 k_m \cos\theta \,\frac{(2a^2 + 2r^2 - ar\sin\theta)}{[r^2 + a^2 - 2ar\sin\theta]^{5/2}} \,.$$

These functions are singular for points on the ring ($\theta = \pi/2$ and $r = a$). However, because of the approximation used, they should not be employed for such points. The variation of these fields with θ, which are shown in Fig. 5.16, is typical for points with $r > a$. For points with $a > r$ the field B_θ has a similar variation but with negative values. With the additional approximation

$$\frac{1}{\left[r^2 + a^2 - 2ra\sin\theta\right]^{n/2}} = \frac{1}{r^n} + \ldots$$

one finds in the asymptotic region $r \gg a$ for the vector potential

$$A_\varphi(r,\theta) \xrightarrow{r \gg a} i k_m \pi\, a^2 \frac{\sin\theta}{r^2}$$

as well as for the magnetic dipole moment $m = ik_f \pi a^2$ of the ring (see p. 257)

$$A_\varphi(r,\theta) \xrightarrow{r \gg a} \frac{m k_m}{k_f}\frac{\sin\theta}{r^2} \,.$$

The variation of the associated components of the field vary with the variable r as $1/r^3$

$$B_r(r,\theta) \xrightarrow{r\gg a} \frac{2m\,k_m}{k_f}\frac{\cos\theta}{r^3}$$

$$B_\theta(r,\theta) \xrightarrow{r\gg a} \frac{m\,k_m}{k_f}\frac{\sin\theta}{r^3}\,.$$

This corresponds to the previous result for points on the z-axis ($\theta = 0$) (5.19).

An alternative approach for the discussion of the problem of the current ring is the multipole expansion of the function of the distance (a more detailed discussion can be found in Detail 5.6.2.4), which provides the results

$$B_\theta = -\frac{1}{r}\frac{\partial(rA(r,\theta))}{\partial r} = i\pi\,a^2 k_m \sum_{n=0}^{\infty}(-1)^n \frac{(2n-1)!!}{2^n(n+1)!}$$

$$\times \left\{\begin{array}{l} (2n+2)\left(\dfrac{r^{2n}}{a^{2n+3}}\right) \\[2ex] -(2n+1)\left(\dfrac{a^{2n+1}}{r^{2n+3}}\right) \end{array}\right\} P^1_{2n+1}(\cos\theta)\left\{\begin{array}{l} r<a \\[2ex] r>a \end{array}\right.$$

and

$$B_r = \frac{1}{r\sin\theta}\frac{\partial \sin\theta A(r,\theta)}{\partial\theta}$$

$$= 2i\pi k_m\left(\frac{a}{r}\right)\sum_{n=0}^{\infty}(-1)^n\frac{(2n+1)!!}{2^n n!}\left(\frac{r_<^{2n+1}}{r_>^{2n+2}}\right)P_{2n+1}(\cos\theta)\,.$$

These results apply to all values of $r \neq a$ (independent of the question whether the convergence of the expansion is slower or faster).

The following qualitative statements can be extracted from these results: The vector potential $\boldsymbol{A}$ is always tangential to circles around the z-axis and has the same value for all tangential points of a circle. This value varies with the distance from the origin of the coordinate system and the opening angle of the cone (Fig. 5.17a). The associated field lines of the magnetic field $\boldsymbol{B}$ in the x-z plane are closed curves about the points of the ring. They are nearly circular in the vicinity of the conductor and go slowly over into a straight line in the vicinity of the z-axis, which is a field line (Fig. 5.17b). The direction of the field corresponds to the right hand rule. A three dimensional impression is obtained by rotating the figure in Fig. 5.17b about the z-axis.

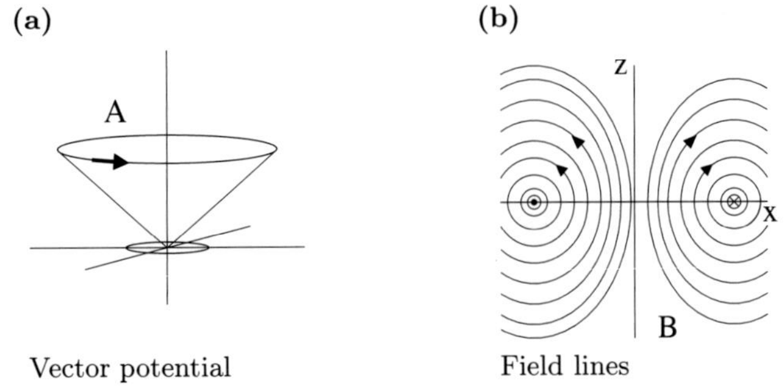

Fig. 5.17 Magnetic field of a circular ring

This example (with relatively simple geometry) will demonstrate that the explicit calculation of magnetic fields can be a rather intricate matter. No additional specific examples will be presented for this reason. but some general aspects will be indicated on the basis of results assembled up to this point.

The following problem is posed for a thin, closed current loop of arbitrary shape (Fig. 5.18): Calculate the vector potential and the magnetic field $\boldsymbol{B}$ for points with a sufficiently large separation from the loop. In order to answer this question one has to expand the function for the separation of two points, which appears in the formula

$$\boldsymbol{A}(\boldsymbol{r}) = i\, k_m \oint \frac{\mathbf{d}\boldsymbol{r}'}{|\boldsymbol{r} - \boldsymbol{r}'|}$$

in an appropriate way. Instead of using the standard multipole expansion (which can be used as well), one can employ a geometric variant

$$|\boldsymbol{r} - \boldsymbol{r}'|^{-1} = \left[r^2 + r'^2 - 2\boldsymbol{r} \cdot \boldsymbol{r}'\right]^{-1/2} .$$

For $r > r'$ one may approximate

$$|\boldsymbol{r} - \boldsymbol{r}'|^{-1} \approx \frac{1}{r}\left[1 - 2\frac{(\boldsymbol{r} \cdot \boldsymbol{r}')}{r^2}\right]^{-1/2}$$

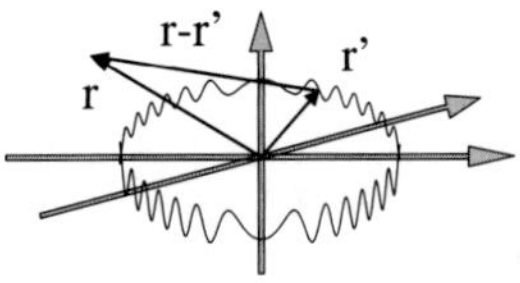

Fig. 5.18 The geometry used for the discussion of the current loop

and expand with the aid of the binomial formula. The result is

$$\approx \frac{1}{r} + \frac{(\boldsymbol{r}\cdot\boldsymbol{r}')}{r^3} + \dots .$$

The result for the vector potential to first order is therefore

$$\boldsymbol{A}(\boldsymbol{r}) = \frac{i\,k_m}{r}\oint \mathbf{d}\boldsymbol{r}' + \frac{i\,k_m}{r^3}\oint (\boldsymbol{r}\cdot\boldsymbol{r}')\,\mathbf{d}\boldsymbol{r}' .$$

The first term vanishes, as the sum of infinitesimal vectors, which form a closed curve, is a null vector. This expresses, once more, the fact, that magnetic charges (monopoles) do not exist.

The second term is the dipole contribution, which can be converted into a more useful form by use of the following formula for a double vector product

$$(\boldsymbol{r}'\times\mathbf{d}\boldsymbol{r}')\times\boldsymbol{r} = (\boldsymbol{r}\cdot\boldsymbol{r}')\,\mathrm{d}\boldsymbol{r}' - (\boldsymbol{r}\cdot\mathbf{d}\boldsymbol{r}')\,\boldsymbol{r}' . \tag{5.27}$$

The first term on the right hand side is the term obtained by the expansion. The term on the left side allows an explicit separation of the variables of integration and the field points. In order to find a useful form for the second term on the right side one uses the total differential

$$\mathrm{d}'\left[(\boldsymbol{r}\cdot\boldsymbol{r}')\,\boldsymbol{r}'\right] = (\boldsymbol{r}\cdot\boldsymbol{r}')\,\mathrm{d}\boldsymbol{r}' + (\boldsymbol{r}\cdot\mathbf{d}\boldsymbol{r}')\,\boldsymbol{r}' . \tag{5.28}$$

Addition of the equations (5.27) and (5.28) leads to

$$(\boldsymbol{r}\cdot\mathbf{d}\boldsymbol{r}')\,\boldsymbol{r}' = \frac{1}{2}\left\{(\boldsymbol{r}'\times\mathbf{d}\boldsymbol{r}')\times\boldsymbol{r} + \mathrm{d}'\left[(\boldsymbol{r}\cdot\boldsymbol{r}')\,\boldsymbol{r}'\right]\right\} .$$

The integration along a closed curve with vectorial elements gives

$$\oint \mathrm{d}'\left[(\boldsymbol{r}\cdot\boldsymbol{r}')\,\boldsymbol{r}'\right] = \left[(\boldsymbol{r}\cdot\boldsymbol{r}')\,\boldsymbol{r}'\right]_B^{E=B} = 0 .$$

The final result for the vector potential in the limit $r > r'$ in lowest non-vanishing order is therefore

$$\boldsymbol{A}(\boldsymbol{r}) = \left[\frac{i\,k_m}{2}\oint (\boldsymbol{r}'\times\mathbf{d}\boldsymbol{r}')\right]\times\frac{\boldsymbol{r}}{r^3} + \dots \quad . \tag{5.29}$$

The leading term of the vector potential has dipole character $(1/r^2)$. The dots indicate the contributions of higher order multipoles (magnetic quadrupole, octupole, etc.), which occur, if terms with $(r'/r)^2$ and higher powers are not neglected. For

Fig. 5.19 Magnetic moment: plane current loop

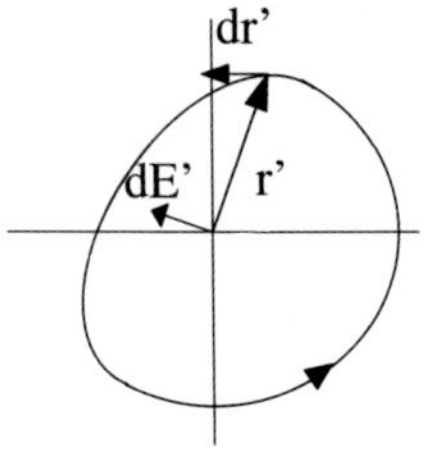

the discussion of the dipole one can define the **magnetic dipole moment** of a thin current loop as (compare p. 257)

$$\boldsymbol{m} = \frac{i k_f}{2} \oint \left(\boldsymbol{r}' \times \mathbf{d}\boldsymbol{r}'\right) . \tag{5.30}$$

The vector potential is then

$$\boldsymbol{A}(\boldsymbol{r}) = \frac{k_m}{k_f} \frac{(\boldsymbol{m} \times \boldsymbol{r})}{r^3} + \dots . \tag{5.31}$$

Every closed current loop looks like a magnetic dipole, if viewed from a sufficient distance. The corresponding $\boldsymbol{B}$-field can be obtained with (5.21)

$$\boldsymbol{B}(\boldsymbol{r}) = \frac{k_m}{k_f} \left(\frac{3(\boldsymbol{m} \cdot \boldsymbol{r})\boldsymbol{r}}{r^5} - \frac{\boldsymbol{m}}{r^3} + \dots \right) . \tag{5.32}$$

The result for a planar but otherwise arbitrary current loop is (Fig. 5.19)

$$\mathbf{d}\boldsymbol{F}' = \frac{1}{2} \left(\boldsymbol{r}' \times \mathbf{d}\boldsymbol{r}'\right) .$$

The integrand in the definition of the magnetic moment is a surface element of the plane loop, traced by the vector $\boldsymbol{r}'$

$$\frac{1}{2} \oint \left(\boldsymbol{r}' \times \mathbf{d}\boldsymbol{r}'\right) = \boldsymbol{F} .$$

The line integral represents the oriented surface of the loop. The orientation is determined by the sense of circulation, that is the direction of the current. The direction of the current and the orientation of the surface are related by the right hand rule. The integral represents the oriented surface of the loop. The line integral produces an effective surface of the current loop (the projection on a median plane), if the surface is not plane. The magnetic moment is

$$\boldsymbol{m} = i\, k_f \boldsymbol{F} ,$$

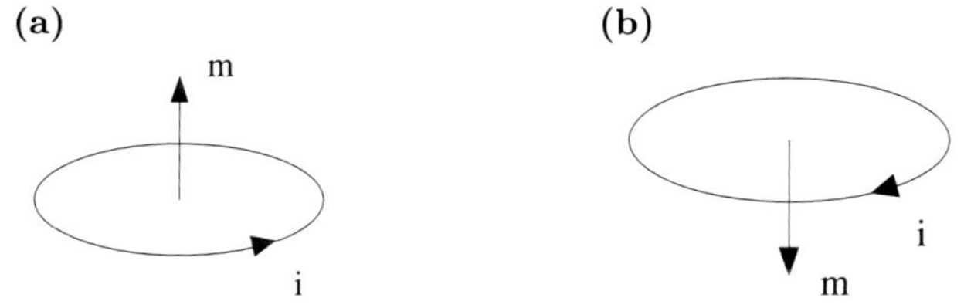

Fig. 5.20 Magnetic moment: orientation

where $\boldsymbol{F}$ is either the true or the effective surface of the current loop. The orientation of the moment is given by the true or an effective direction of the current (Fig. 5.20a, b).

The definition of the magnetic moment can be generalised, if the Biot-Savart form, which has been used thus far, is replaced by the relation between current and current density

$$i\,\mathbf{ds}' \Longrightarrow \iint \boldsymbol{j}(\boldsymbol{r}')\mathrm{d}V' .$$

The magnetic moment of a general distribution of currents is then

$$\boldsymbol{m} = \frac{k_f}{2} \iiint \left(\boldsymbol{r}' \times \boldsymbol{j}(\boldsymbol{r}')\right) \mathrm{d}V' . \tag{5.33}$$

A simple, but important, example is the magnetic moment of a point charge q, which circulates about the origin of a coordinate system on an arbitrary closed path (Fig. 5.21). The current density is in this case (see Detail 5.6.3)

$$\boldsymbol{j}(\boldsymbol{r}') = q\,\boldsymbol{v}(t)\,\delta(\boldsymbol{r}' - \boldsymbol{R}(t)) .$$

The vectors $\boldsymbol{v}(t)$ and $\boldsymbol{R}(t)$ characterise the velocity and the position of the point charge. The magnetic moment of the rotating point charge is

$$\boldsymbol{m} = \frac{q\,k_f}{2}\,(\boldsymbol{R}(t) \times \boldsymbol{v}(t)) .$$

Fig. 5.21 Magnetic moment of a rotating charge

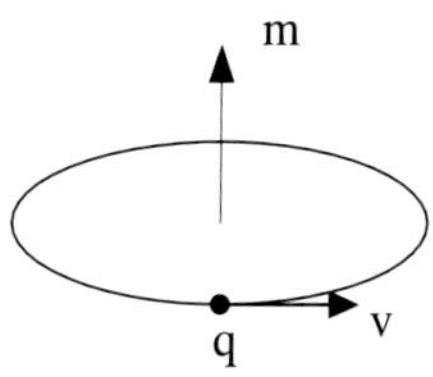

The vector product in this expression is proportional to the angular momentum of the point charge. The magnetic moment is therefore

$$\boldsymbol{m} = \frac{q\,k_f}{2M}\,\boldsymbol{l}(t)\,, \tag{5.34}$$

if the mass of the charge is M. The dipole moment is independent of time, if the angular momentum is a constant of the motion

$$\dot{\boldsymbol{l}} = 0 \Longrightarrow \boldsymbol{m} = \boldsymbol{const.}$$

It is only possible to associate a stationary magnetic moment and a stationary vector potential with the rotating charge in this case. A discussion within the frame of electrodynamics is required, if these conditions are not fulfilled.

The relation between the angular momentum and the magnetic moment is also valid for quantum systems. This could not have been expected, but this correspondence was very useful in helping to understand the properties of such systems. The electrons in an atom move (in a statistical manner) about the nucleus. It is this motion which generates a magnetic moment of each electron,

$$\boldsymbol{m}_i = -\frac{e_0\,k_f}{2m_e}\,\boldsymbol{l}_i(t)\,,$$

which depends on the ratio of charge and mass e_0/m_e. The total angular momentum of all the electrons in the atom is determined by the addition of the individual contributions

$$\boldsymbol{m}_{\text{Atom}} = \sum_{i=1}^{Z} \boldsymbol{m}_i\,.$$

Such relations exist, cum grano salis, on the level of elementary particles. Most particles, which are classified as elementary, possess an inner angular momentum, the spin s. The usual value for Fermions is[6]

$$s = \frac{\hbar}{2}\,.$$

The corresponding magnetic moment is

$$m = \mu_{\text{ref}}\,\frac{g}{2} = \left\{\frac{e_0\hbar}{2Mc}\right\}\frac{g}{2}\,.$$

[6] The quantity $\hbar = h/2\pi$ is related to Planck's constant $h = 6.63 \cdot 10^{-27}$ erg s. Consult a text on quantum mechanics.

Here e_0 is of the elementary charge and M the mass of the elementary particles. The quantity g is the gyromagnetic factor. The reference magnetic moment for electrons is the Bohr magneton

$$\mu_\mathrm{B} = \frac{e_0\hbar}{2m_e c} = 9.274096\cdots\ 10^{-24}\ \frac{\text{Joule}}{\text{Tesla}}$$

$$= 9.274096\cdots\ 10^{-21}\ \frac{\text{erg}}{\text{Gauss}},$$

for protons the nuclear magneton

$$\mu_\mathrm{N} = \frac{e_0\hbar}{2m_p c} = 5.050951\cdots\ 10^{-27}\ \frac{\text{Joule}}{\text{Tesla}}$$

$$= 5.050951\cdots\ 10^{-24}\ \frac{\text{erg}}{\text{Gauss}}.$$

The experimental value of the gyromagnetic factor is

$$g_e = 2.0023\ldots \qquad \text{for electrons}$$

$$g_p = 5.5856\ldots \qquad \text{for protons}.$$

On the basis of a model, that an elementary particle is a rotating spherical charge distribution, one expects $g = 1$ (see Detail 5.6.4). The actual value is therefore an indication of the inner structure of the particular elementary particle. For instance the magnetic moment of the neutron

$$\mu_\mathrm{n} = \frac{e_0}{2M_\mathrm{n}c} g_\mathrm{n} s \qquad \text{with} \quad g_\mathrm{n} = -3.8263\ (!)$$

shows, that there must be a charge distribution inside the neutron, even though it is electrically neutral, if viewed from the outside. The topic of the next chapter will address the question of how the discussion of classical magnetostatics has to be modified, if matter is present in the space surrounding the current carrying conductors.

5.4 Matter in a Magnetic Field

The presence of matter in a magnetic field requires, as in the case of electrostatics, the investigation of the response of the material on a microscopic level. The situation is definitely more complicated as the situation for dielectrics in electric fields. This is the reason for the restriction to rather simple models on the basis of simple experiments. These models can serve as the foundation for a formal theory of

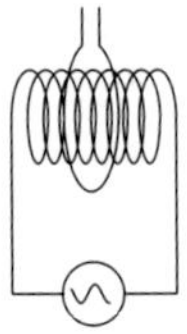

Fig. 5.22 Experiment: Matter in a magnetic field

macroscopic magnetostatics in terms of three magnetic fields (in a certain analogy to the situation for electric fields).

5.4.1 Global Models of Magnetisation

The present topic is also introduced by a simple experiment. A relatively uniform magnetic field can be produced by the current in the interior of a sufficiently long coil (see Chap. 5.2.3, (5.20)). A material is inserted into this coil and the magnetic flux across the cross section of the coil

$$\Phi = \oiint \boldsymbol{B} \cdot \mathbf{d}\boldsymbol{f}$$

is compared for the situation

- current i, field in vacuum: $\boldsymbol{B}_0$
- current i, field with matter: $\boldsymbol{B}_{\text{with}}$.

One would notice in such an experiment, that

$$\oiint \boldsymbol{B}_{\text{with}} \cdot \mathbf{d}\boldsymbol{f} \qquad \neq \qquad \oiint \boldsymbol{B}_0 \cdot \mathbf{d}\boldsymbol{f}\,.$$

The measurement of the flux requires the use of an alternating current and the measurement of the induced voltage in a second coil around the first[7] (see Fig. 5.22).

As the geometry has not changed in this experiment, one can conclude: It is the magnetic field, which is different in the two situations. This can be expressed in the tentatively form

$$\boldsymbol{B}_{\text{with}} = \mu \boldsymbol{B}_0\,.$$

The dimensionless quantity μ is the **magnetic permeability**. The value of μ is used to distinguish three groups of materials.

[7] This remark will be explained in Dreizler and Lüdde, 2024, Electrodynamics and Special Theory of Relativity (Springer Berlin Heidelberg), Chap. 1.1 about induction.

- If $\mu < 1$, the material is **diamagnetic**. The magnetic field is weakened by placing such materials into the field. This effect can be observed e.g. for Bi, Au and H_2O, but the value of μ is small. A typical value is

$$\mu_{\,\text{Bi}} = (10^{-5} - 2 \cdot 10^{-5})\,.$$

- **Paramagnetic** materials as Al, W, Ti are characterised by $\mu > 1$. The effect is also small. The order of magnitude is

$$\mu_{\,\text{Ti}} = (1 + 7 \cdot 10^{-5})\,.$$

- **Ferromagnetic** materials e.g. the metals Fe, Co, Ni, the rare earths Gd, Dy, as well as alloys of these materials with $\mu \gg 1$. In this case one is faced with complicate details. As an example one can look at an alloy on the basis of iron, which has never been exposed to a magnetic field, and measure the ration $\mu = \boldsymbol{B}_{\text{with}}/\boldsymbol{B}_0$ as a function of the current in the coil. As the current i and the field $\boldsymbol{B}_0$ are proportional with respect to each other, one can use the dependence on the field $\boldsymbol{B}_0$ instead of the dependence on the current. The measured function shows roughly the following pattern: An immense amplification up to a value of $\mu_{\text{max}} \approx 5500$, for the particular alloy. The amplification decreases though and one finds some saturation after $\mu_{\text{sat.}} \approx 1000$ (Fig. 5.23). It is quite apparent that a function

$$\boldsymbol{B}_{\text{with}} = \mu \cdot \boldsymbol{B}_0$$

does not agree with this situation. A more general function

$$\boldsymbol{B}_{\text{with}} = \boldsymbol{\mu}\,(\boldsymbol{B}_0)$$

is needed to reproduce the initial magnetisation plot. For dia- and paramagnetic substances one observes

$$i \longrightarrow 0 \qquad \boldsymbol{B}_{\text{with}} \longrightarrow \boldsymbol{0}\,.$$

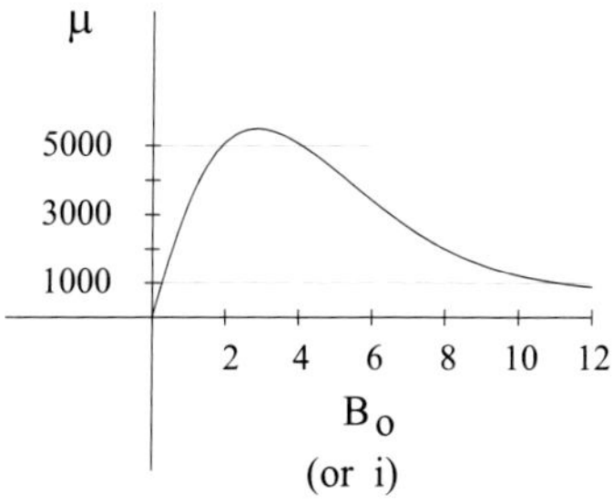

Fig. 5.23 Schematic representation of an initial magnetisation plot of $\mu = \mu(B_0)$ for an iron alloy

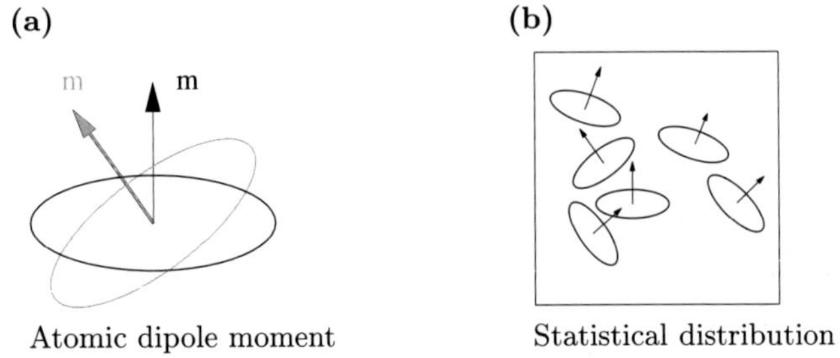

Fig. 5.24 Model of magnetic properties

For ferromagnetic substances the statement is

$$i \longrightarrow 0 \qquad \boldsymbol{B}_{\text{with}} \not\longrightarrow \boldsymbol{0}\,.$$

There remains a permanent magnetic field.

For a correct understanding of matter one has to deal with solid state physics (that is applied quantum mechanics). It is necessary to develop a more naive model of matter, which does not explain the differences between dia-, para- and ferromagnetism, in order to proceed with the discussion of magnetostatics. This naive model still allows a macroscopic treatment of the topic 'matter in a magnetic field' in terms of three magnetic fields.

The naive model for the magnetic properties of matter is the following: Every atom of a material can be characterised by a dipole moment. The dipole moment is produced by circulating electrons and their spin (Fig. 5.24a). There exist also effects of higher multipolarity, but they can be ignored for the present purpose.

The atomic or molecular dipole moments are distributed in a statistical manner, if no external field is present (Fig. 5.24b). The magnetic field $\boldsymbol{B}_0$ in a block of material (the vector sum of all atomic/molecular dipole moments) is normally zero. The atomic dipole moments are oriented, if the block is placed into an external magnetic field. The oriented dipoles produce an additional field,

$$\boldsymbol{B}_{\text{with}} = \boldsymbol{B}_0 + \boldsymbol{B}_{\text{dipole}}\,.$$

The different response of different materials is not taken into account. The differences are, roughly speaking, due to the following effects:

Diamagnetism:	The magnetic dipoles are induced, when the external field is switched on. For this reason they point in the opposite direction.
Paramagnetism:	Individual alignment of exiting dipoles.
Ferromagnetism:	Collective alignment of groups of coupled dipoles.

A magnetic field, which is produced by the sum of all atomic dipoles, would suggest to calculate the corresponding vector potential by (compare (5.31))

$$A_{\text{dipole}}(r) = \frac{k_m}{k_f} \sum_{i=1}^{10^{23}} \frac{m_i(r_i) \times (r - r_i)}{|r - r_i|^3} .$$

Every dipole contributes its share to the dipole moment of a block of material. This ansatz can not be used in this form, as the distribution (orientation and position) of the individual atomic dipoles is not known (and the number of terms too large, about 10^{23} atoms per mol of material). In order to convert it into a usable macroscopic form, one can define a vectorial dipole density $M(r)$

$$M(r) = \frac{1}{\mathrm{d}V} \sum_{i \in \mathrm{d}V} m_i(r)$$

in an infinitesimal volume element $\mathrm{d}V$ and add all atomic dipoles in the volume element. This vector quantity is referred to as the **magnetisation**. It can vary with the position of the volume elements. The volume element is supposed to be infinitesimal. This means, that the definition only makes sense, if (as for the charge density) one understands $M(r)$ as a mean value of atomic dipole densities in the volume $\mathrm{d}V$, which have been averaged over sufficiently large domains. In the vacuum and in the case of a statistical distribution of the atomic dipoles one has

$$M(r) = \mathbf{0} ,$$

in the magnetised state it is

$$M(r) \neq \mathbf{0} .$$

For the transition from the microscopic to the macroscopic ansatz one uses

$$m_i \longrightarrow M \, \mathrm{d}V \qquad \sum_i \longrightarrow \iiint$$

and thus

$$A_{\text{dipole}}(r) \longrightarrow A_{\text{M}}(r) = \frac{k_m}{k_f} \iiint_{\text{V}} \frac{M(r') \times (r - r')}{|r - r'|^3} \mathrm{d}V'$$

$$= \frac{k_m}{k_f} \iiint_{\text{space}} \frac{M(r') \times (r - r')}{|r - r'|^3} \mathrm{d}V' .$$

The integration is over all points $\boldsymbol{r}'$ of the material, or with the assumption that $\boldsymbol{M}=\boldsymbol{0}$ in the exterior region, over the complete space. This expression is transformed with

$$\nabla'\left(\frac{1}{|\boldsymbol{r}-\boldsymbol{r}'|}\right)=\frac{(\boldsymbol{r}-\boldsymbol{r}')}{|\boldsymbol{r}-\boldsymbol{r}'|^3}$$

and

$$\nabla'\times\big(\varphi(\boldsymbol{r}')\boldsymbol{M}(\boldsymbol{r}')\big)=\varphi(\boldsymbol{r}')(\nabla'\times\boldsymbol{M}(\boldsymbol{r}'))+(\nabla'\varphi(\boldsymbol{r}'))\times\boldsymbol{M}(\boldsymbol{r}')\,,$$

so that the final statement is

$$\boldsymbol{A}_{\mathrm{M}}(\boldsymbol{r})=\frac{k_m}{k_f}\iiint\frac{(\nabla'\times\boldsymbol{M}(\boldsymbol{r}'))}{|\boldsymbol{r}-\boldsymbol{r}'|}\,\mathrm{d}V'$$

$$-\frac{k_m}{k_f}\iiint\nabla'\times\left(\frac{\boldsymbol{M}(\boldsymbol{r}')}{|\boldsymbol{r}-\boldsymbol{r}'|}\right)\mathrm{d}V'\,. \tag{5.35}$$

The second integral can be treated with a variant of the integral theorem of Stokes

$$\iiint_{\mathrm{V}}\nabla\boldsymbol{F}(\boldsymbol{r})\,\mathrm{d}V=\iint_{\mathrm{O(V)}}\mathbf{d}\boldsymbol{f}\times\boldsymbol{F}(\boldsymbol{r})$$

rot $\boldsymbol{F}$, which has been quoted in (Vol. 1, Math. Chap. 5.3.3)

$$\nabla\times\boldsymbol{F}(\boldsymbol{r})\,\mathrm{d}V=\sum_i\mathbf{d}\boldsymbol{f}_i\times\boldsymbol{F}(\boldsymbol{r})\,.$$

For the second term in (5.35) one obtains

$$T_2(\boldsymbol{r})=\frac{k_m}{k_f}\iint_{\infty\,\text{sphere}}\frac{\mathbf{d}\boldsymbol{f}'\times\boldsymbol{M}(\boldsymbol{r}')}{|\boldsymbol{r}-\boldsymbol{r}'|}\longrightarrow 0\,,$$

if the block of material has a finite extension. The final result is thus

$$\boldsymbol{A}_{\mathrm{M}}(\boldsymbol{r})=\frac{k_m}{k_f}\iiint\frac{\nabla'\times\boldsymbol{M}(\boldsymbol{r}')}{|\boldsymbol{r}-\boldsymbol{r}'|}\,\mathrm{d}V'\,. \tag{5.36}$$

If one has found (e.g. with quantum statistical models) an expression for the magnetisation of a material, one can use this equation to calculate its macroscopic consequences. The alternative is the use of simpler models for $\boldsymbol{M}$.

If the final expression for $\boldsymbol{A}_{\mathrm{M}}$ is compared with the general representation of a vector potential (5.22)

$$\boldsymbol{A}(\boldsymbol{r}) = k_m \iiint \frac{\boldsymbol{j}(\boldsymbol{r}')}{|\boldsymbol{r} - \boldsymbol{r}'|} \, \mathrm{d}V' ,$$

the definition of a **magnetisation current density** offers itself

$$\boldsymbol{j}_{\mathrm{M}}(\boldsymbol{r}) = \frac{1}{k_f} \left(\nabla \times \boldsymbol{M}(\boldsymbol{r}) \right) . \tag{5.37}$$

This relation can be viewed as a connection between microphysics and macrophysics, as one can obtain $\boldsymbol{j}_{\mathrm{M}}$ from $\boldsymbol{M}$ and in this way calculate the macroscopic vector potential

$$\boldsymbol{A}_{\mathrm{M}}(\boldsymbol{r}) = k_m \iiint \frac{\boldsymbol{j}_{\mathrm{M}}(\boldsymbol{r}')}{|\boldsymbol{r} - \boldsymbol{r}'|} \, \mathrm{d}V' . \tag{5.38}$$

In order to illustrate the final expression one can take a look of a uniformly magnetised material with densely packed, equally oriented ring currents (magnetic moments) (Fig. 5.25). The ring currents can be interpreted as a net of currents. The currents in the interior cancel each other. There remains an effective current (an effective current density) on the macroscopic surface of the material, which is characterised by $\boldsymbol{j}_{\mathrm{M}}$. If the material is not uniform, one finds at the surfaces between uniform regions (this means also in the interior) an effective magnetisation current (Fig. 5.25b). One should, however, note that there exists no direct transport of charges. The electrons move about the nuclei. The global effect of the atomic ring currents (or the atomic magnetic moments) can be interpreted as an effective current density $\boldsymbol{j}_{\mathrm{M}}$. The formal definition of the three magnetic fields in the presence of materials can be effected after this discussion of the global aspects of the model.

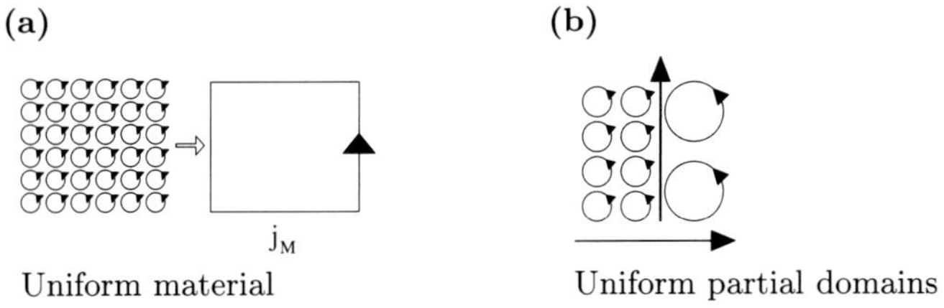

Fig. 5.25 Model of the magnetisation current density

5.4.2 The Three Magnetic Fields

It is useful to start with the extension of the definition of the $\boldsymbol{B}$-field: This field is produced by the real currents (the current, which are due to the actual transport of charges, usually referred to as **true** currents) and the magnetisation currents. The differential equations, which characterise this field, are

$$\nabla \times \boldsymbol{B}(\boldsymbol{r}) = 4\pi k_m \left(\boldsymbol{j}_{\mathrm{w}}(\boldsymbol{r}) + \boldsymbol{j}_{\mathrm{M}}(\boldsymbol{r})\right)$$

$$\nabla \cdot \boldsymbol{B}(\boldsymbol{r}) = 0 . \tag{5.39}$$

The second equation states: The $\boldsymbol{B}$-field has no sources, even in the general case. The two equations define the $\boldsymbol{B}$-field as the field, which is measured in experiments. It is equal to the quantity, which has been denoted provisionally as $\boldsymbol{B}_{\mathrm{with}}$.

If one uses the representation (5.37) of $\boldsymbol{j}_{\mathrm{M}}$ by the magnetisation, one obtains

$$\nabla \times \left(\boldsymbol{B}(\boldsymbol{r}) - 4\pi \frac{k_m}{k_f} \boldsymbol{M}(\boldsymbol{r})\right) = 4\pi k_m \, \boldsymbol{j}_{\mathrm{w}}(\boldsymbol{r}) .$$

This equation suggests the introduction of another field, which is produced solely by the true currents. It is (as the $\boldsymbol{D}$-field) an auxiliary quantity, which is generally denoted by $\boldsymbol{H}$. In order to accommodate the different system of units, one introduces another constant of proportionality k_h at this stage with the values

$$k_{h,\mathrm{SI}} = \frac{1}{4\pi} \qquad\qquad k_{h,\mathrm{CGS}} = \frac{1}{c} .$$

The defining equation of this field is

$$\frac{k_m}{k_h} \boldsymbol{H}(\boldsymbol{r}) = \boldsymbol{B}(\boldsymbol{r}) - 4\pi \frac{k_m}{k_f} \boldsymbol{M}(\boldsymbol{r}) \tag{5.40}$$

with

$$\operatorname{rot} \boldsymbol{H}(\boldsymbol{r}) = 4\pi k_h \, \boldsymbol{j}_{\mathrm{w}}(\boldsymbol{r}) .$$

It is equal to the provisionally notation $\boldsymbol{B}_0$ in Chap. 5.4.1. The official names of the three fields are

$\boldsymbol{B}$ $\longrightarrow$ is the magnetic induction.

$\boldsymbol{H}$ $\longrightarrow$ is called the magnetic field strength.

$\boldsymbol{M}$ $\longrightarrow$ is the magnetisation.

The units in the CGS-system are equal for the three fields and correspond to the unit of the electric field

$$[B]_{CGS} = [H]_{CGS} = [M]_{CGS} = [E]_{CGS} = \frac{\mathrm{g}^{1/2}}{\mathrm{cm}^{1/2}\,\mathrm{s}}\,.$$

The names of the units are different though. They are

$$[B]_{CGS} = \text{Gauss}, \quad [H]_{CGS} = [M]_{CGS} = \text{Ørstedt}\,.$$

The unit of the $\boldsymbol{B}$-field is

$$[B]_{SI} = \frac{\mathrm{kg\,C}}{\mathrm{s}} = \text{Tesla}\,,$$

of the magnetic field strength $\boldsymbol{H}$ and the magnetisation $\boldsymbol{M}$ one uses

$$[H]_{SI} = [M]_{SI} = \frac{\mathrm{C}}{\mathrm{s\,m}} = \mathrm{A/m}\,.$$

Comparison with electrostatics shows common features in general, but differences in detail.

A field, which is accessible in the experiments, exists in both areas

$$\boldsymbol{B}(\boldsymbol{r}) \qquad\qquad \boldsymbol{E}(\boldsymbol{r})\,.$$

The differential equations, which characterise these fields, are so to speak complementary.

$$\operatorname{div}\boldsymbol{B}(\boldsymbol{r}) = \boldsymbol{0} \qquad \operatorname{div}\boldsymbol{E}(\boldsymbol{r}) = 4\pi k_e(\rho_{\mathrm{w}}(\boldsymbol{r}) + \rho_{\mathrm{pol}}(\boldsymbol{r}))$$

$$\operatorname{rot}\boldsymbol{B}(\boldsymbol{r}) = 4\pi k_m(\boldsymbol{j}_{\mathrm{w}}(\boldsymbol{r}) + \boldsymbol{j}_{\mathrm{M}}(\boldsymbol{r})) \qquad \operatorname{rot}\boldsymbol{E}(\boldsymbol{r}) = \boldsymbol{0}\,.$$

In both areas an auxiliary field is introduced

$$\boldsymbol{H}(\boldsymbol{r}) \qquad\qquad \boldsymbol{D}(\boldsymbol{r})\,.$$

These fields are produced by the true charges or currents

$$\operatorname{rot}\boldsymbol{H}(\boldsymbol{r}) = 4\pi k_h\,\boldsymbol{j}_{\mathrm{w}}(\boldsymbol{r}) \qquad \operatorname{div}\boldsymbol{D}(\boldsymbol{r}) = 4\pi k_d\,\rho_{\mathrm{w}}(\boldsymbol{r})\,.$$

The differential equations are the means for the practical calculation of the fields. With the specification of the true charge and current distributions one first calculates the auxiliary fields. When these have been calculated with the respective differential equations and boundary conditions, one can deal with the measurable fields. A correct approach requires knowledge of the response of the material to the presence of external fields (charge separation in atoms in the electric case, orientation of dipoles in the magnetic case). If the treatment of the atomic approach is not possible or available, one can try to use empirical relations (material equations) in order to include response effects.

The simplest material equations are

$$\boldsymbol{B}(\boldsymbol{r}) = \mu \frac{k_m}{k_h} \boldsymbol{H}(\boldsymbol{r}) \qquad \boldsymbol{D}(\boldsymbol{r}) = \varepsilon \frac{k_d}{k_e} \boldsymbol{E}(\boldsymbol{r}) \,.$$

Magnetic fields are treated by expressing the measurable field as a material constant times the auxiliary field. In the electric case the auxiliary field equals a material constant times the measurable field. The proportionalities might have to be replaced by non-uniform, non-isotropic and non-linear relations.

If space is filled with different materials, conditions are needed for the behaviour of the fields on the separating layer of different materials. The material equations and the differential equations allow (provided there are no true charges or currents on the separating layer) the formulation of continuity conditions

$$\boldsymbol{H}_{2,t} - H_{1,t} = 0 \qquad \varepsilon_2 \boldsymbol{D}_{2t} - \varepsilon_1 \boldsymbol{D}_{1t} = 0$$

$$\mu_2 \boldsymbol{H}_{2,n} - \mu_1 \boldsymbol{H}_{1,n} = 0 \qquad \boldsymbol{D}_{2n} - \boldsymbol{D}_{1n} = 0$$

The averaged atomic background is expressed in the quantities

$$\boldsymbol{M}(\boldsymbol{r}) = \frac{k_f}{4\pi}\left(\frac{\boldsymbol{B}(\boldsymbol{r})}{k_m} - \frac{\boldsymbol{H}(\boldsymbol{r})}{k_h}\right) \qquad \boldsymbol{P}(\boldsymbol{r}) = \frac{1}{4\pi}\left(\frac{\boldsymbol{D}(\boldsymbol{r})}{k_d} - \frac{\boldsymbol{E}(\boldsymbol{r})}{k_e}\right) .$$

They are determined by the measurable and the auxiliary fields. The magnetisation current density and the polarisation charge density can be obtained from a macroscopic variant of the atomic structure for these response quantities

$$\boldsymbol{j}_{\mathrm{M}}(\boldsymbol{r}) = \frac{1}{k_f}(\nabla \times \boldsymbol{M}(\boldsymbol{r})) \qquad \rho_{\mathrm{pol}}(\boldsymbol{r}) = -\nabla \cdot \boldsymbol{P}(\boldsymbol{r}) \,.$$

Two points have to be added to the summary of electrostatics and magnetostatics in this section.

5.4.3 The Magnetic Material Equation

The relation $\boldsymbol{B} = \mu k_m \boldsymbol{H}/k_h$ can be used for diamagnetic and paramagnetic materials, as the permeability μ does not differ very much from the vacuum value 1. The explicit relation in the two systems of units is

$$\boldsymbol{B}_{\mathrm{SI}} = \mu\mu_0 \boldsymbol{H} \qquad \boldsymbol{B}_{\mathrm{CGS}} = \mu \boldsymbol{H} \,.$$

In the SI-system one uses often the permeability number μ_r instead of the permeability μ. With respect to the numerical value one finds in any case $\mu = \mu_r$. It is possible to connect the magnetisation in a simple way with the other magnetic fields for such materials

$$\boldsymbol{M}(\boldsymbol{r}) = \frac{k_f}{4\pi}\left(\frac{\boldsymbol{B}(\boldsymbol{r})}{k_m} - \frac{\boldsymbol{H}(\boldsymbol{r})}{k_h}\right) = \frac{k_f}{4\pi k_m}(\mu - 1)\,\boldsymbol{H}(\boldsymbol{r}) \tag{5.41}$$

$$= \frac{k_f}{4\pi k_m}\left(\frac{\mu - 1}{\mu}\right)\boldsymbol{B}(\boldsymbol{r})\,.$$

The quantity $\kappa = (\mu - 1)/\mu$ carries the name **magnetic Susceptibility**.

For ferromagnetic materials the linear relation has to be replaced by a non-linear one

$$\boldsymbol{B}(\boldsymbol{r}) = \mu(\boldsymbol{H}(\boldsymbol{r}))\boldsymbol{H}(\boldsymbol{r}) \quad \text{or} \quad \boldsymbol{B}(\boldsymbol{r}) = \boldsymbol{f}(\boldsymbol{H}(\boldsymbol{r}))\,. \tag{5.42}$$

The vector function $\boldsymbol{f}$ is rather complicated. The experiment, already discussed in Fig. 5.26a, in which an iron core was placed into a coil, involves the following details: The uniform section of the field in the coil can be described in a good approximation by the equation (see (5.20))

$$H = 4\pi k_m\, i\, \frac{N}{L}\,,$$

where N is the number of windings and L the length of the coil. It is therefore possible to convert the measured current in the coil to Ørstedt, the unit for the $\boldsymbol{H}$-field. If one plots the induction $\boldsymbol{B}$ as a function of the magnetic field strength $\boldsymbol{H}$ (in a chosen direction of space, that is $B = f(H)$) one obtains a figure, which is known as a **hysteresis loop** (Fig. 5.26b). For a piece of material, which has never been subjected to a H-field one measure a rise of B with growing H up to a saturation value ($B_{\max} \approx 10^3\ H$ up to $10^5\ H$, depending on the kind of alloy). If the H-field is then reduced, the B-field is also reduced, but does not follow the initial curve. For $H = i = 0$ there remains a permanent rest magnetisation of the material. If now a

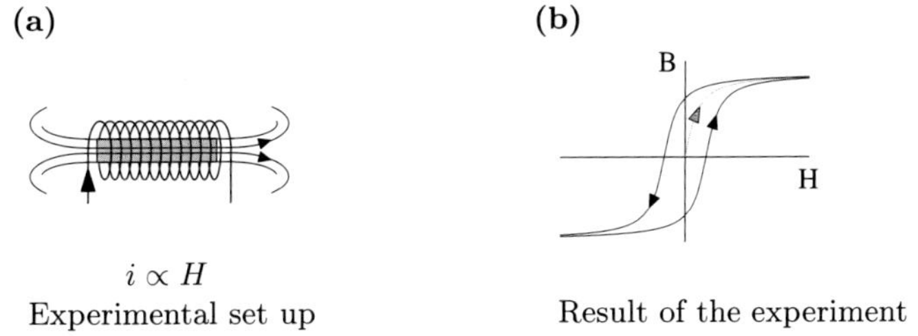

Fig. 5.26 Hysteresis

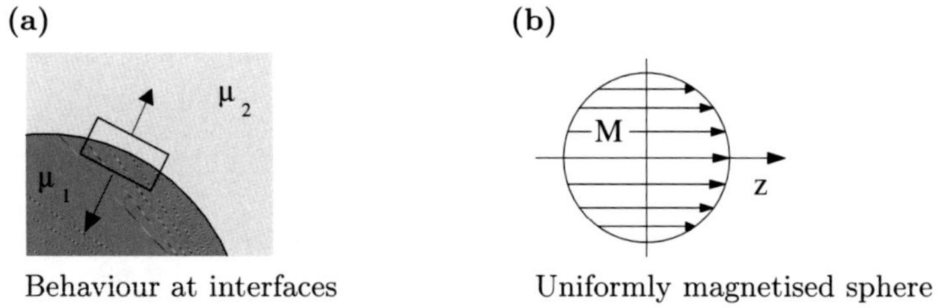

Fig. 5.27 Material in a magnetic field

field in the opposite direction is applied by reversing the direction of the current), B is reduced further until H reaches the value 0. This value of H is called the coercive force (also, the coercivity) H_c. For large, negative values of H saturation of B is reached again (with the same absolute value as for first saturation). If now the process is reverted again, one finds the mirror image of the upper loop. The function $B = f(H)$ is not unique. It also depends on the prehistory of the process.

It is obvious, that, due to the complicated relation between H and B, the calculation of the magnetic fields for ferromagnetic materials is not a simple matter.

5.4.4 The Behaviour of B and H on Interfaces

The question of the behaviour of the two magnetic fields on interfaces can be answered easily, if the linear relation $\boldsymbol{B} = \mu \boldsymbol{H}$ is valid. At a surface separating two materials with the permeabilities μ_i and μ_a (Fig. 5.27a) the equation div $\boldsymbol{B} = 0$ holds. The argument with the Gaussian boxes yields the relation

$$\boldsymbol{B}_{a,n} - \boldsymbol{B}_{i,n} = 0\,. \tag{5.43}$$

The normal components of the $\boldsymbol{B}$-field are continuous. It then follows, that the normal components of the $\boldsymbol{H}$-field are discontinuous

$$\mu_a \boldsymbol{H}_{a,n} - \mu_i \boldsymbol{H}_{i,n} = 0\,. \tag{5.44}$$

For the tangential components of $\boldsymbol{H}$ one finds if there are no true currents in the interface

$$\nabla \times \boldsymbol{H} = \boldsymbol{0} \qquad \text{if} \quad \boldsymbol{j}_{tr} = \boldsymbol{0}\,.$$

The argument with the Stokes lines then leads to

$$\boldsymbol{H}_{a,t} - \boldsymbol{H}_{i,t} = 0 \qquad \text{resp.} \quad \frac{\boldsymbol{B}_{a,t}}{\mu_a} - \frac{\boldsymbol{B}_{i,t}}{\mu_i} = 0\,. \tag{5.45}$$

The tangential components of $\boldsymbol{H}$are continuous, those of B are discontinuous.

5.4.5 Explicit Survey of Matter in Magnetic Fields

It is worthwhile to look at some examples, in order to illustrate such situations in some detail. The first example is the calculation of the $\boldsymbol{B}$-field and the $\boldsymbol{H}$-field in the inner and the outer regions of a sphere (radius R) of uniform magnetisable material (Fig. 5.27b). With the standard choice of the coordinate system the magnetisation $\boldsymbol{M}$ is

$$\boldsymbol{M}(\boldsymbol{r}) = \begin{cases} M_0\,\boldsymbol{e}_z & r \leq R \\ \boldsymbol{0} & r > R\,. \end{cases}$$

It is possible to determine the magnetisation density directly in this example, as the magnetisation itself is specified

$$\boldsymbol{j}_M(\boldsymbol{r}) = \frac{1}{k_f}\,(\nabla \times \boldsymbol{M}(\boldsymbol{r}))$$

and continue to write down the $\boldsymbol{B}$-field

$$\nabla \times \boldsymbol{B} = 4\pi k_m \boldsymbol{j}_M \qquad \text{e.g.} \qquad \boldsymbol{A}_M(\boldsymbol{r}) = k_m \iiint \frac{\boldsymbol{j}_M(\boldsymbol{r}')}{|\boldsymbol{r} - \boldsymbol{r}'|}\mathrm{d}V'\,.$$

True currents do not feature in this problem.

The current density vanishes in the interior and the exterior of the sphere, as $\boldsymbol{M}$ is either a constant or a zero vector. As a consequence of the discontinuity there is, however, a surface contribution. In order to accommodate the discontinuity correctly one has to use the (discontinuous) **Heaviside function**, which is defined by (Fig. 5.28a)

$$\theta(x - a) = \begin{cases} 0 & x \leq a \\ 1 & x > a\,. \end{cases}$$

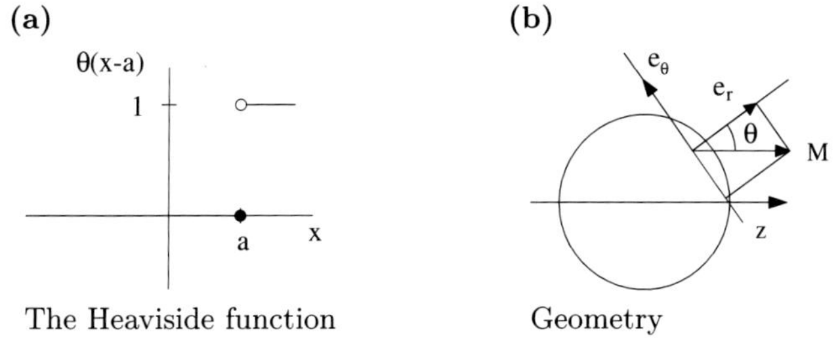

Fig. 5.28 The magnetic field of a uniformly magnetised sphere

The derivative of this function is

$$\frac{\mathrm{d}}{\mathrm{d}x}\,\theta(x-a) = \delta(x-a)\,.$$

It is possible to represent the magnetisation in closed form if one uses the Heaviside function. Because of the symmetry of the object, it is opportune to decompose $\boldsymbol{M}$ in spherical coordinates (Fig. 5.28b)

$$M_r(r,\theta) = M_0\,[1-\theta(r-R)]\cos\theta$$

$$M_\theta(r,\theta) = -M_0\,[1-\theta(r-R)]\sin\theta$$

$$M_\varphi(r,\theta) = 0\,,$$

or

$$\boldsymbol{M}(r,\theta) = M_0\,[1-\theta(r-R)]\,\boldsymbol{e}_z\,.$$

This decomposition allows to write down the components of $\boldsymbol{j}_M$ in spherical coordinates. Using the corresponding decomposition of the operator rot in spherical coordinates, one finds in detail

$$(\nabla\times\boldsymbol{M}(r,\theta))_r = \frac{1}{r\sin\theta}\left\{\frac{\partial}{\partial\theta}\left(\sin\theta M_\varphi(r,\theta)\right) - \frac{\partial M_\theta(r,\theta)}{\partial\varphi}\right\} = 0$$

$$(\nabla\times\boldsymbol{M}(r,\theta))_\theta = \frac{1}{r\sin\theta}\left\{\frac{\partial}{\partial\varphi}M_r(r,\theta) - \sin\theta\frac{\partial}{\partial r}\left(rM_\varphi(r,\theta)\right)\right\} = 0$$

$$\begin{aligned}(\nabla\times\boldsymbol{M}(r,\theta))_\varphi &= \frac{1}{r}\left\{\frac{\partial}{\partial r}\left(rM_\theta(r,\theta)\right) - \frac{\partial}{\partial\theta}\left(M_r(r,\theta)\right)\right\}\\ &= M_0\left\{-\frac{1}{r}\left(1-\theta(r-R)\right)+\delta(r-R)\right.\\ &\qquad\left.+\frac{1}{r}\left(1-\theta(r-R)\right)\right\}\sin\theta\\ &= M_0\delta(r-R)\sin\theta\end{aligned}$$

or in summary

$$\boldsymbol{j}_M(\boldsymbol{r}) = \frac{1}{k_f}M_0\,\delta(r-R)\,\sin\theta\;\boldsymbol{e}_\varphi\,.$$

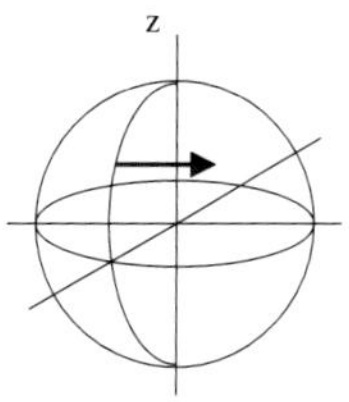

Fig. 5.29 Magnetic field of a magnetised sphere: Magnetisation current $\boldsymbol{j}_{\mathrm{M}}$

The magnetisation current is tangential to the circles of latitude and represents obviously a current layer on the surface (Fig. 5.29).

As the expression for the vector potential is only possible in the Cartesian decomposition, one needs this decomposition of $\boldsymbol{j}_M$ in the next step. With the projections

$$(\boldsymbol{e}_\varphi)_{\mathrm{x}} = -\sin\varphi \qquad (\boldsymbol{e}_\varphi)_{\mathrm{y}} = \cos\varphi \qquad (\boldsymbol{e}_\varphi)_{\mathrm{z}} = 0$$

one finds for the integrals, which have to be evaluated

$$A_{M,x}(r,\,\theta,\,\varphi) = -M_0\frac{k_m}{k_f}\iiint \frac{\delta(r'-R)\sin\theta'\sin\varphi'}{|\boldsymbol{r}-\boldsymbol{r}'|}\,\mathrm{d}V'$$

$$A_{M,y}(r,\,\theta,\,\varphi) = M_0\frac{k_m}{k_f}\iiint \frac{\delta(r'-R)\sin\theta'\cos\varphi'}{|\boldsymbol{r}-\boldsymbol{r}'|}\,\mathrm{d}V'$$

$$A_{M,z}(r,\,\theta,\,\varphi) = 0\,.$$

The result of a rather involved angular integration is given in (Detail 5.6.5).

$$\boldsymbol{A}_M(\boldsymbol{r}) = (-A(r,\theta)\sin\varphi,\ A(r,\theta)\cos\varphi,\ 0)$$

with

$$A(r,\theta) = \begin{cases} \dfrac{4\pi}{3}R^3 M_0\dfrac{k_m}{k_f}\dfrac{\sin\theta}{r^2} & \text{for the exterior} \quad r \geq R \\[2ex] \dfrac{4\pi}{3}M_0\dfrac{k_m}{k_f}\,r\sin\theta & \text{for the interior} \quad r < R\,. \end{cases}$$

In order to determine the associated $\boldsymbol{B}$-field, one needs to calculate $\nabla \times \boldsymbol{A}$. For this purpose one defines the dipole moment of the sphere

$$\boldsymbol{m} = \iiint_{\text{sphere}} \boldsymbol{M}(\boldsymbol{r})\mathrm{d}V = \frac{4\pi}{3}R^3 M_0\boldsymbol{e}_{\mathrm{z}}$$

and writes for the interior ($r < R$)

$$A_{M,i}(\boldsymbol{r}) = \frac{k_m}{k_f\, R^3}\,(\boldsymbol{m} \times \boldsymbol{r})\ .$$

Direct evaluation yields

$$\boldsymbol{B}_i(\boldsymbol{r}) = \frac{2k_m}{k_f\, R^3}\,\boldsymbol{m} = \frac{8\pi k_m}{3k_f}\,M_0\,\boldsymbol{e}_z\,.$$

The calculation for the exterior region can be abbreviated. One notes the for $r \geq R$ the relation

$$A_{M,a}(\boldsymbol{r}) = \frac{k_m}{k_f}\left(\frac{\boldsymbol{m} \times \boldsymbol{r}}{r^3}\right)$$

is correct, so that the $\boldsymbol{B}$-field follows with the formula (5.32) for a magnetic dipole field, which has already been used

$$\boldsymbol{B}_a(\boldsymbol{r}) = \frac{k_m}{k_f}\left(\frac{3(\boldsymbol{m}\cdot\boldsymbol{r})\boldsymbol{r}}{r^5} - \frac{\boldsymbol{m}}{r^3}\right)\,.$$

The field lines are sketched in Fig. 5.30. The field $\boldsymbol{B}$ is constant in the interior and points in the direction of the magnetisation. It is continued in the exterior by a dipole field. The normal component is continuous on the surface of the sphere. The tangential components are discontinuous on the surface, a feature, which is responsible for the kink. The conditions for separation layers are automatically taken care of by use of the formula for $\boldsymbol{A}_M$.

The $\boldsymbol{H}$-field can be obtained most directly from the relation

$$\boldsymbol{H} = k_h\left(\frac{\boldsymbol{B}}{k_m} - \frac{4\pi}{k_f}\,\boldsymbol{M}\right)$$

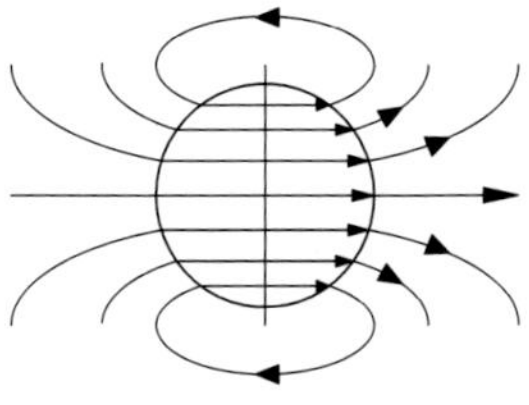

Fig. 5.30 Magnetic field of a magnetised sphere: Magnetic induction $\boldsymbol{B}$

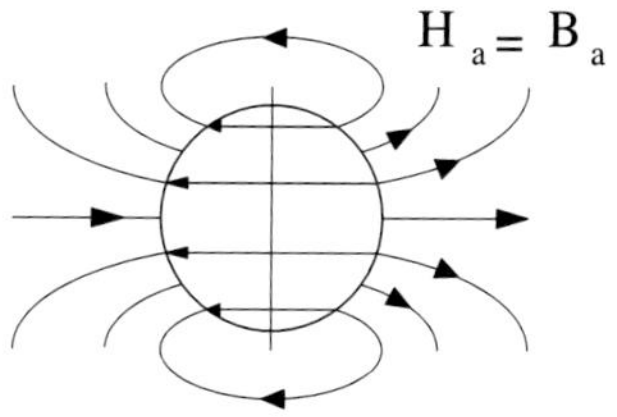

Fig. 5.31 Magnetic field of a magnetised sphere: Magnetic field ***H***

as

$$\boldsymbol{H}_i = -\frac{4\pi\, k_h}{3k_f} M_0 \boldsymbol{e}_z \qquad r < R$$

$$\boldsymbol{H}_a = \frac{k_h}{k_f}\left(\frac{3(\boldsymbol{m}\cdot\boldsymbol{r})\boldsymbol{r}}{r^5} - \frac{\boldsymbol{m}}{r^3}\right) \qquad r \geq R\,.$$

The second line follows as $\boldsymbol{M}_a = \boldsymbol{0}$ for $r \geq R$. The field lines of the $\boldsymbol{H}$-field are displayed in Fig. 5.31. The $\boldsymbol{H}$-field is weaker by a factor of two in the interior compared to the $\boldsymbol{B}$-field and points in the opposite direction. In the exterior one recognises the same dipole field as for $\boldsymbol{B}$.

The result for $\boldsymbol{H}$ could have been obtained without an additional calculation. As there are no true currents, one has $\operatorname{rot}\boldsymbol{H} = \boldsymbol{0}$. On the other side one finds on the surface

$$\nabla\cdot\boldsymbol{H} = \nabla\cdot\left(\frac{k_h\,\boldsymbol{B}}{k_m} - \frac{4\pi}{k_f}\boldsymbol{M}\right) = -\frac{4\pi}{k_f}\nabla\cdot\boldsymbol{M} \neq \boldsymbol{0}\,.$$

The $\boldsymbol{H}$-field has sources on the surface. For this reason one can state: The tangential component $\boldsymbol{H}_t$ is continuous, the normal component H_n is discontinuous.

The results for the uniformly magnetised sphere can be used for the second example, which is: A sphere of magnetisable material is placed into a uniform magnetic field $\boldsymbol{B}_0 = B_0\,\boldsymbol{e}_z$. The task is the calculation of the three magnetic fields (Fig. 5.32).

In order to treat this problem one uses the superposition principle and the fact, that the homogeneous material can be uniformly magnetised. The corresponding ansatz for the solution is

$$\boldsymbol{B} = \boldsymbol{B}_0 + \boldsymbol{B}_M$$

$$\boldsymbol{H} = \frac{k_h\,\boldsymbol{B}_0}{k_m} + \boldsymbol{H}_M\,.$$

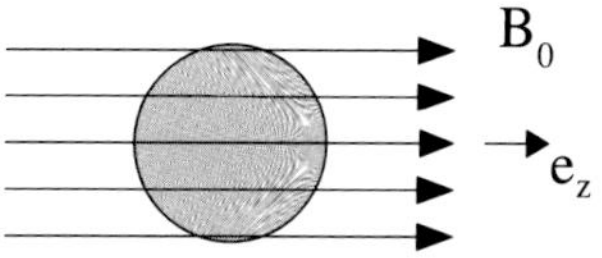

Fig. 5.32 Illustration: Magnetisable sphere in an uniform magnetic field

The fields $\boldsymbol{B}_M$ and $\boldsymbol{H}_M$ are given by the formulae of the first problem, but the magnitude of the magnetisation is not known and has to be calculated. The fact, that both these equations contain $\boldsymbol{B}_0$, is justified with the consideration: If the sphere is removed ($\boldsymbol{M} = \boldsymbol{0}$), then the relation

$$\boldsymbol{H} = k_h \frac{\boldsymbol{B}}{k_m} = k_h \frac{\boldsymbol{B}_0}{k_m}$$

must follow. This statement implies the ansatz for the fields in the interior

$$\boldsymbol{B}_i = \boldsymbol{B}_0 + \frac{8\pi\, k_m}{3k_f} \boldsymbol{M}$$

$$\boldsymbol{H}_i = \frac{k_h\, \boldsymbol{B}_0}{k_m} - \frac{4\pi\, k_h}{3k_f} \boldsymbol{M}\,.$$

Elimination of the unknown magnetisation yields a relation between the fields $\boldsymbol{B}_i$ and $\boldsymbol{H}_i$

$$\boldsymbol{B}_i + \frac{2k_m}{k_h} \boldsymbol{H}_i = 3\, \boldsymbol{B}_0\,.$$

A second vector equation is needed for the determination of the two fields. In the case of dia- or paramagnetic materials one is able to use the material equation

$$\boldsymbol{B}_i = \mu \frac{k_m}{k_h} \boldsymbol{H}_i\,.$$

The solution of the resulting system of vector equations is

$$\boldsymbol{B}_i = \frac{3\mu}{(\mu + 2)} \boldsymbol{B}_0 \qquad \boldsymbol{H}_i = \frac{3k_h}{k_f(\mu + 2)} \boldsymbol{B}_0\,.$$

The magnetisation $\boldsymbol{M}$ can the be obtained as

$$\boldsymbol{M} = \frac{k_f}{4\pi} \left(\frac{\boldsymbol{B}}{k_m} - \frac{\boldsymbol{H}}{k_h} \right) = \frac{3k_f}{4\pi\, k_m} \left(\frac{\mu - 1}{\mu + 2} \right) \boldsymbol{B}_0\,.$$

If $\boldsymbol{M}$ is known, one can also find the exterior field (as in the previous example)

$$\boldsymbol{H}_a = \frac{k_h}{k_m} \boldsymbol{B}_a = \frac{k_h}{k_m} \boldsymbol{B}_0 + \frac{k_h}{k_f} \left(\frac{3(\boldsymbol{m} \cdot \boldsymbol{r})\boldsymbol{r}}{r^5} - \frac{1}{r^3} \boldsymbol{m} \right)$$

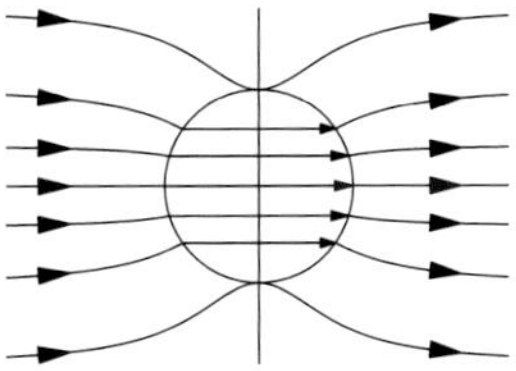

Fig. 5.33 Magnetisable sphere in an external magnetic field: Magnetic induction $\boldsymbol{B}$

with

$$\boldsymbol{m} = \frac{4\pi}{3} R^3 \boldsymbol{M} = \left(\frac{\mu - 1}{\mu + 2}\right) \frac{k_f R^3}{k_m} \boldsymbol{B}_0 .$$

The field lines for $\boldsymbol{B}$ are shown in Fig. 5.33. The $\boldsymbol{B}$-field is constant in the interior and proportional to $\boldsymbol{B}_0$. In the exterior one finds a superposition of the constant and the dipole fields.

The results could be used to check explicitly that the relations

$$\boldsymbol{B}_{a,n} - \boldsymbol{B}_{i,n} = 0 \qquad \boldsymbol{H}_{a,t} - \boldsymbol{H}_{i,t} = 0$$

are satisfied on the surface of the sphere.

The situation for the case of ferromagnetic materials will only be indicated rather than followed to the end. If one ignores the vector character of the fields, the solution of the system of equations for dia- or paramagnetic materials

$$B_i = -\frac{2k_m}{k_h} H_i + 3B_0$$

$$B_i = \mu \frac{k_m}{k_h} H_i$$

is found as the intersection of two straight lines in the H_i-B_i plane (Fig. 5.34a). For ferromagnetic materials the material equation is represented by the hysteresis

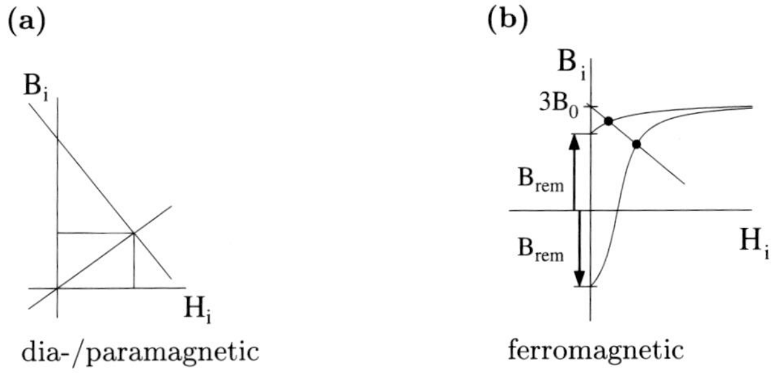

Fig. 5.34 Magnetised spheres in a uniform field: Relation between $\boldsymbol{B}$ and $\boldsymbol{H}$

curve. As this curve is the image of a transcendent function (for a simple case by the hyperbolic tangent) one has to proceed with a numerical evaluation. One finds in this case for a magnetic material with a prehistory two points, for which the hysteresis curve and the straight line intersect (Fig. 5.34b). The two points indicate the possibility, that the pre-magnetised spheres are brought into field with

$$\boldsymbol{B}_{\text{rem}} \text{ parallel to } \boldsymbol{B}_0 \qquad \text{(upper point)}$$

$$\boldsymbol{B}_{\text{rem}} \text{ antiparallel to } \boldsymbol{B}_0 \text{ (lower point)}$$

Naturally it is possible to bring the sphere into the external field with an arbitrary orientation. To discuss this case, one has to face up to the vector character of the system of equations and determine the coordinates of the points of intersection with the equations of the components.

If the points of intersection (for the case that $\boldsymbol{B}_{\text{rem}}$ parallel/antiparallel to $\boldsymbol{B}_0$ is) have been determined graphically or numerically, the discussion proceeds along the same pattern as for dia- or paramagnetic materials up to the determination of the exterior field.

One should realise, that different situations can be found for the determination of the points of intersection, as e.g.

$\boldsymbol{B}_{\text{rem}} > 3\,\boldsymbol{B}_0$. The internal field $\boldsymbol{B}_i$ is stronger for the parallel case. The parallel interior field points in the opposite direction (Fig. 5.35a).

$\boldsymbol{B}_{\text{rem}} < 3\,\boldsymbol{B}_0$. The internal field $\boldsymbol{B}_i$ is weaker. Both fields point in the same direction (Fig. 5.35b).

The orientation of the $\boldsymbol{B}$-field can change its orientation in the case of an initial antiparallel orientation (Fig. 5.35c).

The last section of the chapter on magnetostatics addresses the forces on charges in electric and magnetic fields.

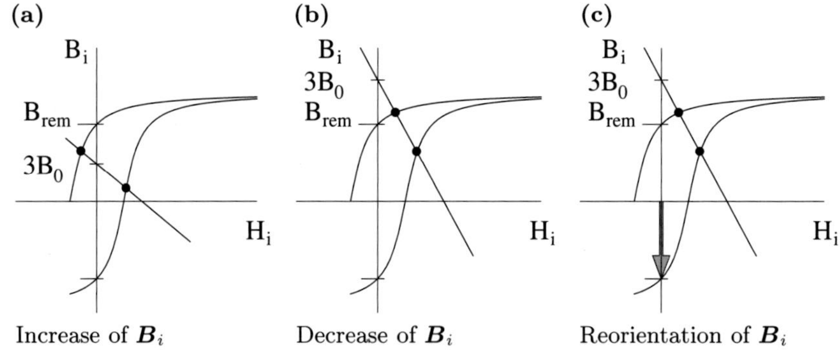

Fig. 5.35 Different effects for ferromagnetic spheres in a uniform field

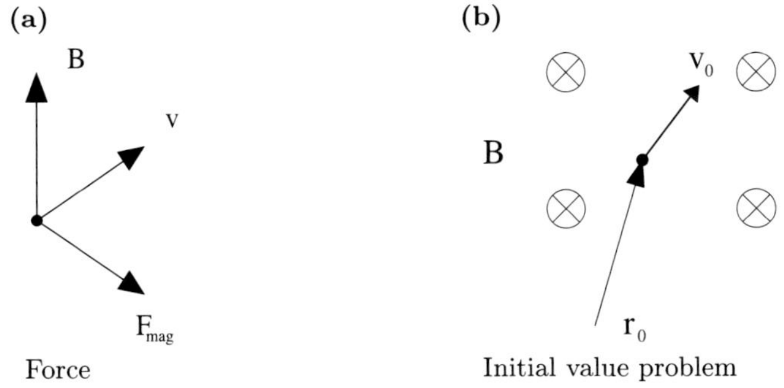

Fig. 5.36 The Lorentz force

5.5 Forces on Charges in Magnetic Fields

The key formula is the **law of Lorentz** for magnetic forces. It has been obtained by experiments investigating the deflection of electrons and ions in magnetic fields. The form of this law is[8]

$$\boldsymbol{F}_{\text{mag}}(\boldsymbol{r}) = qk_f\Big[\boldsymbol{v}(\boldsymbol{r}) \times \boldsymbol{B}(\boldsymbol{r})\Big]. \tag{5.46}$$

A force acts on a charge q in a $\boldsymbol{B}$-field only if the charge is moving. The vector representing the force is perpendicular to the direction of the field and the vector of the velocity of the charge (Fig. 5.36a). The force is proportional to the charge and a factor k_f, which is defined on p. 257. This factor and the units of the magnetic induction and the charge reproduce the correct unit of the force in each of the systems of units (compare Chaps. 5.2.1 and 1.2). The factor v/c, which appears if the CGS-system is used, emphasises the vicinity to the theory of relativity.

The more involved form of the force law indicates that the treatment of problems of motion in magnetic fields is not as easy as the motion in electric fields. The simplest problem is the motion of a point charge (mass m, charge q) in a uniform magnetic field $\boldsymbol{B} = (0, 0, B)$ with the initial conditions $\boldsymbol{r}_0, \boldsymbol{v}_0$ (Fig. 5.36b). The equations of motion for the field pointing in the z-direction are in detail

$$m\ddot{x} = q\,k_f B\dot{y} \qquad m\ddot{y} = -q\,k_f B\dot{x} \qquad m\ddot{z} = 0\,. \tag{5.47}$$

First integration of the third equation gives

$$v_z(t) = v_{0,z}\,.$$

[8] An additional discussion of this force law can be found in Dreizler and Lüdde, 2024, Electrodynamics and Special Theory of Relativity (Springer Berlin Heidelberg), Chap. 3.5.3, which deals with the relation of electrodynamics and the theory of relativity.

It is convenient to introduce the abbreviation

$$\omega = \frac{q\,k_f}{m}\,B$$

for first two equations, so that they take the form

$$\dot{v}_x = \omega\, v_y \qquad \dot{v}_y = -\omega\, v_x\,.$$

Differentiation of the first equation and elimination of v_y with the second leads to a differential equation of the harmonic oscillator

$$\ddot{v}_x + \omega^2 v_x = 0$$

with the solution

$$v_x(t) = v_0 \cos(\omega t + \delta)\,.$$

Insertion into the second equation yields

$$v_y(t) = -v_0 \sin(\omega t + \delta)\,.$$

The integration constants v_0 and δ have to be chosen, so that the velocity components in the two coordinate direction are

$$v_x(0) = v_0 \cos\delta = v_{0,x} \qquad v_y(0) = -v_0 \sin\delta = v_{0,y}\,.$$

The second integration leads to

$$x(t) = x_0' + \frac{v_0}{\omega}\sin(\omega t + \delta) = x_0 + \frac{v_{0,y}}{\omega} + \frac{v_0}{\omega}\sin(\omega t + \delta)$$

$$y(t) = y_0' + \frac{v_0}{\omega}\cos(\omega t + \delta) = y_0 - \frac{v_{0,x}}{\omega} + \frac{v_0}{\omega}\cos(\omega t + \delta)$$

$$z(t) = z_0 + v_{0,z} t\,.$$

The trajectory is a helix in the direction of the field (the z-direction). The projection into the x- y plane is a circle with the radius

$$R = \frac{v_0}{\omega} = \frac{m}{q}\left(\frac{v_0}{k_f B}\right).$$

The radius increases with the initial velocity v_0 and decreases with increasing induction as $1/B$. The time for one revolution is

$$T = \frac{2\pi}{\omega} = \frac{m}{q}\left(\frac{2\pi}{k_f B}\right).$$

The z-component changes during this time by the pitch

$$h = v_{0,z}T = \frac{m}{q}\left(\frac{2\pi\, v_{0,z}}{k_f B}\right).$$

The time for one revolution is independent of the velocity of the charge. It is determined solely by the induction and the ratio q/m. The measurement of the time T in uniform magnetic fields is used for this reason for the determination of the ratio q/m (e.g. e_0/m_e).

One should also point out the special cases $v_{0,z} = 0$, for which one finds a simple motion on a circle, and $v_{0,x} = v_{0,y} = 0$ with a uniform motion along the direction of the field due to $\boldsymbol{F}_{\mathrm{magn}}(t) = \boldsymbol{0}$.

The equation of motion of a charged particle (mass point) in a combined $\boldsymbol{E}$- and $\boldsymbol{B}$-field is obtained by addition of the electric counterpart

$$m\dot{\boldsymbol{v}} = q\boldsymbol{E} + q\,k_f\,[\boldsymbol{v}\times\boldsymbol{B}]\,. \tag{5.48}$$

In order to discuss the situation with respect to the energy, one forms the scalar product of the vector equation of motion (5.48) and the velocity $\boldsymbol{v}$. The result is

$$m\dot{\boldsymbol{v}}\cdot\boldsymbol{v} = \frac{m}{2}\frac{\mathrm{d}}{\mathrm{d}t}(\boldsymbol{v}^2) = q\boldsymbol{E}\cdot\boldsymbol{v} + 0\,,$$

as the triple product with two equal vectors vanishes. The resulting equation shows, that a change of the kinetic energy is solely due to the electric field. In order to obtain the energy theorem one has to apply the chain rule for the potential

$$\frac{\mathrm{d}}{\mathrm{d}t}V(\boldsymbol{r}(t)) = \boldsymbol{\nabla}V\cdot\frac{\mathrm{d}\boldsymbol{r}}{\mathrm{d}t} = -\boldsymbol{E}\cdot\boldsymbol{v}$$

and reformulate $\boldsymbol{E}\cdot\boldsymbol{v}$ in order to obtain

$$\frac{m}{2}v^2 + qV = \text{const.}$$

The $\boldsymbol{B}$-field does not contribute to the energy as the magnetic forces are perpendicular to the direction of motion.

The action of the magnetic force on current carrying conductors can also be discussed in terms the law of Lorentz. One has to use the fact, that an electric current in a given direction corresponds to the motion of the electrons in the

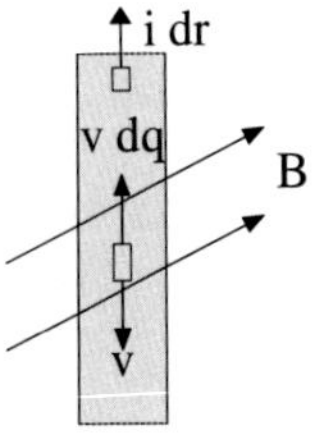

Fig. 5.37 Application of the Biot-Savart approximation for the calculation of the action of the force of a magnetic field on a current carrying conductor

opposite direction (Fig. 5.37). The moving electrons experience a Lorentz force in the magnetic field. The action of the force is transmitted by collisions of the electrons with the ions of the lattice to the material of the conductor. For a volume element $\mathrm{d}V$ of the conductor (Fig. 5.37), one can state

$$\mathrm{d}q\,\boldsymbol{v} = \left(\frac{\mathrm{d}q}{\mathrm{d}t}\right)\mathbf{dr} = i\,\mathbf{dr} = \boldsymbol{j}\,\mathrm{d}V\,.$$

The action of the force on the volume element is

$$\mathrm{d}\boldsymbol{F}_{\mathrm{magn}} = \mathrm{d}q\,k_f\,[\boldsymbol{v}\times\boldsymbol{B}] = k_f\,[\boldsymbol{j}(\boldsymbol{r})\times\boldsymbol{B}(\boldsymbol{r})]\,\mathrm{d}V$$

and the action of the force on the complete conductor is therefore

$$\boldsymbol{F}_{\mathrm{cond}} = k_f \iiint_{\mathrm{V_{cond}}} \Big[\boldsymbol{j}(\boldsymbol{r})\times\boldsymbol{B}(\boldsymbol{r})\Big]\mathrm{d}V\,. \tag{5.49}$$

If the conductor is sufficiently thin, one can use the Biot-Savart approximation $\boldsymbol{j}\,\mathrm{d}V \longrightarrow i\,\mathbf{dr}$ in order to obtain

$$\boldsymbol{F}_{\mathrm{cond}} = i\,k_f \int_{\mathrm{cond}} \Big[\mathbf{dr}\times\boldsymbol{B}(\boldsymbol{r})\Big]\,. \tag{5.50}$$

For the action of a force on a closed current loop in a uniform magnetic field (Fig. 5.38a) one find independent of the geometry

$$\boldsymbol{F} = i\,k_f \oint \Big[\mathbf{dr}\times\boldsymbol{B}\Big] = 0\,. \tag{5.51}$$

In order to prove this statement one can begin by looking at an infinitesimal loop with right angles (Fig. 5.38b) and note, that the contributions of opposite sides cancel in a uniform field. A loop with an arbitrary shape can be viewed as a current net of such rectangular loops (Fig. 5.38c).

For the torque on a closed loop in a uniform magnetic field one finds therefore

$$\boldsymbol{D} = \int \mathbf{d}\boldsymbol{D} = \oint [\boldsymbol{r}\times\mathbf{d}\boldsymbol{F}] = i\,k_f \oint [\boldsymbol{r}\times(\mathbf{dr}\times\boldsymbol{B})]\,.$$

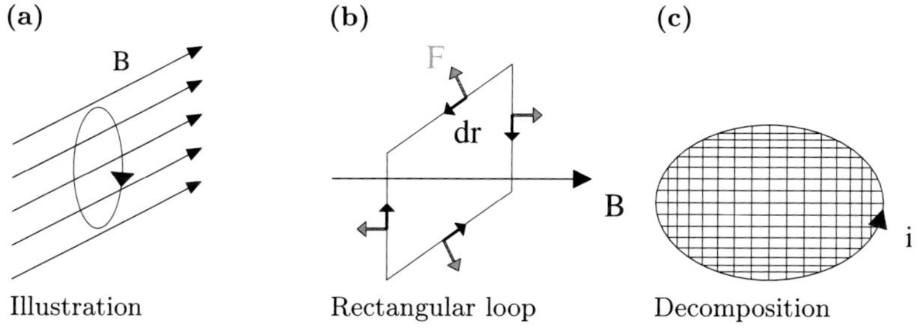

Fig. 5.38 Current loop in a uniform magnetic field

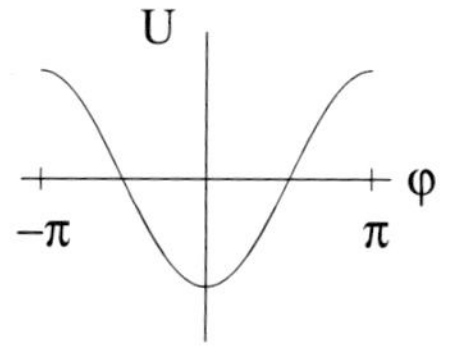

Fig. 5.39 Potential energy of a magnetic dipole in a uniform magnetic field

The triple product of vectors can be rewritten as

$$\boldsymbol{r} \times (\mathbf{d}\boldsymbol{r} \times \boldsymbol{B}) = \frac{1}{2}\mathrm{d}\,(\boldsymbol{r} \times (\boldsymbol{r} \times \boldsymbol{B})) + \frac{1}{2}\,((\boldsymbol{r} \times \mathbf{d}\boldsymbol{r}) \times \boldsymbol{B})\;.$$

The contribution of the total differential vanishes for a closed loop, so that the following expression remains

$$\boldsymbol{D} = \frac{i\,k_f}{2}\oint [\boldsymbol{r} \times \mathbf{d}\boldsymbol{r}] \times \boldsymbol{B} = \boldsymbol{m} \times \boldsymbol{B}\,. \qquad (5.52)$$

A torque acts on a closed current loop in a uniform magnetic field. This torque leads to an alignment of the magnetic dipole moment of the loop in the direction of the field. This statement can be justified by a look at the potential energy

$$\begin{aligned} U &= \int^{\varphi} \boldsymbol{D}(\varphi')\mathrm{d}\,\boldsymbol{\varphi}' = mB\int^{\varphi}\sin\varphi'\,\mathrm{d}\,\varphi' \\ &= -mB\cos\varphi = -\boldsymbol{m}\cdot\boldsymbol{B}\,. \end{aligned} \qquad (5.53)$$

The potential energy is smallest, if the two vectors $\boldsymbol{B}$ and $\boldsymbol{m}$ are parallel. An orientation of $\boldsymbol{m}$ antiparallel to $\boldsymbol{B}$ is not a stable constellation (Fig. 5.39).

The next step is an analysis of the magnetic force between two moving point charges. A snap shot of the situation to be discussed, can be described in the following manner: The two charges q_1 and q_2 are at the position $\boldsymbol{r}_1$ and $\boldsymbol{r}_2$ (Fig. 5.40a). The momentary velocities are $\boldsymbol{v}_1$ and $\boldsymbol{v}_2$. Each of the moving charges

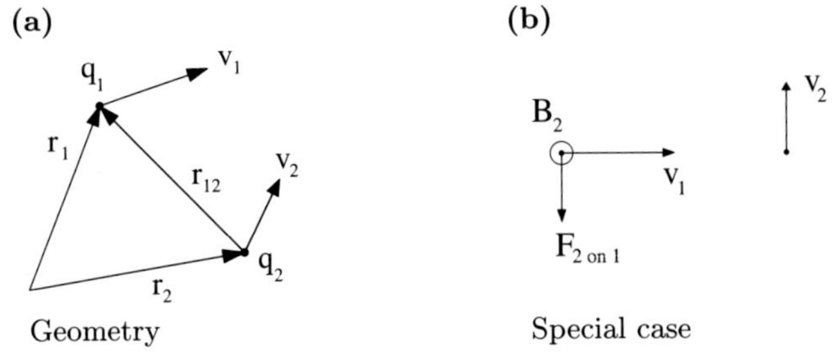

Fig. 5.40 Magnetic forces between two moving charges

produces a magnetic field at the position of the other charge. This field can be given with the formula of Biot-Savart (5.17) using

$$i\mathbf{d}\boldsymbol{r} \qquad \text{instead of} \qquad q\boldsymbol{v}\,.$$

It then follows

$$\boldsymbol{B}_1(\boldsymbol{r}_2) = q_1\,k_m \frac{\boldsymbol{v}_1 \times \boldsymbol{r}_{12}}{r_{12}^3} \qquad \text{(field of } q_1 \text{ on position } \boldsymbol{r}_2\text{)}$$

$$\boldsymbol{B}_2(\boldsymbol{r}_1) = q_2\,k_m \frac{\boldsymbol{v}_2 \times \boldsymbol{r}_{12}}{r_{12}^3} \qquad \text{(field of } q_2 \text{ on position } \boldsymbol{r}_1\text{)}\,.$$

The formula of Lorentz (5.46) can now be used to write down the mutual interaction

$$\begin{aligned} \boldsymbol{F}^{\text{magn}}_{1\,\text{on}\,2} &= -q_1 q_2\,k_f\,k_m\,[\boldsymbol{v}_2 \times (\boldsymbol{v}_1 \times \boldsymbol{r}_{12})]\,\frac{1}{r_{12}^3} \\ \boldsymbol{F}^{\text{magn}}_{2\,\text{on}\,1} &= q_1 q_2\,k_f\,k_m\,[\boldsymbol{v}_1 \times (\boldsymbol{v}_2 \times \boldsymbol{r}_{12})]\,\frac{1}{r_{12}^3}\,. \end{aligned} \tag{5.54}$$

The vector structure of these equations is not very transparent, so that one does not recognise immediately, that one finds in a general situation

$$\boldsymbol{F}^{\text{magn}}_{2\,\text{on}\,1} \neq \boldsymbol{F}^{\text{magn}}_{1\,\text{on}\,2}\,.$$

The magnetic forces between two moving charges do not satisfy the third axiom of Newton.

It is sufficient to look at a possible situation in order to illustrate this statement. The charge q_1 is moving at a given moment on the x-y plane in the x-direction, the charge q_2 is on the x-axis moving in the y-direction (Fig. 5.40b). The forces involved are

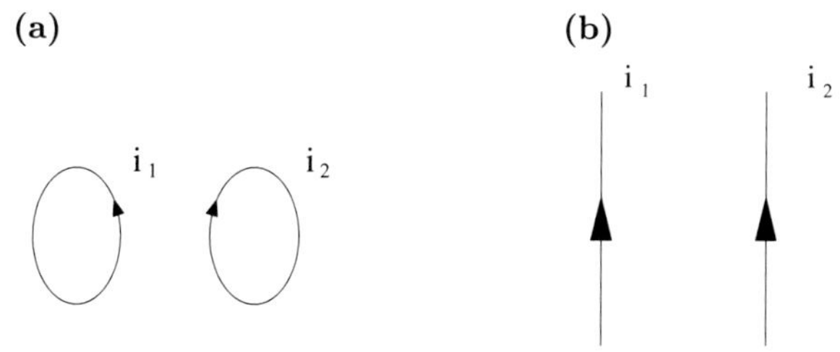

Fig. 5.41 Concerning the validity of the third axiom of Newton for magnetic interactions

(i) The field $\boldsymbol{B}_1$ vanishes $\boldsymbol{B}_1 = \mathbf{0}$, as $\boldsymbol{v}_1$ and $\boldsymbol{r}_{12}$ are anti-parallel. This implies, that the force $\boldsymbol{F}^{\mathrm{magn}}_{1\,\mathrm{on}\,2} = \mathbf{0}$ vanishes as well.
(ii) As the vectors $\boldsymbol{v}_2$ and $\boldsymbol{r}_{12}$ are orthogonal, the field $\boldsymbol{B}_2$ does not vanish, nor does the force $\boldsymbol{F}^{\mathrm{magn}}_{2\,\mathrm{on}\,1} \neq \mathbf{0}$.

With the inclusion of the magnetic interaction it seems that neither the momentum nor the angular momentum of the two particle system are conserved quantities. The discussion in Dreizler and Lüdde, 2024, Electrodynamics and Special Theory of Relativity (Springer Berlin Heidelberg), Chap. 1.4.2 (on the basis of the complete theory of electrodynamics) will, however, demonstrate, that the issue can be sorted by associating a momentum with the fields. The conservation laws are valid for the particles *and* the fields, which are produced by them.[9]

The result (5.54) can be used to discuss the mutual magnetic interaction of two thin, current carrying sections of conducting material (as two current loops, (Fig. 5.41a)). In order to do this, it is necessary to replace $dq\,\boldsymbol{v}$ (resp. $q\,\boldsymbol{v}$) again by $i\,\mathbf{d}\boldsymbol{r}$ and integrate over the infinitesimal elements of the conductors. The force of loop 1 on loop 2 is

$$\boldsymbol{F}^{\mathrm{magn}}_{1\,\mathrm{on}\,2} = -i_1 i_2\, k_f\, k_m \oint\!\!\oint [\mathbf{d}\boldsymbol{r}_1 \times [\mathbf{d}\boldsymbol{r}_2 \times (\boldsymbol{r}_1 - \boldsymbol{r}_2)]]\, \frac{1}{r_{12}^3}$$

and vice versa (5.55)

$$\boldsymbol{F}^{\mathrm{magn}}_{2\,\mathrm{on}\,1} = i_1 i_2\, k_f\, k_m \oint\!\!\oint [\mathrm{d}\boldsymbol{r}_2 \times [\mathbf{d}\boldsymbol{r}_1 \times (\boldsymbol{r}_1 - \boldsymbol{r}_2)]]\, \frac{1}{r_{12}^3}\,.$$

The integrated action of the forces agrees with the third axiom, provided the loops are closed or the length of the wire is infinite (Fig. 5.41b). This statement can be confirmed with the formula

$$[\mathbf{d}\boldsymbol{r}_2 \times (\mathbf{d}\boldsymbol{r}_1 \times \boldsymbol{r}_{12})] = -\,(\mathbf{d}\boldsymbol{r}_2 \cdot \mathbf{d}\boldsymbol{r}_1)\,\boldsymbol{r}_{12} + (\mathbf{d}\boldsymbol{r}_2 \cdot \boldsymbol{r}_{12})\,\mathbf{d}\boldsymbol{r}_1$$

[9] A connection with the interpretation of electromagnetic radiation as an elementary particle (photon) enters at this point.

for the double vector product. The application for $\boldsymbol{F}_{21}^{\text{magn}}$ gives

$$\boldsymbol{F}_{2\,\text{on}\,1}^{\text{magn}} = -i_1 i_2\, k_f\, k_m \oint\!\!\oint (\mathbf{d}\boldsymbol{r}_2 \cdot \mathbf{d}\boldsymbol{r}_1)\, \frac{\boldsymbol{r}_{12}}{r_{12}^3}$$

$$+\, i_1 i_2\, k_f\, k_m \oint \mathbf{d}\boldsymbol{r}_1 \oint \frac{(\mathbf{d}\boldsymbol{r}_2 \cdot \boldsymbol{r}_{12})}{r_{12}^3}\,.$$

The inner integral of the second term is a line integral with an (electric) field of a point charge. This integral vanishes for a closed curve or for infinite limits. In this case only the first term remains. For the force $\boldsymbol{F}_{12}^{\text{magn}}$ one finds

$$\boldsymbol{F}_{1\,\text{on}\,2}^{\text{magn}} = i_1 i_2\, k_f\, k_m \oint\!\!\oint (\mathbf{d}\boldsymbol{r}_1 \cdot \mathbf{d}\boldsymbol{r}_2)\, \frac{\boldsymbol{r}_{12}}{r_{12}^3} + \mathbf{0} \tag{5.56}$$

by interchanging the indices of the differentials. As the sequence of the factors of a scalar product can be interchanged, there follows

$$\boldsymbol{F}_{12}^{\text{magn}} = -\boldsymbol{F}_{21}^{\text{magn}}\,.$$

The expression (5.55) or the expression (5.56) are the basis for the definition of the unit 1 Ampère. This can be seen in Detail 5.6.6.

5.6 Details

5.6.1 Mathematical Properties of the Vector Potential

Mathematical operations with the vector potential $\boldsymbol{A}(\boldsymbol{r})$ are an essential tool for magnetostatics (and electrodynamics). Two of these operations are discussed in this subsection.

5.6.1.1 A Cross Product of Operators

The validity of the relation

$$\nabla \times (\nabla \times \boldsymbol{A}(\boldsymbol{r})) = \nabla\, (\nabla \cdot \boldsymbol{A}(\boldsymbol{r})) - \Delta \boldsymbol{A}(\boldsymbol{r}) \tag{5.57}$$

can be demonstrated by explicit evaluation. One first calculates the rotation of the vector potential

$$\nabla \times \boldsymbol{A}(\boldsymbol{r}) = \begin{vmatrix} \boldsymbol{e}_{\text{x}} & \boldsymbol{e}_{\text{y}} & \boldsymbol{e}_{\text{z}} \\ \dfrac{\partial}{\partial x} & \dfrac{\partial}{\partial y} & \dfrac{\partial}{\partial z} \\ A_{\text{x}} & A_{\text{y}} & A_{\text{z}} \end{vmatrix} = \begin{pmatrix} \dfrac{\partial A_{\text{z}}}{\partial y} - \dfrac{\partial A_{\text{y}}}{\partial z} \\ \dfrac{\partial A_{\text{x}}}{\partial z} - \dfrac{\partial A_{\text{z}}}{\partial x} \\ \dfrac{\partial A_{\text{y}}}{\partial x} - \dfrac{\partial A_{\text{x}}}{\partial y} \end{pmatrix}$$

and then the rotation of the resulting vector

$$\nabla \times (\nabla \times \boldsymbol{A}(\boldsymbol{r})) = \begin{pmatrix} \dfrac{\partial}{\partial y}\left(\dfrac{\partial A_{\mathrm{y}}}{\partial x} - \dfrac{\partial A_{\mathrm{x}}}{\partial y}\right) & - & \dfrac{\partial}{\partial z}\left(\dfrac{\partial A_{\mathrm{x}}}{\partial z} - \dfrac{\partial A_{\mathrm{z}}}{\partial x}\right) \\ \dfrac{\partial}{\partial z}\left(\dfrac{\partial A_{\mathrm{z}}}{\partial y} - \dfrac{\partial A_{\mathrm{y}}}{\partial z}\right) & - & \dfrac{\partial}{\partial x}\left(\dfrac{\partial A_{\mathrm{y}}}{\partial x} - \dfrac{\partial A_{\mathrm{x}}}{\partial y}\right) \\ \dfrac{\partial}{\partial x}\left(\dfrac{\partial A_{\mathrm{x}}}{\partial z} - \dfrac{\partial A_{\mathrm{z}}}{\partial x}\right) & - & \dfrac{\partial}{\partial y}\left(\dfrac{\partial A_{\mathrm{z}}}{\partial y} - \dfrac{\partial A_{\mathrm{y}}}{\partial z}\right) \end{pmatrix} .$$

The correctness of the relation (5.57) can be demonstrated by explicitly evaluating the components on the right hand side of this vector equation. The application of the Laplace operator on the vector function yields

$$\Delta \boldsymbol{A}(\boldsymbol{r}) = \begin{pmatrix} \dfrac{\partial^2}{\partial x^2} A_{\mathrm{x}} & + & \dfrac{\partial^2}{\partial y^2} A_{\mathrm{x}} & + & \dfrac{\partial^2}{\partial z^2} A_{\mathrm{x}} \\ \dfrac{\partial^2}{\partial x^2} A_{\mathrm{y}} & + & \dfrac{\partial^2}{\partial y^2} A_{\mathrm{y}} & + & \dfrac{\partial^2}{\partial z^2} A_{\mathrm{y}} \\ \dfrac{\partial^2}{\partial x^2} A_{\mathrm{z}} & + & \dfrac{\partial^2}{\partial y^2} A_{\mathrm{z}} & + & \dfrac{\partial^2}{\partial z^2} A_{\mathrm{z}} \end{pmatrix} ,$$

the gradient of the divergence (going over to a short hand notation) is

$$\begin{aligned} \nabla (\nabla \cdot \boldsymbol{A}(\boldsymbol{r})) &= \nabla \left(\partial_{\mathrm{x}} A_{\mathrm{x}} + \partial_{\mathrm{y}} A_{\mathrm{y}} + \partial_{\mathrm{z}} A_{\mathrm{z}}\right) \\ &= \begin{pmatrix} \partial_{\mathrm{x}}\partial_{\mathrm{x}} A_{\mathrm{x}} & + & \partial_{\mathrm{x}}\partial_{\mathrm{y}} A_{\mathrm{y}} & + & \partial_{\mathrm{x}}\partial_{\mathrm{z}} A_{\mathrm{z}} \\ \partial_{\mathrm{y}}\partial_{\mathrm{x}} A_{\mathrm{x}} & + & \partial_{\mathrm{y}}\partial_{\mathrm{y}} A_{\mathrm{y}} & + & \partial_{\mathrm{x}}\partial_{\mathrm{z}} A_{\mathrm{z}} \\ \partial_{\mathrm{z}}\partial_{\mathrm{x}} A_{\mathrm{x}} & + & \partial_{\mathrm{z}}\partial_{\mathrm{y}} A_{\mathrm{y}} & + & \partial_{\mathrm{z}}\partial_{\mathrm{z}} A_{\mathrm{z}} \end{pmatrix} . \end{aligned}$$

One can then check, whether both sides of (5.57) match term by term, e.g. for the x-component

$$\begin{aligned} &\partial_{\mathrm{y}} \left(\partial_{\mathrm{x}} A_{\mathrm{y}} - \partial_{\mathrm{y}} A_{\mathrm{x}}\right) - \partial_{\mathrm{z}} \left(\partial_{\mathrm{z}} A_{\mathrm{x}} - \partial_{\mathrm{x}} A_{\mathrm{z}}\right) \\ &\quad = \partial_{\mathrm{x}} \left(\partial_{\mathrm{x}} A_{\mathrm{x}} + \partial_{\mathrm{y}} A_{\mathrm{y}} + \partial_{\mathrm{z}} A_{\mathrm{z}}\right) - \left(\partial_{\mathrm{x}}^2 + \partial_{\mathrm{y}}^2 + \partial_{\mathrm{z}}^2\right) A_{\mathrm{x}} . \end{aligned}$$

If one wants to find a direct representation of the components of $\nabla \times (\nabla \times \boldsymbol{A}(\boldsymbol{r}))$, the following consideration is useful: Add to and subtract from each component the complement of the second derivative with respect to the same coordinate, e.g. for the x-component

$$\begin{aligned} &\partial_{\mathrm{y}}\partial_{\mathrm{x}} A_{\mathrm{y}} - \partial_{\mathrm{y}}\partial_{\mathrm{y}} A_{\mathrm{x}} - \partial_{\mathrm{z}}\partial_{\mathrm{z}} A_{\mathrm{x}} + \partial_{\mathrm{z}}\partial_{\mathrm{x}} A_{\mathrm{z}} = \\ &\quad -\partial_{\mathrm{x}}^2 A_{\mathrm{x}} - \partial_{\mathrm{y}}^2 A_{\mathrm{x}} - \partial_{\mathrm{z}}^2 A_{\mathrm{x}} + \partial_{\mathrm{x}}\partial_{\mathrm{x}} A_{\mathrm{x}} + \partial_{\mathrm{y}}\partial_{\mathrm{x}} A_{\mathrm{y}} + \partial_{\mathrm{z}}\partial_{\mathrm{x}} A_{\mathrm{z}} . \end{aligned}$$

One recognises a term of the form $-\Delta A_{\mathrm{x}}$. The remaining terms contain all a partial derivative with respect to x, which can be 'factored' out

$$\partial_{\mathrm{x}}\partial_{\mathrm{x}}A_{\mathrm{x}} + \partial_{\mathrm{y}}\partial_{\mathrm{x}}A_{\mathrm{y}} + \partial_{\mathrm{z}}\partial_{\mathrm{x}}A_{\mathrm{z}} = \partial_{\mathrm{x}}(\nabla \cdot \boldsymbol{A})\,.$$

The same procedure is possible for all components, so that the validity of the assertion is demonstrated.

It should be realised, that a quantity as e.g. $\Delta \boldsymbol{A}$ can not be expressed in a simple manner in terms of spherical or other curvilinear coordinates. It is necessary to express the Cartesian unit vectors (Vol. 1, Chap. 2.4)

$$\boldsymbol{A}(\boldsymbol{r}) = A_x(\boldsymbol{r})\,\boldsymbol{e}_x + A_y(\boldsymbol{r})\,\boldsymbol{e}_y + A_z(\boldsymbol{r})\,\boldsymbol{e}_z$$

in terms of curvilinear coordinates, so that one can apply the Laplace operator (in curvilinear coordinates).

5.6.1.2 Check of $\nabla \cdot A = 0$ for a Special Case

The solution of the differential equation of Ampère with simple boundary conditions is

$$\boldsymbol{A}(\boldsymbol{r}) = k_m \iiint \boldsymbol{j}(\boldsymbol{r}')\left(\frac{1}{|\boldsymbol{r}-\boldsymbol{r}'|}\right)\,\mathrm{d}V'\,,$$

where the current density vector is

$$\boldsymbol{j}(\boldsymbol{r}) \xrightarrow{r\to\infty} \boldsymbol{0}\,.$$

The differentiation and the integration can be interchanged, for the calculation of the divergence of the vector potential, as the operator acts on the unprimed coordinates (and the integral is not divergent). The scalar character of the equation is preserved, if one uses a scalar product of the current density operator and the ***nabla***-operator

$$\nabla \cdot \boldsymbol{A}(\boldsymbol{r}) = k_m \iiint \boldsymbol{j}(\boldsymbol{r}')\cdot\nabla\left(\frac{1}{|\boldsymbol{r}-\boldsymbol{r}'|}\right)\,\mathrm{d}V'\,.$$

The action of the operator on the function for the separation of two points is (as can be checked easily)

$$\nabla\left(\frac{1}{|\boldsymbol{r}-\boldsymbol{r}'|}\right) = -\nabla'\left(\frac{1}{|\boldsymbol{r}-\boldsymbol{r}'|}\right)\,.$$

The gradient acting on the unprimed coordinates can be replaced by the gradient acting on the primed coordinates

$$\nabla \cdot \boldsymbol{A}(\boldsymbol{r}) = -k_m \iiint \boldsymbol{j}(\boldsymbol{r}') \cdot \nabla' \left(\frac{1}{|\boldsymbol{r} - \boldsymbol{r}'|} \right) \mathrm{d}V \, .$$

In order to rewrite the resulting expression, the relation (see Append. C.3)

$$\nabla' \cdot \left[\varphi(\boldsymbol{r}')(\boldsymbol{j}(\boldsymbol{r}')\right] = \varphi(\boldsymbol{r}') \left[\nabla' \cdot \boldsymbol{j}(\boldsymbol{r}')\right] + \left[\nabla' \varphi(\boldsymbol{r}')\right] \cdot \boldsymbol{j}(\boldsymbol{r}') \, ,$$

which follows from the product rule, can be used to obtain the result

$$\nabla \cdot \boldsymbol{A}(\boldsymbol{r}) = -k_m \iiint \mathrm{d}V' \left\{ \nabla' \cdot \left(\boldsymbol{j}(\boldsymbol{r}') \frac{1}{|\boldsymbol{r} - \boldsymbol{r}'|} \right) \right. \\ \left. - \frac{1}{|\boldsymbol{r} - \boldsymbol{r}'|} \nabla' \cdot \boldsymbol{j}(\boldsymbol{r}') \right\} .$$

The first term can be rewritten with the divergence theorem

$$\iiint \nabla' \cdot \left(\boldsymbol{j}(\boldsymbol{r}') \frac{1}{|\boldsymbol{r} - \boldsymbol{r}'|} \right) \mathrm{d}V' = \oiint_{\infty \text{ sphere}} \left\{ \boldsymbol{j}(\boldsymbol{r}') \frac{1}{|\boldsymbol{r} - \boldsymbol{r}'|} \right\} \cdot \mathrm{d}\boldsymbol{f}' = 0$$

and yields zero, as the current density vector $\boldsymbol{j}(\boldsymbol{r})$ becomes a null vector on the surface of an infinite sphere. The second term vanishes also, as the statement $\nabla \cdot \boldsymbol{j} = 0$ is valid for a situation with a stationary current distribution.

5.6.2 The Magnetic Field of a Circular Current

The calculation of a magnetic field of a circular ring with a stationary current for any point in space is a lengthy exercise. The following shorter problems are to be discussed here as a complement to Chap. 5.3.

(a) Calculation of the vector potential with the Biot-Savart approximation.
(b) Calculation of the magnetic field by $\boldsymbol{B} = \nabla \times \boldsymbol{A}$.
(c) Discussion of the magnetic field for points at large distances from the circular ring.
(d) Calculation of the magnetic field by a multipole expansion.

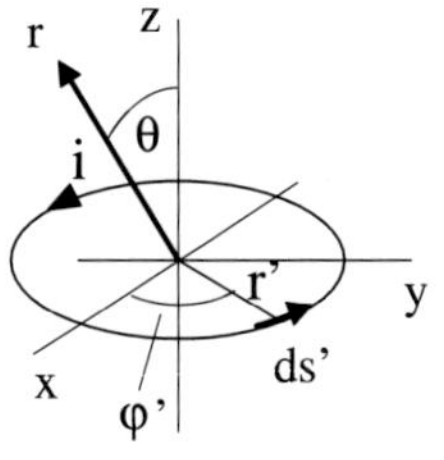

Fig. 5.42 Magnetic field of a circular current: Geometry

5.6.2.1 The Vector Potential

Use the following quantities

- Radius of the circular ring a.
- Spherical coordinates $r,\ \varphi,\ \theta$ for a ring in the x- y plane about the origin of the coordinate system.
- Motion of the current in the x- z plane. The symmetry allows the calculation of the potential/field in arbitrary planes by rotation about the z-axis.

The answer to the first question requires the evaluation of the integral $A_{\mathrm{y}}(r,\theta) \equiv A_{\varphi}(r,\theta)$ for points with the coordinates $r,\ \theta,\ \varphi = 0$ (Fig. 5.42). This integral

$$A_{\mathrm{y}}(r,\theta) = i\,ak_m \int_0^{2\pi} \frac{\cos\varphi'}{\left[r^2 + a^2 - 2ar\sin\theta\cos\varphi'\right]^{1/2}}\,\mathrm{d}\varphi'$$

can be written in the form

$$A_{\mathrm{y}}(r,\theta) = i\,ak_m \int_0^{2\pi} \frac{\cos\varphi'}{[B - C\cos\varphi']^{1/2}}\,\mathrm{d}\varphi' \equiv i\,ak_m\, I\,,$$

if one introduces the quantities

$$B = r^2 + a^2 \qquad C = 2ar\sin\theta\,.$$

The substitution $\varphi = 2\eta$ with

$$\cos 2\eta = \cos^2\eta - \sin^2\eta = 1 - 2\sin^2\eta$$

then leads to the integral

$$I = 2\int_0^{\pi} \frac{1 - 2\sin^2\eta}{\left[B - C + 2C\sin^2\eta\right]^{1/2}}\mathrm{d}\eta\,,$$

which can be sorted further by extracting a factor from the square root

$$I = \frac{2}{[B-C]^{1/2}} \int_0^{\pi} \mathrm{d}\eta \frac{1 - 2\sin^2\eta}{\left[1 + \dfrac{2C\sin^2\eta}{B-C}\right]^{1/2}}$$

and defining the parameter

$$\kappa^2 = \frac{2C}{C-B} = \frac{-4ar\sin\theta}{r^2 + a^2 - 2ar\sin\theta},$$

so that one arrives at

$$I = \frac{2}{[B-C]^{1/2}} \int_0^{\pi} \mathrm{d}\eta \left[\frac{1}{\left[1 - \kappa^2\sin^2\eta\right]^{1/2}} - \frac{2\sin^2\eta}{\left[1 - \kappa^2\sin^2\eta\right]^{1/2}} \right]. \tag{5.58}$$

The first integral is, except for the upper limit of integration, a complete elliptic integral. In order to go over to the complete form one uses the substitution $\eta = \pi - \eta'$, which leads to a replacement of the integrand $f(\eta)$ in equation (5.58) by $f(\eta) = f(\pi - \eta)$, so that the following sequence of steps

$$\begin{aligned}
\int_0^{\pi} f(\sin^2\eta)\,\mathrm{d}\eta &= \int_0^{\pi/2} f(\sin^2\eta)\,\mathrm{d}\eta + \int_{\pi/2}^{\pi} f(\sin^2\eta)\,\mathrm{d}\eta \\
&= \int_0^{\pi/2} f(\sin^2\eta)\,\mathrm{d}\eta + \int_{\pi/2}^{0} f(\sin^2\eta')\,(-\mathrm{d}\eta') \\
&= 2\int_0^{\pi/2} f(\sin^2\eta)\,\mathrm{d}\eta
\end{aligned}$$

results in a complete elliptic integral.

The second integral in (5.58) can also be converted to a complete elliptic integral. Here one uses the replacement

$$\sin^2\eta = -\frac{1}{\kappa^2}\left(1 - \kappa^2\sin^2\eta - 1\right)$$

and finds for the integrand of the second integral

$$\frac{-2\sin^2\eta}{\left[1-\kappa^2\sin^2\eta\right]^{1/2}} = \frac{2}{\kappa^2\left[1-\kappa^2\sin^2\eta\right]^{1/2}}\left(1-\kappa^2\sin^2\eta-1\right)$$

$$= \frac{2}{\kappa^2}\left[1-\kappa^2\sin^2\eta\right]^{1/2} - \frac{2}{\kappa^2\left[1-\kappa^2\sin^2\eta\right]^{1/2}} .$$

Together with the necessary change of the upper limits one thus arrives at the (deceptively) compact representation for the vector potential

$$I = \frac{4}{[B-C]^{1/2}}\int_0^{\pi/2} \mathrm{d}\eta \left[\left(1-\frac{2}{\kappa^2}\right)\frac{1}{\left[1-\kappa^2\sin^2\eta\right]^{1/2}} + \frac{2}{\kappa^2}\left[1-\kappa^2\sin^2\eta\right]^{1/2}\right] \tag{5.59}$$

with complete elliptic integrals of the first and the second kind. The notation for these integrals is K (used in physics) (or F, used in mathematics)

$$\text{1. kind: } K(\kappa) = \int_0^{\pi/2} \mathrm{d}\eta \left[1-\kappa^2\sin^2\eta\right]^{-1/2} = \frac{\pi}{2} F\left(\frac{1}{2}, \frac{1}{2}, 1, \kappa^2\right)$$

$$\text{2. kind: } E(\kappa) = \int_0^{\pi/2} \mathrm{d}\eta \left[1-\kappa^2\sin^2\eta\right]^{1/2} = \frac{\pi}{2} F\left(-\frac{1}{2}, \frac{1}{2}, 1, \kappa^2\right) .$$

The elliptic integrals can be expressed by hypergeometric functions $F(a, b, c, z)$, for which the series expansion

$$F(a, b, c, z) = \frac{\Gamma(c)}{\Gamma(a)\Gamma(b)}\sum_{n=0}^{\infty}\frac{\Gamma(a+n)\Gamma(b+n)}{\Gamma(c+n)}\frac{z^n}{n!}$$

is possible. The final result for the vector potential is therefore

$$A_y(r,\theta) = \frac{4\,i\,a\,k_m}{[B-C]^{1/2}}\left[\left(1-\frac{2}{\kappa^2}\right)K(\kappa) + \frac{2}{\kappa^2}E(\kappa)\right] \tag{5.60}$$

$$= \frac{2\pi\,i\,a\,k_m}{[B-C]^{1/2}}\left\{\left(1-\frac{2}{\kappa^2}\right)F\left(\frac{1}{2}, \frac{1}{2}, 1; \kappa^2\right) + \frac{2}{\kappa^2}F\left(-\frac{1}{2}, \frac{1}{2}, 1; \kappa^2\right)\right\} . \tag{5.61}$$

The constants involved are

$$B - C = r^2 + a^2 - 2ar\sin\theta \qquad \kappa^2 = \frac{2C}{C-B} = \frac{-4ar\sin\theta}{r^2+a^2-2ar\sin\theta}\,.$$

5.6.2.2 The Magnetic Field

For the calculation of the magnetic induction one can use Cartesian or spherical coordinates. As the vector potential has been obtained in terms of spherical coordinates

$$\boldsymbol{A}(r,\theta) = (A_r,\ A_\theta,\ A_\varphi) = (0,\ 0,\ A(r,\theta))\,, \tag{5.62}$$

this option is the more direct one. The same result can also be found with Cartesian coordinates, as will be shown.

- As the induction is orthogonal to the vector potential, one should have a magnetic field with a radial- and an azimuthal angular component. Application of the operator rot in spherical coordinates yields with (5.62)

$$\boldsymbol{B}(r,\theta) = (B_r,\ B_\theta,\ B\varphi) = \nabla \times \boldsymbol{A}$$

$$= \left(\frac{1}{r\sin\theta}\frac{\partial}{\partial\theta}\left(\sin\theta A(r,\theta)\right),\ -\frac{1}{r}\frac{\partial}{\partial r}\left(rA(r,\theta)\right),\ 0\right),$$

where $A(r,\theta)$ is given by (5.60) as

$$A(r,\theta) = \frac{4\,i\,a\,k_m}{[B-C]^{1/2}}\left[\left(1-\frac{2}{\kappa^2}\right)K(\kappa) + \frac{2}{\kappa^2}\,E(\kappa)\right].$$

The calculation of the components of the $\boldsymbol{B}$-field with the chain rule is rather tedious. One first needs the derivative

$$\frac{\mathrm{d}}{\mathrm{d}\kappa^2}F\left(a,\,b,\,c;\,\kappa^2\right) = \frac{a\,b}{c}F\left(a+1,\,b+1,\,c+1;\,\kappa^2\right)$$

and obtains

$$\frac{\mathrm{d}}{\mathrm{d}\kappa^2}\left\{\left(1-\frac{2}{\kappa^2}\right)F\left(\frac{1}{2},\,\frac{1}{2},\,1;\,\kappa^2\right) + \frac{2}{\kappa^2}F\left(-\frac{1}{2},\,\frac{1}{2},\,1;\,\kappa^2\right)\right\} =$$

$$\frac{2}{\kappa^4}F\left(\frac{1}{2},\,\frac{1}{2},\,1;\,\kappa^2\right) + \frac{1}{4}\left(1-\frac{2}{\kappa^2}\right)F\left(\frac{3}{2},\,\frac{3}{2},\,2;\,\kappa^2\right)$$

$$-\frac{2}{\kappa^4}F\left(-\frac{1}{2},\,\frac{1}{2},\,1;\,\kappa^2\right) + \frac{1}{2\kappa^2}F\left(\frac{1}{2},\,\frac{3}{2},\,2;\,\kappa^2\right).$$

The derivatives of κ^2 lead to additional factors

$$\frac{\partial \kappa^2}{\partial r} = \frac{4a(r^2 - a^2)\sin\theta}{(r^2 + a^2 - 2ar\sin\theta)^2}$$

$$\frac{\partial \kappa^2}{\partial \theta} = \frac{-4ar(a^2 + r^2)\cos\theta}{(r^2 + a^2 - 2ar\sin\theta)^2}\,.$$

The derivatives of the prefactors, which depend on r and θ are

$$\frac{\partial [B - C]^{-1/2}}{\partial r} = -(r - a\sin\theta)[B - C]^{-3/2}$$

$$\frac{\partial [B - C]^{-1/2}}{\partial \theta} = (ar\cos\theta)[B - C]^{-3/2}\,.$$

All these 'ingredients' have to be incorporated into the explicit expressions for the components of the B-field

$$B_r = \frac{\cos\theta}{r\sin\theta}A + \frac{1}{r}\frac{\partial A}{\partial \theta} \qquad\qquad B_\theta = -\frac{A}{r} - \frac{\partial A}{\partial r}\,,$$

which need, however, not be done.

- Spherical coordinates are adapted to local situations. This implies that one may just differentiate in a straight forward manner. Such a statement is not possible for Cartesian coordinates, so that one has to sort the procedure beforehand. In order to calculate the magnetic field for any point of space, one has to calculate first the vector potential in this point. This requires (see Fig. 5.43a) a rotation of the coordinate system by an angle $-\varphi$ about the z-axis, as well as (see Fig. 5.43b)

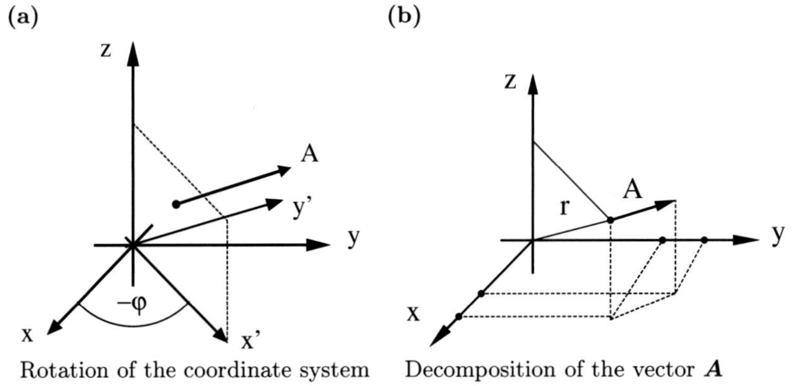

Fig. 5.43 Calculation of the ***B***-field in Cartesian coordinates

a corresponding adjustment of the radial coordinate

$$r = [x^2 + z^2]^{1/2} \longrightarrow r = [x^2 + y^2 + z^2]^{1/2}$$

and the azimuthal angle

$$\theta = \arctan\frac{x}{z} \longrightarrow \theta = \arctan\frac{[x^2 + y^2]^{1/2}}{z} .$$

The vector potential in the rotated coordinate system has in any point of space (see Fig. 5.43b) an x- as well as a y-component

$$\boldsymbol{A}(r, \theta, \varphi) = (A_x,\ A_y,\ A_z) = (-\sin\varphi A(r, \theta),\ \cos\varphi A(r, \theta),\ 0)\,,$$

where each of the spherical coordinates has to be expressed in terms of the Cartesian coordinates. The components of the magnetic field are therefore

$$\boldsymbol{B}(x, y, z) = \boldsymbol{\nabla} \times \boldsymbol{A}(x, y, z) = (B_x,\ B_y,\ B_z)$$

$$= \left(-\frac{\partial A_y}{\partial z},\ \frac{\partial A_x}{\partial z},\ \frac{\partial A_y}{\partial x} - \frac{\partial A_x}{\partial y}\right) .$$

As the components of $\boldsymbol{A}$ are quite involved functions of the Cartesian coordinates

$$A_i = A_i(r(x, y, z), \theta(x, y, z), \varphi(x, y, z))\,,$$

it is necessary to calculate the derivatives involved with the chain rule as e.g.

$$B_x = -\frac{\partial A_y}{\partial z} = -\left(\frac{\partial A_y}{\partial r}\frac{\partial r}{\partial z} + \frac{\partial A_y}{\partial \theta}\frac{\partial \theta}{\partial z} + \frac{\partial A_y}{\partial \varphi}\frac{\partial \varphi}{\partial z}\right)$$

$$= -\cos\varphi\cos\theta\,\frac{\partial A(r, \theta)}{\partial r} + \frac{\cos\varphi\sin\theta}{r}\frac{\partial A(r, \theta)}{\partial \theta} .$$

The other components are

$$B_y = -\sin\varphi\cos\theta\,\frac{\partial A(r, \theta)}{\partial r} + \frac{\sin\varphi\sin\theta}{r}\frac{\partial A(r, \theta)}{\partial \theta}$$

$$B_z = \sin\theta\,\frac{\partial A(r, \theta)}{\partial r} + \frac{\cos\theta}{r}\frac{\partial A(r, \theta)}{\partial \theta} + \frac{A(r, \theta)}{r\sin\theta} .$$

This decomposition of the magnetic field will also not be displayed, but the connection with the magnetic field in spherical coordinates will be indicated.

For this purpose one has to insert the representation of the Cartesian components of the $\boldsymbol{B}$-field

$$\boldsymbol{B} = B_x \boldsymbol{e}_x + B_y \boldsymbol{e}_y + B_z \boldsymbol{e}_z \,,$$

into the relations

$$\begin{aligned}
\boldsymbol{e}_x &= \cos\varphi \sin\theta \boldsymbol{e}_r + \cos\varphi \cos\theta \boldsymbol{e}_\theta - \sin\varphi \boldsymbol{e}_\varphi \\
\boldsymbol{e}_y &= \sin\varphi \sin\theta \boldsymbol{e}_r + \sin\varphi \cos\theta \boldsymbol{e}_\theta + \cos\varphi \boldsymbol{e}_\varphi \\
\boldsymbol{e}_z &= \cos\theta \boldsymbol{e}_r - \sin\theta \boldsymbol{e}_\theta \,.
\end{aligned}$$

The result is

$$\boldsymbol{B} = \left(\frac{\cos\theta}{r\,\sin\theta} A + \frac{1}{r}\frac{\partial A}{\partial \theta} \right) \boldsymbol{e}_r - \left(\frac{A}{r} + \frac{\partial A}{\partial r} \right) \boldsymbol{e}_\theta \,.$$

5.6.2.3 The Far Field

The magnetic field of a current ring leads, as seen in the last subsection, to a rather complicated set of expressions for the components of the field. The results are simpler if the parameter κ^2 is small enough, for the following situations

- for points about the axis of the ring with $\sin\theta \ll 1$,
- for points about the origin, which are characterised by $r \ll a$ and
- for points which are sufficiently distant from the ring with $r \gg a$.

In this case only the lowest order expansion of the elliptic integrals are needed. They can be obtained from the power series of the hypergeometric functions

$$K(\kappa) = \frac{\pi}{2} \left(1 + \frac{\kappa^2}{4} + \frac{9\kappa^4}{64} + \ldots \right)$$

$$E(\kappa) = \frac{\pi}{2} \left(1 - \frac{\kappa^2}{4} - \frac{3\kappa^4}{64} + \ldots \right) .$$

For the combination of these functions in (5.60) one obtains to order κ^2

$$\begin{aligned}
\left(1 - \frac{2}{\kappa^2} \right) K(\kappa) + \frac{2}{\kappa^2} E(\kappa) &= \frac{\pi}{2} \left(1 + \frac{\kappa^2}{4} - \frac{2}{\kappa^2} - \frac{1}{2} - \frac{9\kappa^2}{32} \right. \\
&\qquad \left. + \frac{2}{\kappa^2} - \frac{1}{2} - \frac{3\kappa^2}{32} + \ldots \right) \\
&= -\frac{\pi}{16}\kappa^2 + \ldots \,,
\end{aligned}$$

so that the vector potential in that order is

$$A(r,\theta) \approx \frac{i\pi a^2 r k_m \sin\theta}{[r^2+a^2-2ar\sin\theta]^{3/2}} \, .$$

The components of the magnetic induction to that order are then

$$B_\theta = -\frac{1}{r}\frac{\partial(rA(r,\theta))}{\partial r} = -i\pi a^2 k_m \sin\theta \frac{(2a^2-r^2-ar\sin\theta)}{[r^2+a^2-2ar\sin\theta]^{5/2}}$$

$$B_r = \frac{1}{r\sin\theta}\frac{\partial \sin\theta A(r,\theta)}{\partial\theta} = i\pi a^2 k_m \cos\theta \frac{(2a^2+2r^2-ar\sin\theta)}{[r^2+a^2-2ar\sin\theta]^{5/2}} \, .$$

These results are simplify further for the far field, for which it is possible to use the approximation in lowest order in $1/r$

$$\frac{1}{\left[r^2+a^2-2ra\sin\theta\right]^{n/2}} = \frac{1}{r^n} + \dots \, .$$

In this approximation one obtains for the $A_\varphi(r,\theta)$-component of the vector potential

$$A_\varphi(r,\theta) \xrightarrow{r\gg a} i\pi\, a^2 k_m \frac{\sin\theta}{r^2}$$

as well as for the $\boldsymbol{B}$-field

$$B_r(r,\theta) \xrightarrow{r\gg a} 2i\pi a^2 k_m \frac{\cos\theta}{r^3} \qquad B_\theta(r,\theta) \xrightarrow{r\gg a} i\pi a^2 k_m \frac{\sin\theta}{r^3} \, .$$

The prefactor can be simplified with the introduction of the magnetic moment of the spherical ring

$$m = i\pi a^2 k_f \, .$$

5.6.2.4 Multipole Expansion

The starting point is the relation (5.25) of Chap. 5.3

$$\boldsymbol{A}(\boldsymbol{r}) = ik_m \int \frac{\mathrm{d}\boldsymbol{s}'}{|\boldsymbol{r}-\boldsymbol{r}'|} \, .$$

Only the y-component contributes to the line integral, which has been written for the evaluation with the multipole expansion in the form

$$\mathrm{d}s'_y = a\cos\varphi'\mathrm{d}\varphi' = a\cos\varphi'\mathrm{d}\varphi' \iint \mathrm{d}\cos\theta'\, r'^2\mathrm{d}r'\, \delta(\cos\theta')\frac{\delta(r'-a)}{a^2} \, .$$

The multipole expansion of the separation of two points (see Chap. 3.3 (3.34))

$$\frac{1}{|\boldsymbol{r}-\boldsymbol{r}'|} = 4\pi \sum_{l,m} \frac{1}{(2l+1)} \left(\frac{r_<^l}{r_>^{l+1}}\right) Y_{lm}^*(\Omega')Y_{lm}(\Omega)$$

is needed for $\theta' = \pi/2$ (the ring is in the x- y plane) and for $\varphi = 0$ (calculation of the vector potential in the x- y plane). If this is inserted into the initial form of the φ-component of the vector potential

$$A_\varphi(r,\theta) = i\, ak_m \iiint r'^2 \mathrm{d}r' \mathrm{d}\Omega' \frac{\cos\varphi'}{|\boldsymbol{r}-\boldsymbol{r}'|} \delta(cos\theta') \frac{\delta(r'-a)}{a^2}$$

and if the simple integration over r' and θ' are executed, one finds

$$A_\varphi(r,\theta) = 4\pi i\, ak_m \sum_{l,m} \frac{1}{(2l+1)} \left(\frac{r_<^l}{r_>^{l+1}}\right) Y_{lm}(\theta,0)$$
$$\int \mathrm{d}\varphi' \cos\varphi' Y_{lm}^*(\pi/2,\varphi')\,,$$

where $r_>$ respectively $r_<$ is the larger respectively the smaller of the variables r and a. For the integration over φ', which is still to be executed, one uses the definition of the spherical harmonics

$$Y_{lm}^*(\theta',\varphi') = \left[\frac{(2l+1)(l-m)!}{4\pi(l+m)!}\right]^{1/2} P_l^m(\cos\theta')\,\mathrm{e}^{-im\varphi'} \tag{5.63}$$

and

$$\int \mathrm{d}\varphi' \cos\varphi'\,\mathrm{e}^{-im\varphi'} = \int \mathrm{d}\varphi' \frac{1}{2}\left(\mathrm{e}^{-\mathrm{i}(m-1)\varphi'} + \mathrm{e}^{-\mathrm{i}(m+1)\varphi'}\right)$$
$$= \pi(\delta_{m,-1} + \delta_{m,1})\,.$$

If the function $Y_{lm}(\theta,0)$ is also used as in the definition (5.63) and if the factors are sorted, there remains for the component of the vector potential

$$A_\varphi(r,\theta) = i\pi\, ak_m \sum_l \left(\frac{r_<^l}{r_>^{l+1}}\right) \left(\frac{P_l^1(\cos\theta)P_l^1(0)}{l(l+1)}\right.$$
$$\left. + l(l+1)P_l^{-1}(\cos\theta)P_l^{-1}(0)\right)\,.$$

At this point one has to use the values of the associated Legendre polynomials at the position $\cos\theta = 0$. One first uses the relation

$$P_l^{-m}(\cos\theta) = (-1)^m \frac{(l-m)!}{(l+m)!} P_l^m(\cos\theta)$$

and finds that the second term in the bracket is equal to the first. If one uses $(2n+1)!! = 1 \cdot 3 \cdots (2n-1) \cdot (2n+1)$ and the definition $(-1)!! = 1$ (see also M. Abramovitz, I. Stegun: 'Handbook of Mathematical Functions' Appendix A.5)

$$P_l^1(0) = \begin{cases} 0 & \text{for} \quad l = \text{ even} \\ (-1)^{n+1} \dfrac{(2n+1)!!}{2^n n! \sqrt{\pi}} & \text{for} \quad l = \text{odd} = 2n+1 \quad n = 0, 1, \ldots \end{cases}$$

and collects all factors, one obtains the multipole expansion of the vector potential

$$A_\varphi(r,\theta) = -i\pi\, a k_m \sum_{n=0}^{\infty} (-1)^n \frac{(2n-1)!!}{2^n (n+1)!} \left(\frac{r_<^{2n+1}}{r_>^{2n+2}} \right) P_{2n+1}^1(\cos\theta)\,,$$

which can be inserted for $r > a$ as well as for $r < a$. The first terms of this expansion are

$$A_\varphi(r,\theta) = -i\pi\, a k_m \left\{ \left(\frac{r_<}{r_>^2} \right) P_1^1(\cos\theta) - \frac{1}{4} \left(\frac{r_<^3}{r_>^4} \right) P_3^1(\cos\theta) \right.$$

$$\left. + \frac{1}{8} \left(\frac{r_<^5}{r_>^6} \right) P_5^1(\cos\theta) - \ldots \right\}$$

respectively for the case $r > a$

$$A_\varphi(r,\theta) = -i\pi\, a^2 k_m \frac{\sin\theta}{r^2} \left\{ 1 - \frac{3}{8} \left(\frac{a}{r} \right)^2 (5\cos^2\theta - 1) \right.$$

$$\left. + \frac{15}{64} \left(\frac{a}{r} \right)^4 (21\cos^4\theta - 14\cos^2\theta + 1) - \ldots \right\}.$$

The convergence of the multipole expansion is illustrated in Fig. 5.44a–c. In order to show the azimuthal symmetry, the range of the angle θ, that is $0 \geq \theta \geq \pi$, is extended up to 2π. The following three figures show the variation of the vector potential with the angles $\theta \equiv x$ for different values of (a/r). The multipole expansion up to the order $(a/r)^2$ (grey) is compared with the expansion up to order $(a/r)^4$ (black). For the case with $(a/r) = 0.2$ there is no significant difference if one considers an expansion to order $(a/r)^2$ or to order $(a/r)^4$. The two expansions

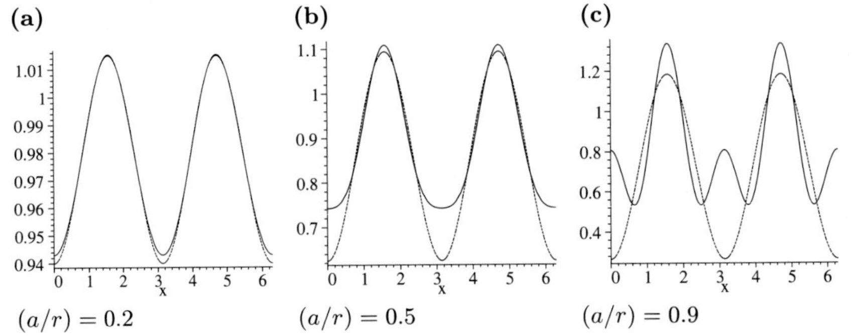

Fig. 5.44 The multipole expansion up to the order $(a/r)^2$ (grey) compared with the expansion up to order $(a/r)^4$ (black)

differ already more for $(a/r) = 0.5$ at the values $\theta = 0, \pi$. For the larger value $(a/r) = 0.9$ the expansion up to order $(a/r)^2$ is apparently not sufficient.

The component of the field B_θ can be calculated directly, but two cases have to be distinguished

$$B_\theta = -\frac{1}{r}\frac{\partial(rA(r,\theta))}{\partial r} = i\pi\, a^2 k_m \sum_{n=0}^{\infty}(-1)^n \frac{(2n-1)!!}{2^n(n+1)!}$$

$$\times \left\{ \begin{array}{c} (2n+2)\left(\dfrac{r^{2n}}{a^{2n+3}}\right) \\ \\ -(2n+1)\left(\dfrac{a^{2n+1}}{r^{2n+3}}\right) \end{array} \right\} P^1_{2n+1}(\cos\theta) \left\{ \begin{array}{c} r < a \\ \\ r > a \end{array} \right. .$$

For the calculation of the component B_r a formula for the derivative of the associated Legendre polynomial P_l^1 is needed

$$\frac{1}{\sin\theta}\frac{\partial}{\partial\theta}\left(\sin\theta\, P_l^1(\cos\theta)\right) = -\frac{\mathrm{d}}{\mathrm{d}\cos\theta}\left([1-\cos^2\theta]^{1/2} P_l^1(\cos\theta)\right)$$
$$= -l(l+1)P_l(\cos\theta)\,,$$

which follows from the definition of the function $P_l^1(x)$

$$P_l^1(x) = [1-x^2]^{1/2}\frac{\mathrm{d}}{\mathrm{d}x}P_l(x)$$

and the differential equation for $P_l(x)$

$$\frac{\mathrm{d}}{\mathrm{d}x}\left((1-x^2)\frac{\mathrm{d}}{\mathrm{d}x}P_l(x)\right)+l(l+1)P_l(x)=0\,.$$

This can be used directly for the calculation of

$$\begin{aligned} B_r &= \frac{1}{r\sin\theta}\frac{\partial \sin\theta A(r,\theta)}{\partial\theta} \\ &= 2i\pi k_m\left(\frac{a}{r}\right)\sum_{n=0}^{\infty}(-1)^n\frac{(2n+1)!!}{2^n n!}\left(\frac{r_<^{2n+1}}{r_>^{2n+2}}\right)P_{2n+1}(\cos\theta)\,. \end{aligned}$$

5.6.3 The Current Density of a Circulating Point Charge

A point charge q is moving with the velocity $\boldsymbol{V(t)}(t)$ on a circular trajectory $\boldsymbol{R}(t)$ (Fig. 5.45).

In order to demonstrate that the time dependent current density

$$\rho(\boldsymbol{r},t)=q\,\delta^{(3)}(\boldsymbol{r}-\boldsymbol{R}(t))=q\,\delta(x-R_x(t))\,\delta(y-R_y(t))\,\delta(z-R_z(t))$$

corresponds to the time dependent charge density

$$\boldsymbol{j}(\boldsymbol{r})=q\,\dot{\boldsymbol{R}}(t)\delta^{(3)}(\boldsymbol{r}-\boldsymbol{R}(t))\,,$$

one has to show, that these quantities satisfy the equation of continuity. For this purpose it is necessary to calculate the derivatives

$$\frac{\partial\rho(\boldsymbol{r},t)}{\partial t}\qquad\text{and}\qquad\nabla\cdot\boldsymbol{j}(\boldsymbol{r},t)\,.$$

One finds for instance

$$\begin{aligned} \frac{\partial}{\partial t}\delta(x-R_x(t)) &= \frac{\partial\delta(x-R_x(t))}{\partial(x-R_x(t))}\cdot\frac{\partial(x-R_x(t))}{\partial t} \\ &= -\dot{R}_x(t)\frac{\partial\delta(x-R_x(t))}{\partial(x-R_x(t))} \end{aligned}$$

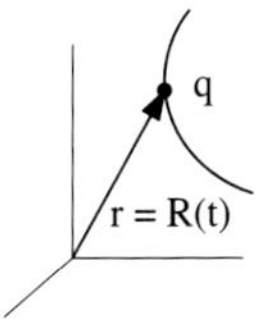

Fig. 5.45 Point charge q moving on a circular trajectory $\boldsymbol{R}(t)$

and

$$\frac{\partial}{\partial x}\delta(x - R_x(t)) = \frac{\partial \delta(x - R_x(t))}{\partial(x - R_x(t))} \cdot \frac{\partial(x - R_x(t))}{\partial x}$$

$$= \frac{\partial \delta(x - R_x(t))}{\partial(x - R_x(t))} \, .$$

Corresponding results are valid for the derivatives of $\delta(y - R_y(t))$ and $\delta(z - R_z(t))$ with respect to t, y and z.
There follows

$$\frac{\partial}{\partial x}\rho(\boldsymbol{r}, t)) = -q \left\{ \dot{R}_x(t) \frac{\partial \delta(x - R_x(t))}{\partial(x - R_x(t))} \delta(y - R_y(t)) \delta(z - R_z(t)) \right.$$
$$+ \dot{R}_y(t) \delta(x - R_x(t)) \frac{\partial \delta(y - R_y(t))}{\partial(x - R_x(t))} \delta(z - R_z(t))$$
$$\left. + \dot{R}_z(t) \delta(x - R_x(t)) \delta(y - R_y(t)) \frac{\partial \delta(z - R_z(t))}{\partial(x - R_x(t))} \right\}$$

$$\nabla \cdot \boldsymbol{j}(\boldsymbol{r}, t) = q \left\{ \dot{R}_x(t) \frac{\partial \delta(x - R_x(t))}{\partial(x - R_x(t))} \delta(y - R_y(t)) \delta(z - R_z(t)) \right.$$

$$+ \dot{R}_y(t) \delta(x - R_x(t)) \frac{\partial \delta(y - R_y(t))}{\partial(x - R_x(t))} \delta(z - R_z(t))$$

$$\left. + \dot{R}_z(t) \delta(x - R_x(t)) \delta(y - R_y(t)) \frac{\partial \delta(z - R_z(t))}{\partial(x - R_x(t))} \right\}$$

and thus

$$\frac{\partial \rho(\boldsymbol{r}, t)}{\partial t} + \nabla \cdot \boldsymbol{j}(\boldsymbol{r}, t) = 0 \, .$$

The magnetic moment of a moving point charge is

$$\boldsymbol{m} = \frac{k_f}{2} \iiint \left(\boldsymbol{r}' \times \boldsymbol{j}(\boldsymbol{r}')\right) \mathrm{d}V'$$

$$= \frac{q k_f}{2} \iiint \left(\boldsymbol{r}' \times \dot{\boldsymbol{R}}(t)\right) \delta^{(3)}(\boldsymbol{r}' - \boldsymbol{R}(t)) \, \mathrm{d}V'$$

$$= \frac{q k_f}{2} \left(\boldsymbol{R}(t) \times \dot{\boldsymbol{R}}(t)\right) = \frac{q k_f}{2} \left(\boldsymbol{R}(t) \times \boldsymbol{v}(t)\right) \, ,$$

if one uses $\boldsymbol{v}(t)$ instead of $\dot{\boldsymbol{R}}(t)$.

5.6.4 The Gyromagnetic Factor of a Uniformly Rotating Homogeneous Charge Distribution

The gyromagnetic factor of elementary particles differs generally from the expected classical value $g = 1$. This can be regarded as an indication that the particle has a structured charge distribution. It will be demonstrated in this section, that a uniformly rotating uniform charge distribution in the form of a sphere has indeed the classical value $g = 1$.

A uniformly rotating mass distribution in the shape of a sphere (radius R) with a uniform density ϱ_m) can be characterised by the following properties: The total mass of the sphere is $M = 4\pi R^3 \varrho_m/3$, the moment of inertia is $I = 2MR^2/5$ and the relation between angular momentum $\boldsymbol{l}$ and the angular velocity is $\boldsymbol{\omega} = \omega\, \boldsymbol{e}_z$. The relation

$$\boldsymbol{l} = \left(\frac{4}{3}\pi R^3 \varrho_m\right) \frac{2}{5} R^2\, \boldsymbol{\omega} = \frac{2}{5} M R^2\, \boldsymbol{\omega}$$

is valid for ω =const. This formula can be obtained in the following way: One considers the contribution of a mass element dm at the position $\boldsymbol{r}$, which moves with the velocity $\boldsymbol{v}$ (see Fig. 5.46). The contribution of a mass element dm to the angular momentum is

$$\mathbf{d}\boldsymbol{l} = \mathrm{d}m\, (\boldsymbol{r} \times \boldsymbol{v})\,.$$

The addition of the contributions of all mass elements yields the total angular momentum

$$\boldsymbol{l} = \varrho_m \iiint (\boldsymbol{r} \times \boldsymbol{v})\, \mathrm{d}V\,, \tag{5.64}$$

which is the basis of the formula given above.

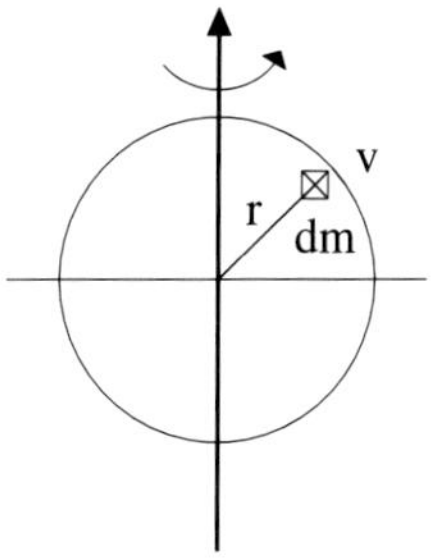

Fig. 5.46 Mass element dm at the position $\boldsymbol{r}$ moving with the velocity $\boldsymbol{v}$

The magnetic moment of a uniformly rotating, uniform charge distribution with the charge density ϱ_q is according to (5.33) given as

$$\boldsymbol{m}_{\text{total}} = \frac{k_f}{2} \iiint (\boldsymbol{r} \times \boldsymbol{j}(\boldsymbol{r}))\, \mathrm{d}V \, .$$

The current density of a charge element

$$\mathrm{d}q = \varrho_q \, \mathrm{d}V \, ,$$

which moves with the velocity $\boldsymbol{v}$ is

$$\boldsymbol{j}(\boldsymbol{r})\, \mathrm{d}V = \varrho_q \boldsymbol{v}\, \mathrm{d}V = \mathrm{d}q\, \boldsymbol{v} \, .$$

Therefore one finds

$$\boldsymbol{m}_{\text{total}} = \frac{k_f}{2} \varrho_q \iiint (\boldsymbol{r} \times \boldsymbol{v})\, \mathrm{d}V \, .$$

The integral can (with (5.64)) be replaced by the angular momentum

$$\boldsymbol{m}_{\text{total}} = \frac{k_f}{2} \frac{\varrho_q}{\varrho_m} \boldsymbol{l} \, .$$

If the volume of the sphere is introduced in this expression in order to replace the density by the total mass M and the charge density by the total charge Q, one finds the result

$$\boldsymbol{m}_{\text{total}} = \frac{k_f Q}{2M} \boldsymbol{l} = \left\{ \frac{Q}{2Mc}\, g\, \boldsymbol{l} \right\}_{\text{CGS}} ,$$

which corresponds to the gyromagnetic factor $g = 1$ ($k_f = 1/c$ in the CGS-system).

Addendum The contribution of a mass element $\mathrm{d}m$ is

$$\mathbf{d}\boldsymbol{l} = \mathrm{d}m\, (\boldsymbol{r} \times \boldsymbol{v}) \, .$$

The vector $\boldsymbol{v}$ is perpendicular to the vector $\boldsymbol{r}$ (see Fig. 5.47) and has the value

$$v = \omega\, \rho_D \, ,$$

where $\rho_D = r \sin\theta$ is the distance from the axis of rotation (of the z-axis). The vector product is therefore

$$\boldsymbol{r} \times \boldsymbol{v} = -r\, \rho_D\, \omega\, \boldsymbol{e}_\theta \, .$$

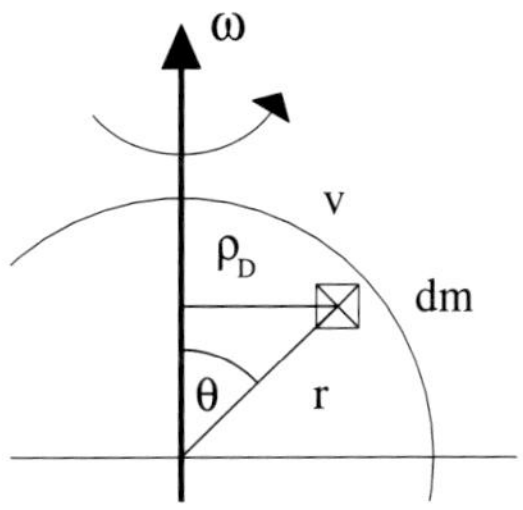

Fig. 5.47 Mass element dm at the position $\boldsymbol{r}$ moving with the velocity $\boldsymbol{v}$ (spherical coordinates)

The unit vector $\boldsymbol{e}_\theta$ of the spherical coordinates can be expressed by the Cartesian unit vectors as

$$\boldsymbol{e}_\theta = \cos\theta\cos\varphi\,\boldsymbol{e}_\mathrm{x} + \cos\theta\sin\varphi\,\boldsymbol{e}_\mathrm{y} - \sin\theta\,\boldsymbol{e}_\mathrm{z}.$$

The total angular momentum follows with integration over all mass elements, as $\mathrm{d}m = \varrho_m\,\mathrm{d}V$, thus

$$\boldsymbol{l} = \varrho_m\omega\int_0^{2\pi}\mathrm{d}\varphi\int_0^{\pi}\sin\theta\,\mathrm{d}\theta\int_0^R r^2\,\mathrm{d}r\left\{-r^2\left(\cos\theta\sin\theta\cos\varphi\,\boldsymbol{e}_\mathrm{x}\right.\right.$$

$$\left.\left.+\cos\theta\sin\theta\sin\varphi\,\boldsymbol{e}_\mathrm{y} - \sin^2\theta\,\boldsymbol{e}_\mathrm{z}\right)\right\}.$$

The integrals over the angle φ give

$$\int_0^{2\pi}\sin\varphi\,\mathrm{d}\varphi = \int_0^{2\pi}\cos\varphi\,\mathrm{d}\varphi = 0\,.$$

The angular momentum points, as expected in the direction of the axis of rotation (here the z-axis). The evaluation of the last angular integral gives

$$\int_0^{\pi}\sin^3\theta\,\mathrm{d}\theta = \left(-\cos\theta + \frac{1}{3}\cos^3\theta\right)\Bigg|_0^{\pi} = \frac{4}{3}\,.$$

The total angular momentum of the uniformly rotating homogeneous sphere is therefore

$$\boldsymbol{l} = 2\pi\,\varrho_m\omega\frac{4}{3}\frac{R^5}{5}\,\boldsymbol{e}_\mathrm{z} = \frac{2}{5}\left(\frac{4}{3}\pi R^3\varrho_m\right)R^2\omega\,\boldsymbol{e}_\mathrm{z}$$

$$= \frac{2}{5}MR^2\omega\,\boldsymbol{e}_\mathrm{z} = I_\mathrm{sphere}\,\boldsymbol{\omega}\,.$$

5.6.5 The Magnetic Field of a Uniformly Magnetised Sphere

One example for the calculation of the magnetic field (actually the magnetic induction) of a macroscopic, magnetised object is the calculation of the magnetic field of a uniformly magnetised sphere. The solution begins with the formula for the vector potential

$$A_M(\boldsymbol{r}) = \frac{k_m}{k_f} \iiint \frac{(\nabla' \times \boldsymbol{M}(\boldsymbol{r}'))}{|\boldsymbol{r} - \boldsymbol{r}'|} \mathrm{d}V' ,$$

which has, as a consequence of the simple form of the magnetisation (and an appropriate choice of the coordinate system), only two components, which are given by the integrals (see Chap. 5.4.5)

$$A_{M,x}(r, \theta, \varphi) = -M_0 \frac{k_m}{k_f} \iiint \frac{\delta(r' - R) \sin\theta' \sin\varphi'}{|\boldsymbol{r} - \boldsymbol{r}'|} \mathrm{d}V'$$

$$A_{M,y}(r, \theta, \varphi) = M_0 \frac{k_m}{k_f} \iiint \frac{\delta(r' - R) \sin\theta' \cos\varphi'}{|\boldsymbol{r} - \boldsymbol{r}'|} \mathrm{d}V' .$$

The evaluation of the integrals requires the steps:

(a) Integrate over the (primed) radial coordinate.
(b) Expand the function $1/|\boldsymbol{r} - \boldsymbol{r}'|$ for $r > R$ and $r < R$ in terms of spherical harmonics.
(c) Express the factors $\sin\theta' \sin\varphi'$ and $\sin\theta' \cos\varphi'$ by spherical harmonics.
(d) Work out the angular integrals using the orthogonality relations of the spherical harmonics.

(a) The integration over the primed radial coordinate can be done directly

$$A_{M,x}(r, \theta, \varphi) = -M_0 \frac{k_m}{k_f} \int_0^R r'^2 \, \mathrm{d}r' \int_0^{2\pi} \mathrm{d}\varphi$$

$$\int_{-1}^{1} \mathrm{d}\cos\theta \, \frac{\delta(r' - R) \sin\theta' \sin\varphi'}{|\boldsymbol{r} - \boldsymbol{r}'|}$$

$$= -M_0 \frac{k_m}{k_f} \int_0^{2\pi} \int_{-1}^{1} \left. \frac{R^2}{|\boldsymbol{r} - \boldsymbol{r}'|} \right|_{r'=R} \sin\theta' \sin\varphi' \, \mathrm{d}\varphi \, \mathrm{d}\cos\theta .$$

The expansion of the function $1/|\boldsymbol{r}-\boldsymbol{r}'|$ is (see Chap. 3.3, (3.34))

$$\left.\frac{1}{|\boldsymbol{r}-\boldsymbol{r}'|}\right|_{r'=R} = \sum_{l=0}^{\infty}\sum_{m=-l}^{m=l} \frac{4\pi}{(2l+1)}\frac{R^l}{r^{l+1}} \mathrm{Y}_{l,m}(\theta,\varphi)\mathrm{Y}^*_{l,m}(\theta',\varphi') \qquad \text{for} \quad r > R$$

and

$$\left.\frac{1}{|\boldsymbol{r}-\boldsymbol{r}'|}\right|_{r'=R} = \sum_{l,m} \frac{4\pi}{(2l+1)}\frac{r^l}{R^{l+1}} \mathrm{Y}_{l,m}(\theta,\varphi)\mathrm{Y}^*_{l,m}(\theta',\varphi') \quad \text{for} \quad r < R\,.$$

(c) The functions $\sin\theta'\cos\varphi'$ and $\sin\theta'\sin\varphi'$ can be expressed in terms of spherical harmonics. Any list of these functions will give

$$\mathrm{Y}_{1,1}(\Omega) = -\sqrt{\frac{3}{8\pi}}\,\sin\theta\,\mathrm{e}^{\mathrm{i}\varphi} \qquad \mathrm{Y}_{1,-1}(\Omega) = \sqrt{\frac{3}{8\pi}}\,\sin\theta\,\mathrm{e}^{-\mathrm{i}\varphi}\,,$$

so that

$$\mathrm{Y}_{1,1}(\Omega) - \mathrm{Y}_{1,-1}(\Omega) = \sqrt{\frac{3}{8\pi}}\,\sin\theta\big[-\mathrm{e}^{\mathrm{i}\varphi} + \mathrm{e}^{\mathrm{i}\varphi}\big] = -\sqrt{\frac{3}{2\pi}}\,\sin\theta\,\cos\varphi\,,$$

respectively

$$\sin\theta\,\cos\varphi = -\sqrt{\frac{2\pi}{3}}\,\big[\mathrm{Y}_{1,1}(\Omega) - \mathrm{Y}_{1,-1}(\Omega)\big]\,.$$

In the same fashion one finds

$$\sin\theta\,\sin\varphi = \mathrm{i}\sqrt{\frac{2\pi}{3}}\,\big[\mathrm{Y}_{1,1}(\Omega) + \mathrm{Y}_{1,-1}(\Omega)\big]\,.$$

(d) Insert the expression for $\sin\theta'\sin\varphi'$ into the integral $A_{M,x}$ and use the orthogonality relations of the spherical harmonics. The result for $r > R$ is

$$A_{M,x}(r,\,\theta,\,\varphi) = -M_0\frac{k_m}{k_f}\int_0^{2\pi}\!\!\int_{-1}^{1} \left.\frac{R^2}{|\boldsymbol{r}-\boldsymbol{r}'|}\right|_{r'=R} \sin\theta'\sin\varphi'\,\mathrm{d}\varphi\,\mathrm{d}\cos\theta$$

$$= -M_0\frac{k_m}{k_f}R^2\int_0^{2\pi}\!\!\int_{-1}^{1}\sum_{l,m}\frac{4\pi}{(2l+1)}\,\mathrm{i}\sqrt{\frac{2\pi}{3}}\,\frac{R^l}{r^{l+1}}\mathrm{Y}_{l,m}(\Omega)$$

$$*\Big[\mathrm{Y}^*_{l,m}(\Omega')\mathrm{Y}_{1,1}(\Omega') + \mathrm{Y}^*_{l,m}(\Omega')\mathrm{Y}_{1,-1}(\Omega')\Big]\mathrm{d}\Omega'$$

$$= -M_0 \frac{k_m}{k_f} R^2 \sum_{l,m} \frac{4\pi}{(2l+1)}\, \mathrm{i} \sqrt{\frac{2\pi}{3}}\, \frac{R^l}{r^{l+1}} \mathrm{Y}_{l,m}(\Omega)$$

$$* \Big[\delta_{l,1}\, \delta_{m,1} + \delta_{l,1}\, \delta_{m,-1} \Big]$$

$$= -M_0 \frac{k_m}{k_f} \frac{R^3}{r^2} \frac{4\pi}{3}\, \mathrm{i} \sqrt{\frac{2\pi}{3}} \Big[\mathrm{Y}_{11}(\Omega) + \mathrm{Y}_{1-1}(\Omega) \Big]$$

$$= -M_0 \frac{k_m}{k_f} \frac{R^3}{r^2} \frac{4\pi}{3} \sin\theta \sin\varphi\, .$$

For $r < R$ one obtains in the same way

$$A_{M,x}(r,\, \theta,\, \varphi) = -M_0 \frac{k_m}{k_f} \int_0^{2\pi} \int_{-1}^{1} \frac{R^2}{|\boldsymbol{r} - \boldsymbol{r}'|} \bigg|_{r'=R} \sin\theta' \sin\varphi' \,\mathrm{d}\varphi \,\mathrm{d}\cos\theta$$

$$= -M_0 \frac{k_m}{k_f} R^2 \int_0^{2\pi} \int_{-1}^{1} \sum_{l,m} \frac{4\pi}{(2l+1)}\, \mathrm{i} \sqrt{\frac{2\pi}{3}}\, \frac{r^l}{R^{l+1}} \mathrm{Y}_{l,m}(\Omega)$$

$$* \Big[\delta_{l,1}\, \delta_{m,1} - \delta_{l,1}\, \delta_{m,-1} \Big]$$

$$= -M_0 \frac{k_m}{k_f}\, r\, \frac{4\pi}{3} \sin\theta \sin\varphi\, .$$

The integral $A_{M,y}(r,\, \theta,\, \varphi)$ can be calculated in the same fashion. Combining the results of the two calculations, one finds for the total vector potential

$$\boldsymbol{A}_M(\boldsymbol{r}) = (-A(r,\theta) \sin\varphi,\; A(r,\theta)\cos\varphi,\; 0)$$

with

$$A(r,\theta) = \begin{cases} \dfrac{4\pi}{3} R^3 M_0 \dfrac{\sin\theta}{r^2} & \text{for the outer region} \qquad r \geq R \\[2ex] \dfrac{4\pi}{3} M_0\, r \sin\theta & \text{for the inner region} \qquad r < R. \end{cases}$$

The result for the corresponding magnetic induction can be found in Chap. 5.4.5.

5.6.6 Definition of the Unit Ampère

The Ampère, the unit for the electric current, is defined in the SI-system as

$$1 \text{ Ampère} = \frac{1 \text{ Coulomb}}{1 \text{ second}} .$$

This formal definition does, however, not address the question, whether there is a current of one or of x Ampère in the conductor. A definite rule for the strength of the current has to be supplied. The quantity, which is used for this purposes is the force between two parallel wires. The official definition is based on the formula of Biot-Savart. It states: 1 Ampère is the strength of a stationary current in each of two straight, thin parallel wires in vacuum, if there is a distance of 1 meter between them and if there exists a force of $2 \cdot 10^{-7}$ Newton per 1 meter of the length of the wires.

This definition is based on the evaluation of Eqs. (5.56)

$$\boldsymbol{F}^{\text{mag}}_{1\,\text{on}\,2} = i_1 i_2 k_f k_m \iint (\mathbf{d}\boldsymbol{r}_1 \cdot \mathbf{d}\boldsymbol{r}_2) \frac{\boldsymbol{r}_{12}}{r_{12}^3}$$

$$\boldsymbol{F}^{\text{mag}}_{2\,\text{on}\,1} = -i_1 i_2 k_f k_m \iint (\mathbf{d}\boldsymbol{r}_1 \cdot \mathbf{d}\boldsymbol{r}_2) \frac{\boldsymbol{r}_{12}}{r_{12}^3} .$$

The quantities involved are: $\boldsymbol{r}_{12} = \boldsymbol{r}_1 - \boldsymbol{r}_2$ is the vector from the line element $\mathbf{d}\boldsymbol{r}_2$ to the line element $\mathbf{d}\boldsymbol{r}_1$ (Fig. 5.48a). As the wires are parallel, one can state

$$\mathbf{d}\boldsymbol{r}_1 \cdot \mathbf{d}\boldsymbol{r}_2 = \mathrm{d}r_1\, \mathrm{d}r_2 .$$

It is therefore necessary to evaluate e.g. the integral (Fig. 5.48b)

$$\boldsymbol{F}^{\text{mag}}_{2\,\text{on}\,1} = -i_1 i_2 k_f k_m \int_{-\infty}^{\infty} \mathrm{d}r_1 \int_{-\infty}^{\infty} \mathrm{d}r_2 \frac{\boldsymbol{r}_{12}}{r_{12}^3} .$$

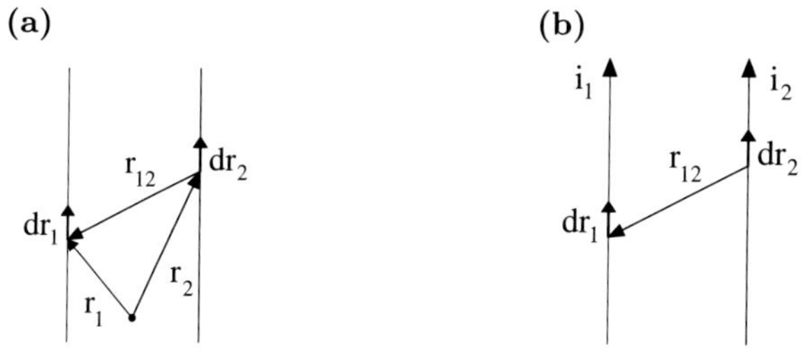

Fig. 5.48 Geometry used for the definition of the unit 1 Ampère

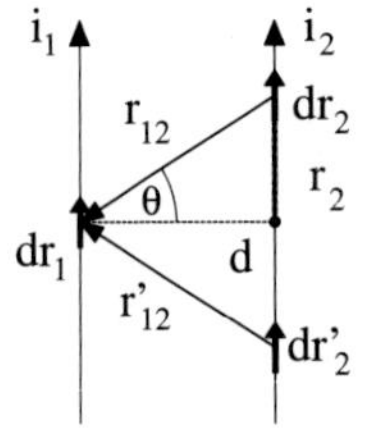

Fig. 5.49 Measurement of the strength of currents

In order to execute the integration over dr_2, one should consider two line elements $\mathbf{dr_2}$ and $\mathbf{dr'_2}$, which are diametric with respect to dr_1 (Fig. 5.49). The contribution of these infinitesimal elements to the force

$$\mathbf{d}\boldsymbol{F} = -i_1 i_2 k_f k_m \frac{\boldsymbol{r}_{12} + \boldsymbol{r}'_{12}}{r_{12}^3} \mathrm{d}r_1 \mathrm{d}r_2$$

is a vector perpendicular to the line element $\mathbf{d}\boldsymbol{r}_1$. The action of this force is attractive (in the direction towards the wire), if the currents move in the same direction, it is repulsive (the wires repel each other) for currents in different directions. The geometry of the arrangement yields

$$\boldsymbol{r}_{12} + \boldsymbol{r}'_{12} = 2r_{12} \cos\theta \, \boldsymbol{e}_1 = 2d \, \boldsymbol{e}_1$$

and

$$r_{12} = \left[r_2^2 + d^2\right]^{1/2} ,$$

if the separation of the wires is d.

The integral over r_2 leads to

$$I = \int_0^\infty \mathrm{d}r_2 \frac{2d}{\left[r_2^2 + d^2\right]^{3/2}} = 2d \left. \frac{r_2}{d^2 \left[r_2^2 + d^2\right]^{1/2}} \right|_0^\infty = \frac{2}{d} .$$

The remaining integral over r_1 is infinite, but one can extract the force acting on wire 1 per unit of length as

$$\frac{\boldsymbol{F}^{\mathrm{mag}}_{2\,\mathrm{on}\,1}}{l_1} = -i_1 i_2 k_f k_m \frac{2}{d} .$$

If the necessary quantities are inserted in SI units, ($d = 1\text{m}$, $k_m = 10^{-7}(\text{kg m})/\text{C}^2$, $k_f = 1$) one obtains for the magnitude of the force per unit length

$$\left|\frac{\boldsymbol{F}^{\text{mag}}_{2\,\text{on}\,1}}{l_1}\right| = \left|\frac{\boldsymbol{F}^{\text{mag}}_{1\,\text{on}\,2}}{l_2}\right|$$

$$= \frac{\text{C}^2}{\text{s}^2} 1 \cdot 10^{-7} \frac{\text{kg m}}{\text{C}^2} \frac{2}{1\,\text{m}} = 2 \cdot 10^{-7} \frac{\text{kg}}{\text{s}^2} = 2 \cdot 10^{-7} \frac{\text{N}}{\text{m}},$$

as specified in the definition.

Literature

A

A.1 Books and Literature Quoted in the Text

- L.P. Fulcher
  Phys. Rev. A33 (1986)
  p. 759
- S.J. Plimpton and W.E. Lawton
  Phys. Rev. 50 (1936)
  p. 1066
- Edwin Hall
  Amer. Journ. Math. **2** (1879)
  p. 287
- P. Moon and D.E. Spencer
  Field Theory Handbook
  Springer (1961)
- P. Morse and H. Feshbach
  Methods of Theoretical Physics
  Mc Graw-Hill (1953)
- M.E. Rose
  Elementary theory of angular momentum
  Dover (1995)

A.2 Introductory Texts

- R.P. Feynman and R.B. Leighton and M.L. Sands
  The Feynman lectures on physics: definitive edition
  Pearson Addison-Wesley (2006)

© The Author(s), under exclusive license to Springer-Verlag GmbH, DE, part of Springer Nature 2024

R. M. Dreizler, C. S. Lüdde, *Electrostatics and Magnetostatics*,
https://doi.org/10.1007/978-3-662-69933-1

- D.J. Morin and E.M. Purcell
  Electricity and Magnetism
  Cambridge University Press (2013)
- E.M. Purcell
  Berkeley physics course Electricity and Magnetism
  Mc Graw-Hill (1966)

A.3 Electrodynamics

- V.D. Barger and M.G. Olsson
  Classical Electricity and Magnetism
  Allyn & Bacon (1987)
- D.J. Griffith
  Introduction to Electrodynamics
  Cambridge University Press (2023)
- J.D. Jackson
  Classical Electrodynamics
  Wiley (2021)
- K. Milton and J. Schwinger
  Classical Electrodynamics: Second edition
  CRC Press (2024)
- G.L. Pollack and D.R. Stump
  Electromagnetism
  Addison-Wesley (2002)
- J. Schwinger and K.A. Milton et al.
  Classical Electrodynamics
  Westview Press (1998)

A.4 Mathematics

- H. Cartan
  The Elementary Theory of Analytic Functions of One or Several Complex Variables
  Dover Pub. (1995)
- F.J. Flanigan
  Complex Variables: Harmonics and Analytic Functions
  Dover Pub. (1983)
- K. Kodaira
  Complex Analysis
  Cambridge University Press (2007)
- A. Kyrala
  Applied Functions of a Complex Variable
  Wiley-Interscience (1972)

A.5 Special Functions and Handbooks

- M. Abramovitz and I. Stegun
  Handbook of Mathematical Functions
  National Bureau of Standards applied mathematics series
  Martino Publishing (2014)
- I.S. Gradstejn and I.M. Ryzik
  Table of Integrals, Series and Products
  Elsevier (2007)
- A. Jeffrey and H. Dai
  Handbook of Mathematical Formulas and Integrals
  Elsevier (2004)
- A.F. Nikiforov and V.B. Uvarov
  Special Functions of Mathematical Physics
  Springer Basel (2013)
- A.D. Polyanin and V. Nazaikinskii
  Handbook of Linear Partial Differential Equations for Engineers and Scientists, 2nd Edition
  Chapman and Hall/CRC (2015)
- A.D. Polyanin and A.I. Chernoutsan
  A Concise Handbook of Mathematics, Physics, and Engineering Sciences
  CRC Press LLC (2010)
- N.M. Temme
  Special Functions: An Introduction to the Functions of Mathematical Physics
  Wiley (1996)
- M. Tenenbaum and H. Pollard
  Ordinary differential equations: an elementary textbook for students of mathematics, engineering, and the sciences
  Dover Publ. (1985)

Systems of Units in Electrodynamics

B

B.1 The Systems

Three parameters, which can in principle be selected freely, are needed to define the units for all quantities of electrodynamics in the vacuum. These constants are usually chosen to be the units of the electric field E, the magnetic induction B and the law of induction (Vol. II, part 2), which contains these two experimentally accessible quantities. The equations, which are relevant for this choice, are:

- The electric field E of a point charge q at a distance r from this charge

$$E(r) = k_e \, \frac{q}{r^2} \, .$$

- The magnetic induction B of a long, thin, straight conducting wire with a stationary current i at a distance r from the axis of the wire

$$B(r) = 2 \, k_m \, \frac{i}{r} \, .$$

  The extracted factor 2 is a useful but not a compelling option.
- The law of induction in the differential form

$$\nabla \times \boldsymbol{E}(\boldsymbol{r}, t) + k_f \frac{\partial}{\partial t} \boldsymbol{B}(\boldsymbol{r}, t) = 0$$

  with the restriction

$$\left[\frac{k_e}{k_m k_f} \right]^{1/2} = c = 2.997925... \cdot 10^{10} \, \frac{\text{cm}}{\text{s}} = 2.997925... \cdot 10^{8} \, \frac{\text{m}}{\text{s}} \, .$$

© The Author(s), under exclusive license to Springer-Verlag GmbH, DE, part of Springer Nature 2024

R. M. Dreizler, C. S. Lüdde, *Electrostatics and Magnetostatics*,
https://doi.org/10.1007/978-3-662-69933-1

This implies, that only two of the three constants can be chosen freely.

The explicit specification of two constants by measurement (traditionally k_e and k_m are chosen) calls for a somewhat longer consideration.
The constant k_e is introduced first in the Coulomb law

$$F_{12} = k_e \frac{q_1 q_2}{r_{12}^2} \,,$$

which addresses the magnitude of the force between two point charges q_1 and q_2 with a separation of r_{12}. With this choice[1] of the dimensionless constant

$$k_{e,\mathrm{CGS}} = 1$$

one connects the unit for the charge in the CGS-system with mechanical units

$$[q]_{\mathrm{CGS}} = \frac{\mathrm{g}^{1/2}\,\mathrm{cm}^{3/2}}{\mathrm{s}} = \mathrm{statcoul}\,.$$

The unit for the electrical field is therefore, because of $E_{q_1} = F_{12}/q_2$,

$$[E]_{\mathrm{CGS}} = \frac{\mathrm{g}^{1/2}}{\mathrm{cm}^{1/2}\,\mathrm{s}} = \frac{\mathrm{statvolt}}{\mathrm{cm}}\,.$$

The choice of the units of the magnetic fields (more precisely magnetic induction) is based on the relation

$$\boldsymbol{D} = \boldsymbol{m} \times \boldsymbol{B}\,.$$

This relation, which connects the magnetic moment $\boldsymbol{m}$ of a test dipole and the magnetic induction $\boldsymbol{B}$ with a torque $\boldsymbol{D}$ shows, that the units of the magnetic moment are connected with the units for the magnetic induction. In the CGS-system one chooses for the constant k_m the value

$$k_{m,\mathrm{CGS}} = \frac{1}{c} = 0.333564\ldots \cdot 10^{-10}\,\frac{\mathrm{s}}{\mathrm{cm}}\,,$$

so that

$$k_{f,\mathrm{CGS}} = \frac{k_{e,\mathrm{CGS}}}{c^2\,k_{m,\mathrm{CGS}}} \qquad \Longrightarrow k_{f,\mathrm{CGS}} = \frac{1}{c}$$

[1] Only the Gaussian CGS-system and the rationalised MKSA-system will be considered in detail. For two additional systems, the constants are given in Table B.1.

follows. The magnetic induction is measured in the units

$$[B]_{\mathrm{CGS}} = \frac{g^{1/2}}{\mathrm{cm}^{1/2}\,\mathrm{s}} = [E]_{\mathrm{CGS}}\,,$$

the magnetic moment in the units

$$[m]_{\mathrm{CGS}} = \frac{g^{1/2}\mathrm{cm}^{5/2}}{\mathrm{s}} = [p]_{\mathrm{CGS}}\,.$$

The electric and the magnetic fields as well as the electric and magnetic dipole moments are associated with the same unit in the CGS-system.

An additional unit, the unit of the strength of the current *Ampère* (short A), is introduced in the rationalised MKSA-system. The definition of the electric current allows an alternative unit for the charge, the *Coulomb* (short C)

$$i = \frac{\mathrm{d}q}{\mathrm{d}t} \qquad \Longleftrightarrow \qquad 1\ \mathrm{A} = 1\ \frac{\mathrm{C}}{s}\,.$$

The technical definition of the unit Ampère includes the measurement of the action of the magnetic forces:

> The flow of a current of the same strength in two long, thin, straight parallel conducting wires in the vacuum is produced by a current of the strength of 1 Ampère, if the wires attract each other with a force F of $2 \cdot 10^{-7}$ N per meter of the length of the wires.

The Detail 5.6.6 shows, that the appropriate equation is

$$i^2 = \frac{|F|d}{2k_f\, k_m\, l}\,.$$

If the values $F = 2 \cdot 10^{-7}\,\mathrm{N}$, $i = 1\ \mathrm{A}$ and $d = l = 1\ \mathrm{m}$ are inserted one finds for the constant k_f

$$k_{f,\mathrm{SI}}\, k_{m,\mathrm{SI}} = 10^{-7}\,\frac{\mathrm{kg\,m}}{\mathrm{C}^2}\,.$$

The constant $k_{f,\mathrm{SI}}$ is usually chosen, so that this quantity is dimensionless

$$k_{f,\mathrm{SI}} = 1\,.$$

The unit *Ampère* corresponds therefore to the constant

$$k_{m,\mathrm{SI}} = 10^{-7}\,\frac{\mathrm{kg\,m}}{\mathrm{C}^2} = \frac{\mu_0}{4\pi}\,,$$

which is often replaced by the permeability of the vacuum

$$\mu_0 = 4\pi \cdot 10^{-7} \frac{\text{kg m}}{\text{C}^2} = 1.25663... \cdot 10^{-6} \frac{\text{kg m}}{\text{C}^2} .$$

The additional restrictive condition for the constants yields therefore

$$k_{e,\text{SI}} = 10^{-7} c^2 \frac{\text{kg m}}{\text{C}^2} = 8.98755.. \cdot 10^9 \frac{\text{kg m}^3}{\text{C}^2\,\text{s}^2} = \frac{1}{4\pi\varepsilon_0} ,$$

so that the dielectric constant of the vacuum is

$$\varepsilon_0 = 8.85418... \cdot 10^{-12} \frac{\text{C}^2\text{s}^2}{\text{kg m}^3} .$$

For the units of the two fields and the magnetic moment in the SI-system one finds

$$[E]_{\text{SI}} = \frac{\text{kg m}}{\text{C s}^2} = \frac{\text{Volt}}{\text{cm}}$$

$$[B]_{\text{SI}} = \frac{\text{kg}}{\text{C s}}$$

$$[m]_{\text{SI}} = \frac{\text{C m}^2}{\text{s}} .$$

It is usual to employ

$$m_{\text{loop,SI}} = i \cdot \text{surface}$$

for the magnetic moment of a plane, current carrying, conducting loop in the SI-system. The unit, which is defined by this definition, agrees with the initially stated unit of the magnetic moment. This is not the case for the CGS-system. The definition in the two system of units is compatible with the general definition of the magnetic moment, if one uses

$$m_{\text{loop}} = k_f (i \cdot \text{surface}) .$$

The discussion of the fields in materials demands the inclusion of the auxiliary fields $\boldsymbol{D}$ and $\boldsymbol{H}$, the dielectric displacement and the magnetic field strength, as well as the quantities, which address the averaged properties of the material, the polarisation $\boldsymbol{P}$ and the magnetisation $\boldsymbol{M}$. The $\boldsymbol{D}$-field is produced by the true charges. This implies the relation

$$\nabla \cdot \boldsymbol{D} = 4\pi k_d \rho_{tr}$$

if the question of the units for this field is left open for the moment. The polarisation is generated by the distribution of the polarisation charges, which suggests the (not compelling) definition

$$\nabla \cdot \boldsymbol{P} = -\rho_{\text{pol}} \, .$$

The fact, that the electric field is produced by the true and the polarisation charges, leads to the relation

$$\boldsymbol{D} = \frac{k_d}{k_e}\, \boldsymbol{E} - 4\pi k_d\, \boldsymbol{P} \, .$$

In the CGS-system the constant k_d, which can still be chosen, is given the value 1

$$k_{d,\text{CGS}} = k_{e,\text{CGS}} = 1 \, .$$

The resulting equation

$$\boldsymbol{D} = \boldsymbol{E} - 4\pi \boldsymbol{P}$$

states, that all three electric fields are measured in the same units in this system.

This constant is also without a dimension in the SI-system

$$k_{d,\text{SI}} = \frac{1}{4\pi} \, ,$$

so that the relation between the three fields is

$$\boldsymbol{D} = \varepsilon_0 \boldsymbol{E} - \boldsymbol{P} \, .$$

The units of the two auxiliary fields are equal, but differ from that of the $\boldsymbol{E}$ -field.[2]

The corresponding statements for magnetic materials are: The $\boldsymbol{H}$ - field is produced by the true currents

$$\nabla \times \boldsymbol{H} = 4\pi k_h \boldsymbol{j}_{tr} \, .$$

The magnetisation $\boldsymbol{M}$ incorporates the averaging process over the loop-like magnetisation currents in the surface of the material

$$\nabla \times \boldsymbol{M} = k_f \boldsymbol{j}_{tr} \, .$$

[2] The name *rationalised* MKSA-system indicates the elimination of the factor 4π in the basic equations.

The statement, that the $\boldsymbol{B}$ -field is produced by both current densities, is the reason for the combination

$$\boldsymbol{H} = \frac{k_h}{k_m}\boldsymbol{B} - 4\pi\frac{k_h}{k_f}\boldsymbol{M}\,.$$

If one wants, that all magnetic fields are measured in the same units - as for the CGS-system - one has to choose

$$k_{h,\mathrm{CGS}} = \frac{1}{c}$$

and finds

$$\boldsymbol{H} = \boldsymbol{B} - 4\pi\boldsymbol{M}\,.$$

The rationalised SI-system is realised by the choice

$$k_{h,\mathrm{SI}} = \frac{1}{4\pi}\,.$$

This leads to

$$\boldsymbol{H} = \frac{1}{\mu_0}\boldsymbol{B} - \boldsymbol{M}\,.$$

The simple material equations for isotropic materials with linear response are used in both systems of units in the form

$$\boldsymbol{D} = \varepsilon\boldsymbol{E} \qquad\qquad \boldsymbol{H} = \frac{1}{\mu}\boldsymbol{B}\,,$$

with a vacuum limit given by

$$\left.\begin{array}{l}\varepsilon_{\mathrm{CGS}}\\[1ex] \mu_{\mathrm{CGS}}\end{array}\right\} \longrightarrow \quad 1$$

and

$$\varepsilon_{\mathrm{SI}} \longrightarrow \varepsilon_0 \qquad \mu_{\mathrm{SI}} \longrightarrow \mu_0\,.$$

The relative dielectric constant and permeability in the SI-system agree with the dielectric constant and permeability in the CGS-system

$$\left(\frac{\varepsilon}{\varepsilon_0}\right)_{\mathrm{SI}} = \varepsilon_{\mathrm{CGS}} \qquad \left(\frac{\mu}{\mu_0}\right)_{\mathrm{SI}} = \mu_{\mathrm{CGS}} \,.$$

The coefficients

$$k_e,\ k_m,\ k_f,\ k_d,\ k_h$$

- for the two most used system of units and the electrostatic (electrostatic units - esu) and the electromagnetic system of units (electromagnetic units - emu) are assembled in Table B.1. The associated numerical values are given in Table B.2.
- Tables B.3 and B.4 contain a list of the physical quantities, which play a role in electrodynamics.

  The formulae, which are used to introduce these quantities, are included the list in order to provide a better orientation besides the units in the SI- and the CGS-system.
- Tables B.5 and B.6 show the factors, which are needed for the conversion of the physical quantities in the SI-system and the CGS-system, which are listed in Tables B.3 and B.4.

B.2 Tables

Table B.1 Definition of constants

Name	SI	CGS	esu	emu
k_e	$\frac{1}{4\pi\,\varepsilon_0}$	1	1	c^2
k_d	$\frac{1}{4\pi}$	1	1	1
k_m	$\frac{\mu_0}{4\pi}$	$\frac{1}{\mathrm{c}}$	$\frac{1}{\mathrm{c}^2}$	1
k_f	1	$\frac{1}{\mathrm{c}}$	1	1
k_h	$\frac{1}{4\pi}$	$\frac{1}{\mathrm{c}}$	1	1

Table B.2 Numerical values for Table B.1

Name	Value (SI)
c	$= 2.997925 \cdot 10^8 \ \frac{\mathrm{m}}{\mathrm{s}}$
e_0	$= 1.602192 \cdot 10^{-19}\ \mathrm{C}$
ε_0	$= 8.85418 \cdot 10^{-12} \ \frac{\mathrm{C}^2}{\mathrm{Nm}^2}$
μ_0	$= 4\pi \cdot 10^{-7} \ \frac{\mathrm{kg\,m}}{\mathrm{C}^2}$
$\frac{1}{4\pi\,\varepsilon_0}$	$= 8.98755 \cdot 10^9 \ \frac{\mathrm{N\,m}^2}{\mathrm{C}^2}$
$\frac{1}{4\pi}$	$= 7.957747 \cdot 10^{-2}$
$\frac{\mu_0}{4\pi}$	$= 1 \cdot 10^{-7} \ \frac{\mathrm{kg\,m}}{\mathrm{C}^2}$

$\mathrm{e}_{0,\mathrm{CGS}} = 4.803250 \cdot 10^{-10}$ esu

Table B.3 Systems of units: Definition and dimension of physical quantities I

Name		Definition	SI	CGS
Charge	q	$q = r\sqrt{\frac{F}{k_e}}$	C (Coulomb)	$\frac{\mathrm{g}^{1/2}\mathrm{cm}^{3/2}}{\mathrm{s}} = \mathrm{statcoul} = \mathrm{esu}$
E-field	E	$E = \frac{F}{q}$	$\frac{\mathrm{N}}{\mathrm{C}} = \frac{\mathrm{kg\,m}}{\mathrm{s}^2\mathrm{C}} = \frac{\mathrm{V}}{\mathrm{m}}$	$\frac{\mathrm{g}^{1/2}}{\mathrm{cm}^{1/2}\mathrm{s}} = \frac{\mathrm{dyn}}{\mathrm{statcoul}} = \frac{\mathrm{statvolt}}{\mathrm{cm}}$
Electr. flux	Φ_e	$\Phi_e = \iint \boldsymbol{E} \cdot \mathbf{d}\boldsymbol{f}$	$\frac{\mathrm{kg\,m}^3}{\mathrm{s}^2\mathrm{C}} = \mathrm{V\,m}$	$\frac{\mathrm{g}^{1/2}\mathrm{cm}^{3/2}}{\mathrm{s}} = \mathrm{statcoul}$
Electr. dipole moment	p	$p = 2aq$	C m	$\frac{\mathrm{g}^{1/2}\mathrm{cm}^{5/2}}{\mathrm{s}} = \mathrm{statcoul\,cm}$
Potential	V	$V = \int \boldsymbol{E} \cdot \mathbf{d}\boldsymbol{s}$	$\frac{\mathrm{Nm}}{\mathrm{C}} = \frac{\mathrm{kg\,m}^2}{\mathrm{s}^2\mathrm{C}} = \mathrm{V}$	$\frac{\mathrm{g}^{1/2}\mathrm{cm}^{1/2}}{\mathrm{s}} = \mathrm{statvolt} = \frac{\mathrm{erg}}{\mathrm{statcoul}}$
Voltage	U	$U = V_2 - V_1$	V (Volt)	statvolt
Capacity	C	$C = \frac{q}{U}$	$\frac{\mathrm{C}^2\mathrm{s}^2}{\mathrm{kg\,m}^2} = \frac{\mathrm{C}}{\mathrm{V}} = \mathrm{F}$ (Farad)	cm
Dielectric displacement	$\boldsymbol{D}$	$\oint\!\!\oint \boldsymbol{D} \cdot \mathbf{d}\boldsymbol{f} = 4\pi\, k_d q_{tr}$	$\frac{\mathrm{C}}{\mathrm{m}^2}\, \frac{\mathrm{g}^{1/2}}{\mathrm{cm}^{1/2}\mathrm{s}}$	$= \frac{\mathrm{statvolt}}{\mathrm{cm}}$
Polarisation	$\boldsymbol{P}$	$\oint\!\!\oint \boldsymbol{P} \cdot \mathbf{d}\boldsymbol{f} = q_{\mathrm{pol}}$	$\frac{\mathrm{C}}{\mathrm{m}^2}$	$\frac{\mathrm{g}^{1/2}}{\mathrm{cm}^{1/2}\mathrm{s}} = \frac{\mathrm{statvolt}}{\mathrm{cm}}$
Energy density	w_{el}	$w_{\mathrm{el}} = \frac{1}{8\pi k_d} \boldsymbol{E} \cdot \boldsymbol{D}$	$\frac{\mathrm{Nm}}{\mathrm{m}^3} = \frac{\mathrm{J}}{\mathrm{m}^3}$	$\frac{\mathrm{dyn\,cm}}{\mathrm{cm}^3} = \frac{\mathrm{erg}}{\mathrm{cm}^3}$
Current	i	$= \frac{\mathrm{d}q}{\mathrm{d}t}$	$\frac{\mathrm{C}}{\mathrm{s}} = \mathrm{A}$ (Ampère)	$\frac{\mathrm{g}^{1/2}\mathrm{cm}^{3/2}}{\mathrm{s}^2} = \mathrm{statamp}$
Current density	$\boldsymbol{j}$	$i = \iint \boldsymbol{j} \cdot \mathbf{d}\boldsymbol{f}$	$\frac{\mathrm{C}}{\mathrm{m}^2\mathrm{s}} = \frac{\mathrm{A}}{\mathrm{m}^2}$	$\frac{\mathrm{g}^{1/2}}{\mathrm{cm}^{1/2}\mathrm{s}^2} = \frac{\mathrm{statamp}}{\mathrm{cm}^2}$

Table B.4 Systems of units: Definition and dimension of physical quantities II

Name		Definition	SI	CGS
Magnet. induction	$\boldsymbol{B}$	$\int \boldsymbol{B} \cdot \mathbf{d}\boldsymbol{s} = 4\pi\, i\, k_m$	$\frac{\text{kg}}{\text{C s}} = \frac{\text{V s}}{\text{m}^2} = \text{T (Tesla)}$	$\frac{\text{g}^{1/2}}{\text{cm}^{1/2}\text{s}} = \text{G (Gauss)}$
Magn. moment	m	$m = \frac{F \cdot r}{B}$	$\frac{\text{C m}^2}{\text{s}} = \text{A m}^2$	$\frac{\text{g}^{1/2}\text{cm}^{5/2}}{\text{s}} = \text{G cm}^3 = \text{emu}$
Vector potential	$\boldsymbol{A}$	$\boldsymbol{B} = \nabla \times \boldsymbol{A}$	$\frac{\text{kg m}}{\text{C s}} = \text{T m}$	$\frac{\text{g}^{1/2}\text{cm}^{1/2}}{\text{s}} = \text{G cm}$
Magn. flux	Φ_m	$\Phi_m = \iint \boldsymbol{B} \cdot \mathbf{d}\boldsymbol{f}$	$\frac{\text{kg m}^2}{\text{C s}} = \text{V s} = \text{Wb (Weber)}$	$\frac{\text{g}^{1/2}\text{cm}^{3/2}}{\text{s}} = \text{G cm}^2 = \text{Mx (Maxwell)}$
Magnetisation	$\boldsymbol{M}$	$\boldsymbol{j}_\text{M} = \frac{1}{k_f}(\nabla \times \boldsymbol{M})$	$\frac{\text{C}}{\text{m s}} = \frac{\text{A}}{\text{m}}$	$\frac{\text{g}^{1/2}}{\text{cm}^{1/2}\text{s}} = \text{Oe (Ørstedt)}$
Magn. field strength	$\boldsymbol{H}$	$\text{rot}\, \boldsymbol{H} = 4\pi k_h \boldsymbol{j}_\text{w}$	$\frac{\text{C}}{\text{m s}} = \frac{\text{A}}{\text{m}}$	$\frac{\text{g}^{1/2}}{\text{cm}^{1/2}\text{s}} = \text{Oe}$
Induction coefficient	L	$U = -L \frac{\text{d}i}{\text{d}t}$	$\frac{\text{kg m}^2}{\text{C}^2} = \frac{\text{V s}}{\text{A}} = \text{H (Henry)}$	$\frac{\text{s}^2}{\text{cm}} = \frac{\text{statvolt s}}{\text{statamp}}$
Magn. energy density	w_m	$w_m = \frac{k_f}{8\pi\, k_h} \boldsymbol{B} \cdot \boldsymbol{H}$	$\frac{\text{kg}}{\text{m s}^2} = \frac{\text{J}}{\text{m}^3}$	$\frac{\text{g}}{\text{cm s}^2} = \frac{\text{erg}}{\text{cm}^3}$
Resistance	R	$U = R \cdot i$	$\frac{\text{kg m}^2}{\text{s C}^2} = \frac{\text{V}}{\text{A}} = \Omega \text{ (Ohm)}$	$\frac{\text{s}}{\text{cm}} = \frac{\text{statvolt}}{\text{statamp}}$
Power	P	$P = U\, I$	$\frac{\text{m}^2\,\text{kg}}{\text{s}^3} = \frac{\text{J}}{\text{s}} = \text{W (Watt)}$	$\frac{\text{cm}^2\text{g}}{\text{s}^3}$

Table B.5 Conversion factors I

Name		SI	= factor · CGS
Charge	q	$1\,\mathrm{C}$	$\frac{\tilde{c}}{10} \cdot \mathrm{statcoul}$
E-field	E	$1\,\frac{\mathrm{N}}{\mathrm{C}} = 1\,\frac{\mathrm{V}}{\mathrm{m}}$	$\frac{10^6}{\tilde{c}} \cdot \frac{\mathrm{statvolt}}{\mathrm{cm}}$
Electr. flux	Φ_e	$1\,\mathrm{V\,m}$	$\frac{\tilde{c}}{10} \cdot \mathrm{statcoul}$
Electr. dipole moment	p	$1\,\mathrm{C\,m}$	$10\,\tilde{c} \cdot \mathrm{statcoul\,cm}$
Potential	V	$1\,\frac{\mathrm{Nm}}{\mathrm{C}} = 1\,\mathrm{V}$ (Volt)	$\frac{10^8}{\tilde{c}} \cdot \mathrm{statvolt}$
Voltage	U	$1\,\mathrm{V}$	$\frac{10^8}{\tilde{c}} \cdot \mathrm{statvolt}$
Capacity	C	$1\,\frac{\mathrm{C}}{\mathrm{V}} = 1\,\mathrm{F}$ (Farad)	$\frac{\tilde{c}^2}{10^9} \cdot \mathrm{cm}$
Dielectric displacement	$\boldsymbol{D}$	$1\,\frac{\mathrm{C}}{\mathrm{m}^2}$	$\frac{4\pi\,\tilde{c}}{10^5} \cdot \frac{\mathrm{statvolt}}{\mathrm{cm}}$
Polarisation	$\boldsymbol{P}$	$1\,\frac{\mathrm{C}}{\mathrm{m}^2}$	$\frac{\tilde{c}}{10^5} \cdot \frac{\mathrm{statvolt}}{\mathrm{cm}}$
Energy density	w_{el}	$1\,\frac{\mathrm{Nm}}{\mathrm{m}^3} = 1\,\frac{\mathrm{J}}{\mathrm{m}^3}$	$10 \cdot \frac{\mathrm{erg}}{\mathrm{cm}^3}$
Current	i	$1\,\frac{\mathrm{C}}{\mathrm{s}} = 1\,\mathrm{A}$ (Ampère)	$\frac{\tilde{c}}{10} \cdot \mathrm{statamp}$
Current density	$\boldsymbol{j}$	$1\,\frac{\mathrm{C}}{\mathrm{m}^2\mathrm{s}} = 1\,\frac{\mathrm{A}}{\mathrm{m}^2}$	$\frac{\tilde{c}}{10^5} \cdot \frac{\mathrm{statamp}}{\mathrm{cm}^2}$

$\tilde{c}$ is the numerical value of the velocity of light in the CGS-system

$\tilde{c} = 2.997925 \cdot 10^{10}$

Table B.6 Conversion factors II

Name		SI	= factor · CGS
Magnetic induction	$\boldsymbol{B}$	$1\,\frac{\text{kg}}{\text{C s}} = 1\,\text{T}$	$10^4 \cdot \text{G}$
Magn. moment	m	$1\,\text{A m}^2$	$10^3 \cdot \text{G cm}^3 = \text{emu}$
Vector potential	$\boldsymbol{A}$	$1\,\frac{\text{kg m}}{\text{C s}} = 1\,\text{T m}$	$10^6 \cdot \text{G cm}$
Magn. flux	Φ_m	$1\,\text{V s} = 1\,\text{Wb}$	$10^8 \cdot \text{G cm}^2 = \text{Mx}$
Magnetisation	$\boldsymbol{M}$	$1\,\frac{\text{A}}{\text{m}}$	$\frac{\tilde{c}}{10^3} \cdot \text{Oe}$
Magn. field strength	$\boldsymbol{H}$	$1\,\frac{\text{A}}{\text{m}}$	$\frac{4\pi}{10^3} \cdot \text{Oe}$
Induction coefficient	L	$1\,\frac{\text{V s}}{\text{A}} = 1\,\text{H}$	$\frac{10^9}{\tilde{c}^2} \cdot \frac{\text{statvolt s}}{\text{statamp}}$
Magn. energy density	w_m	$1\,\frac{\text{J}}{\text{m}^3}$	$10 \cdot \frac{\text{erg}}{\text{cm}^3}$
Resistance	R	$1\,\frac{\text{V}}{\text{A}} = 1\,\Omega$	$\frac{10^9}{\tilde{c}^2} \cdot \frac{\text{statvolt}}{\text{statamp}}$
Power	P	$1\,\frac{\text{J}}{\text{s}} = 1\,\text{W}$	$10^7 \cdot \frac{\text{erg}}{\text{s}}$

$\tilde{c}$ is the numerical value of velocity of light in the CGS-system

$\tilde{c} = 2.997925 \cdot 10^{10}$

Additional Mathematical Topics

C

C.1 Equations of Vector Analysis

In this collection of formulae one finds some multiple products with vectors (C.2), a set of formulae for the application of the ∇ -operator on products of scalar and vector functions (in formal and explicit notation C.3), expressions for the multiple applications of the ∇ -operator (C.4) and the four differential operators ∇, $\nabla\cdot$, $\nabla\times$ and Δ in spherical and cylinder coordinates (C.5). In this chapter one also finds the derivation of the formulae quoted for the two sets of coordinates.

The notation is

$$\begin{aligned} \boldsymbol{a},\ \boldsymbol{b},\ \ldots & \qquad \text{vectors} \\ \varphi(\boldsymbol{r}) & \qquad \text{scalar function} \\ \boldsymbol{A}(\boldsymbol{r}),\ \boldsymbol{B}(\boldsymbol{r}) & \qquad \text{vector functions}\,. \end{aligned}$$

C.2 Multiple Products of Vectors

$$\boldsymbol{a}\cdot(\boldsymbol{b}\times\boldsymbol{c}) = \boldsymbol{b}\cdot(\boldsymbol{c}\times\boldsymbol{a}) = \boldsymbol{c}\cdot(\boldsymbol{a}\times\boldsymbol{b})$$

$$\boldsymbol{a}\times(\boldsymbol{b}\times\boldsymbol{c}) = (\boldsymbol{a}\cdot\boldsymbol{c})\,\boldsymbol{b} - (\boldsymbol{a}\cdot\boldsymbol{b})\,\boldsymbol{c}$$

$$(\boldsymbol{a}\times\boldsymbol{b})\cdot(\boldsymbol{c}\times\boldsymbol{d}) = (\boldsymbol{a}\cdot\boldsymbol{c})\,(\boldsymbol{b}\cdot\boldsymbol{d}) - (\boldsymbol{a}\cdot\boldsymbol{d})\,(\boldsymbol{b}\cdot\boldsymbol{c})$$

© The Author(s), under exclusive license to Springer-Verlag GmbH, DE, part of Springer Nature 2024

R. M. Dreizler, C. S. Lüdde, *Electrostatics and Magnetostatics*,
https://doi.org/10.1007/978-3-662-69933-1

C.3 Product Rules for the Application of the ∇ -Operator

$$\boldsymbol{\nabla} \cdot (\varphi \boldsymbol{A}) = \varphi(\boldsymbol{\nabla} \cdot \boldsymbol{A}) + \boldsymbol{\nabla}\varphi \cdot \boldsymbol{A}$$

$$\operatorname{div}(\varphi \boldsymbol{A}) = \varphi \operatorname{div} \boldsymbol{A} + \operatorname{grad} \varphi \cdot \boldsymbol{A}$$

$$\boldsymbol{\nabla} \times (\varphi \boldsymbol{A}) = \varphi(\boldsymbol{\nabla} \times \boldsymbol{A}) + \boldsymbol{\nabla}\varphi \times \boldsymbol{A}$$

$$\operatorname{rot}(\varphi \boldsymbol{A}) = \varphi \operatorname{rot} \boldsymbol{A} + \operatorname{grad} \varphi \times \boldsymbol{A}$$

$$\boldsymbol{\nabla} \cdot (\boldsymbol{A} \times \boldsymbol{B}) = \boldsymbol{B} \cdot (\boldsymbol{\nabla} \times \boldsymbol{A}) - \boldsymbol{A} \cdot (\boldsymbol{\nabla} \times \boldsymbol{B})$$

$$\operatorname{div}(\boldsymbol{A} \times \boldsymbol{B}) = \boldsymbol{B} \cdot \operatorname{rot} \boldsymbol{A} - \boldsymbol{A} \cdot \operatorname{rot} \boldsymbol{B}$$

$$\boldsymbol{\nabla} \times (\boldsymbol{A} \times \boldsymbol{B}) = \boldsymbol{A}(\boldsymbol{\nabla} \cdot \boldsymbol{B}) - \boldsymbol{B}(\boldsymbol{\nabla} \cdot \boldsymbol{A}) + (\boldsymbol{B} \cdot \boldsymbol{\nabla})\boldsymbol{A} - (\boldsymbol{A} \cdot \boldsymbol{\nabla})\boldsymbol{B}$$

$$\operatorname{rot}(\boldsymbol{A} \times \boldsymbol{B}) = \boldsymbol{A}(\operatorname{div} \boldsymbol{B}) - \boldsymbol{B}(\operatorname{div} \boldsymbol{A}) + (\boldsymbol{B} \cdot \operatorname{grad})\boldsymbol{A} - (\boldsymbol{A} \cdot \operatorname{grad})\boldsymbol{B}$$

C.4 Double Application of ∇

$$\boldsymbol{\nabla} \cdot (\boldsymbol{\nabla} \times \boldsymbol{A}) = \operatorname{div}(\operatorname{rot} \boldsymbol{A}) = 0$$

$$\boldsymbol{\nabla} \times (\boldsymbol{\nabla}\varphi) = \operatorname{rot}(\operatorname{grad} \varphi) = 0$$

$$\boldsymbol{\nabla} \times (\boldsymbol{\nabla} \times \boldsymbol{A}) = \boldsymbol{\nabla}(\boldsymbol{\nabla} \cdot \boldsymbol{A}) - \Delta \boldsymbol{A}$$

$$\operatorname{rot}(\operatorname{rot} \boldsymbol{A}) = \operatorname{grad}(\operatorname{div} \boldsymbol{A}) - (\operatorname{div} \operatorname{grad})\boldsymbol{A}$$

(The last formula is only valid for a representation of $\boldsymbol{A}$ in Cartesian coordinates!)

C.5 Differential Operators in Spherical and Cylinder Coordinates

C.5.1 Spherical Coordinates

$$\nabla V = \frac{\partial V}{\partial r}\,\boldsymbol{e}_r + \frac{1}{r}\frac{\partial V}{\partial \theta}\,\boldsymbol{e}_\theta + \frac{1}{r\sin\theta}\frac{\partial V}{\partial \varphi}\,\boldsymbol{e}_\varphi$$

$$\nabla \cdot \boldsymbol{A} = \frac{1}{r^2}\frac{\partial}{\partial r}\left(r^2 A_r\right) + \frac{1}{r\sin\theta}\frac{\partial}{\partial \theta}\left(\sin\theta\, A_\theta\right) + \frac{1}{r\sin\theta}\frac{\partial A_\varphi}{\partial \varphi}$$

$$\nabla \times \boldsymbol{A} = \frac{1}{r\sin\theta}\left[\frac{\partial}{\partial \theta}\left(\sin\theta\, A_\varphi\right) - \frac{\partial A_\theta}{\partial \varphi}\right]\boldsymbol{e}_r$$
$$+\left[\frac{1}{r\sin\theta}\frac{\partial A_r}{\partial \varphi} - \frac{1}{r}\frac{\partial}{\partial r}\left(r A_\varphi\right)\right]\boldsymbol{e}_\theta + \frac{1}{r}\left[\frac{\partial}{\partial r}\left(r A_\theta\right) - \frac{\partial A_r}{\partial \theta}\right]\boldsymbol{e}_\varphi$$

$$\boldsymbol{\Delta} V = \frac{1}{r^2}\frac{\partial}{\partial r}\left(r^2\frac{\partial V}{\partial r}\right) + \frac{1}{r^2\sin\theta}\frac{\partial}{\partial \theta}\left(\sin\theta\frac{\partial V}{\partial \theta}\right) + \frac{1}{r^2\sin^2\theta}\frac{\partial^2 V}{\partial \varphi^2}$$

C.5.2 Cylinder Coordinates

$$\nabla V = \frac{\partial V}{\partial \rho}\,\boldsymbol{e}_\rho + \frac{1}{\rho}\frac{\partial V}{\partial \varphi}\,\boldsymbol{e}_\varphi + \frac{\partial V}{\partial z}\,\boldsymbol{e}_z$$

$$\nabla \cdot \boldsymbol{A} = \frac{1}{\rho}\frac{\partial}{\partial \rho}\left(\rho A_\rho\right) + \frac{1}{\rho}\frac{\partial A_\varphi}{\partial \varphi} + \frac{\partial A_z}{\partial z}$$

$$\nabla \times \boldsymbol{A} = \left(\frac{1}{\rho}\frac{\partial A_z}{\partial \varphi} - \frac{\partial A_\varphi}{\partial z}\right)\boldsymbol{e}_\rho + \left(\frac{\partial A_\rho}{\partial z} - \frac{\partial A_z}{\partial \rho}\right)\boldsymbol{e}_\varphi$$
$$+\frac{1}{\rho}\left(\frac{\partial}{\partial \rho}\left(\rho A_\varphi\right) - \frac{\partial A_\rho}{\partial \varphi}\right)\boldsymbol{e}_z$$

$$\boldsymbol{\Delta} V = \frac{1}{\rho}\frac{\partial}{\partial \rho}\left(\rho\frac{\partial V}{\partial \rho}\right) + \frac{1}{\rho^2}\frac{\partial^2 V}{\partial \varphi^2} + \frac{\partial^2 V}{\partial z^2}$$

C.6 Angular Functions

This section of the appendix contains often used formulae with the Legendre polynomials, the Legendre functions and the spherical harmonics.

C.6.1 Legendre Polynomials $P_l(x)$

- Differential equation:

$$(1-x^2)\,\frac{\mathrm{d}^2 S_l(x)}{\mathrm{d}x^2} - 2x\,\frac{\mathrm{d}S_l(x)}{\mathrm{d}x} + l(l+1)S_l(x) = 0$$

$$x = \cos\theta \qquad l = 0,\ 1\,2,\ \ldots$$

- Fundamental system:

$$S_l(x) \longrightarrow P_l(x) = \sum_{n=0}^{l/2} a_{2n,l}\, x^{2n} \quad l = \text{even}$$

$$P_l(x) = \sum_{n=0}^{(l-1)/2} a_{2n+1,l}\, x^{2n+1} \quad l = \text{odd}$$

$$Q_l(x) = \sum_{n=0}^{\infty} a_{2n+1,l}\, x^{2n+1} \quad l = \text{even}$$

$$Q_l(x) = \sum_{n=0}^{\infty} a_{2n,l}\, x^{2n} \quad l = \text{odd}$$

- Recursion formula for the expansion coefficients:

$$a_{n+2,l} = \frac{n(n+1) - l(l+1)}{(n+1)(n+2)}\, a_{n,l}$$

- Symmetry:

$$P_l(x) = (-1)^l\, P_l(-x)$$

- Polynomials in lowest order, (normalisation $P_l(1) = 1$):

$$P_0(x) = 1$$

$$P_1(x) = x$$

$$P_2(x) = \frac{1}{2}\left(3\,x^2 - 1\right)$$
$$P_3(x) = \frac{1}{2}\left(5\,x^3 - 3\,x\right)$$
$$P_4(x) = \frac{1}{8}\left(35\,x^4 - 30\,x^2 + 3\right)$$
$$P_5(x) = \frac{1}{8}\left(63\,x^5 - 70\,x^3 + 15x\right)$$

- Generating function:

$$\frac{1}{\left[1 - 2hx + h^2\right]^{1/2}} = \sum_{l=0}^{\infty} h^l\,P_l(x)$$

- Recursion relations, selection:

$$(l+1)\,P_{l+1}(x) = (2l+1)x\,P_l(x) - l\,P_{l-1}(x)$$

$$x\,\frac{\mathrm{d}P_l(x)}{\mathrm{d}x} = \frac{\mathrm{d}P_{l-1}(x)}{\mathrm{d}x} + l\,P_l(x)$$

$$\frac{\mathrm{d}P_{l+1}(x)}{\mathrm{d}x} = \frac{\mathrm{d}P_{l-1}(x)}{\mathrm{d}x} + (2l+1)\,P_l(x)$$

- The formula of Rodriguez:

$$P_l(x) = \frac{1}{2^l\,l!}\,\frac{\mathrm{d}^l}{\mathrm{d}x^l}\left[(x^2-1)^l\right]$$

- Integrals with Legendre polynomials, selection: A basic formula:

$$\int_{-1}^{1}\mathrm{d}x\ f(x)\ P_l(x) = \frac{(-1)^l}{2^l\,l!}\int_{-1}^{1}\mathrm{d}x\ \frac{\mathrm{d}^l f(x)}{\mathrm{d}x^l}\ (x^2-1)^l$$

Some special cases:

$$\int_{-1}^{1}\mathrm{d}x\ P_l(x)\ P_m(x) = \delta_{l,m}\,\frac{2}{(2l+1)}$$

$$\int_{-1}^{1} \mathrm{d}x\, x^m\, P_l(x) = \begin{cases} 0 & l > m \\ 2\dfrac{m!}{(m-l)!}\dfrac{(m-l-1)!!}{(m+l+1)!!} & (m-l)\text{ even, } m \geq l \\ 0 & (m-l)\text{ odd} \end{cases}$$

$$\int_{0}^{1} \mathrm{d}x\, P_0(x) = 1 \qquad \int_{0}^{1} \mathrm{d}x\, P_1(x) = \frac{1}{2}$$

$$\int_{0}^{1} \mathrm{d}x\, P_l(x) = \begin{cases} 0 & l > 0, \text{ even} \\ (-1)^{(l-1)/2}\dfrac{(l-2)!!}{2^{l+1)/2}(l+1)/2)!} & l > 1, \text{ odd} \end{cases}$$

- Legendre series:
  $\longrightarrow$ Representation of a function $f(x)$ in the interval $[-1, 1]$

$$f(x) = \sum_{l=0}^{\infty} A_l P_l(x) \qquad A_l = \frac{(2l+1)}{2} \int_{-1}^{1} f(x) P_l(x)\, \mathrm{d}x$$

C.6.2 The Functions $Q_l(x)$

- Functions in low order:

$$Q_0(x) = \frac{1}{2} \ln\left(\frac{1+x}{1-x}\right) = \sum_{n=0}^{\infty} \frac{x^{2n+1}}{(2n+1)}$$

$$Q_1(x) = \frac{1}{2} x \ln\left(\frac{1+x}{1-x}\right) - 1$$

- Recursion formula:
  The functions Q_l satisfy the same recursion formulae as the polynomials P_l, e.g.

$$Q_{l+1}(x) = (2l+1) x Q_l(x) - l Q_{l-1}(x)\,.$$

C.6.3 Associated Legendre Functions

- Differential equation:

$$(1-x^2)\,\frac{\mathrm{d}^2 P_l^m(x)}{\mathrm{d}x^2} - 2x\,\frac{\mathrm{d}P_l^m(x)}{\mathrm{d}x} + \left(l(l+1) - \frac{m^2}{(1-x^2)}\right) P_l^m(x) = 0$$

- Fundamental system:

$$P_l^m(x) = (-1)^m (1-x^2)^{m/2}\,\frac{\mathrm{d}^m P_l(x)}{\mathrm{d}x^m} \qquad m \geq 0$$

$$Q_l^m(x) = (-1)^m (1-x^2)^{m/2}\,\frac{\mathrm{d}^m Q_l(x)}{\mathrm{d}x^m} \qquad m \geq 0$$

- Functions in lowest order:

$$P_1^1(x) = -(1-x^2)^{1/2} = -\sin\theta$$

$$P_2^1(x) = -3x(1-x^2)^{1/2} = -\frac{3}{2}\sin 2\theta$$

$$P_2^2(x) = 3(1-x^2) = \frac{3}{2}(1-\cos 2\theta)$$

$$P_3^1(x) = -\frac{3}{2}(5x^2-1)(1-x^2)^{1/2} = -\frac{3}{8}(\sin\theta + 5\cos 3\theta)$$

$$P_3^2(x) = 15x(1-x^2) = \frac{15}{4}(\cos\theta - \cos 3\theta)$$

$$P_3^3(x) = -15(1-x^2)^{3/2} = -\frac{15}{4}(3\sin\theta - \sin 3\theta)$$

In addition one has

$$P_l^0(x) = P_l(x)$$

- Recursion formulae, selection:

$$(l+1)P_{l+1}^m(x) = (2l+1)[x\,P_l^m(x) - m\sqrt{1-x^2}\,P_l^{m-1}(x)] - l\,P_{l-1}^m(x)$$

$$x\,P_l^m(x) = P_{l-1}^m(x) - (l-m+1)\sqrt{1-x^2}\,P_l^{m-1}(x)$$

$$(l-m+1)P_{l+1}^m(x) - (2l+1)x\,P_l^m(x) + (l+m)P_{l-1}^m(x) = 0$$

- The formula of Rodriguez:

$$P_l^m(x) = \frac{(-1)^m}{2^l l!}(1-x^2)^{m/2}\frac{\mathrm{d}^{l+m}}{\mathrm{d}x^{l+m}}\left((x^2-1)^l\right)$$

This formula is only valid for $-l \leq m \leq l$. The functions with $+|m|$ and $-|m|$ are linearly dependent. The symmetry relation

$$P_l^{-m}(x) = (-1)^m \frac{(l-m)!}{(l+m)!} P_l^m(x) \qquad m > 0$$

is valid.

- An integral with P_l^m:

$$\int_{-1}^{1} \mathrm{d}x\, P_l^m(x)\, P_{l'}^m(x) = \frac{(l+m)!}{(l-m)!}\int_{-1}^{1} \mathrm{d}x\, P_l(x)\, P_{l'}(x)$$

$$= \delta_{l,l'}\,\frac{(l+m)!}{(l-m)!}\frac{2}{(2l+1)}$$

Note:

$$\int_{-1}^{1} \mathrm{d}x\, P_l^m(x)\, P_{l'}^{m'}(x) \neq \delta_{l,l'}\,\delta_{m,m'}\, I(l,m)$$

C.6.4 Spherical Harmonics

- Definition:

$$Y_{l,m}(\theta,\varphi) \equiv Y_{l,m}(\Omega) = \left[\frac{(2l+1)}{4\pi}\frac{(l-m)!}{(l+m)!}\right]^{1/2} P_l^m(\cos\theta)\, \mathrm{e}^{\mathrm{i}\,m\varphi}$$

in the region

$$0 \leq \theta \leq \pi \qquad 0 \leq \varphi \leq 2\pi\,.$$

Note: The real form of these functions

$$P_l^m(\cos\theta)\cos m\,\varphi \quad \text{and} \quad P_l^m(\cos\theta)\cos m\,\varphi, \quad m \geq 0$$

can also be used

- Special cases:

$$Y_{l,0}(\theta,\varphi) = \left[\frac{(2l+1)}{4\pi}\right]^{1/2} P_l(\cos\theta)$$

$$Y_{l,m}(0,\varphi) = \left[\frac{(2l+1)}{4\pi}\right]^{1/2} \delta_{m,0}$$

- Symmetry relation:

$$Y_{l,-m}(\theta,\varphi) = (-1)^m Y^*_{l,m}(\theta,\varphi)$$

The spherical harmonics

$$Y_{l,m}(\theta,\varphi) \quad \text{and} \quad Y_{l,-m}(\theta,\varphi)$$

are linearly independent!

- Basic integral:

$$\iint \mathrm{d}\Omega\, Y^*_{l,m}(\theta,\varphi) Y_{l',m'}(\theta,\varphi)$$

$$= \int_0^{2\pi} \mathrm{d}\varphi \int_0^{\pi} \sin\theta\ \mathrm{d}\theta\, Y^*_{l,m}(\theta,\varphi) Y_{l',m'}(\theta,\varphi)$$

$$= \delta_{l,l'}\delta_{m,m'}$$

- Differential equation:

$$\frac{1}{\sin\theta}\frac{\partial}{\partial\theta}\left(\sin\theta\frac{\partial Y_{l,m}(\theta,\varphi)}{\partial\theta}\right) + \frac{1}{\sin^2\theta}\frac{\partial^2 Y_{l,m}(\theta,\varphi)}{\partial\varphi^2}$$

$$+ \qquad l(l+1)Y_{l,m}(\theta,\varphi) = 0$$

- Functions in low order:

$$Y_{0,0} = \sqrt{\frac{1}{4\pi}}$$

$$Y_{1,-1} = \sqrt{\frac{3}{8\pi}} \sin\theta\, \mathrm{e}^{-\mathrm{i}\varphi} \qquad Y_{1,0} = \sqrt{\frac{3}{4\pi}} \cos\theta$$

$$Y_{1,1} = -\sqrt{\frac{3}{8\pi}} \sin\theta\, \mathrm{e}^{\mathrm{i}\varphi} \qquad Y_{2,-2} = \sqrt{\frac{15}{32\pi}} \sin^2\theta\, \mathrm{e}^{-2\mathrm{i}\varphi}$$

$$Y_{2,-1} = \sqrt{\frac{15}{8\pi}} \sin\theta \cos\theta\, \mathrm{e}^{-\mathrm{i}\varphi} \qquad Y_{2,0} = \sqrt{\frac{5}{16\pi}} (3\cos^2\theta - 1)$$

$$Y_{2,1} = -\sqrt{\frac{15}{8\pi}} \sin\theta \cos\theta\, \mathrm{e}^{\mathrm{i}\varphi} \qquad Y_{2,2} = \sqrt{\frac{15}{32\pi}} \sin^2\theta\, \mathrm{e}^{2\mathrm{i}\varphi}$$

- Addition theorem:

$$\begin{aligned} P_l(\cos\alpha) &= \frac{4\pi}{(2l+1)} \sum_{m=-l}^{l} Y_{l,m}(\theta,\varphi) Y^*_{l,m}(\theta',\varphi') \\ &= \frac{4\pi}{(2l+1)} \sum_{m=-l}^{l} Y^*_{l,m}(\theta,\varphi) Y_{l,m}(\theta',\varphi') \end{aligned}$$

with the angle α between the two vectors $\boldsymbol{r}$ and $\boldsymbol{r}'$

$$\cos\alpha = \cos(\varphi - \varphi')\ \sin\theta\ \sin\theta' + \cos\theta \cos\theta' .$$

C.7 Radial Functions

The following two functions can represent the radial solution of the Poisson equation, after the substitution $x = \text{const.}\ r$, as many elementary and not so elementary functions are related to them. The properties of these functions are collected here without the related proofs, which can be found e.g. in M. Abramovitz, I. Stegun: 'Handbook of Mathematical Functions' Appendix A.5.

C.7.1 The Hypergeometric Functions $F(a, b, c; x)$

The solutions of the differential equation

$$x(1-x)\frac{\mathrm{d}^2 F(x)}{\mathrm{d}x^2} + [c - (a+b+1)x]\frac{\mathrm{d}F(x)}{\mathrm{d}x} - abF(x) = 0 \tag{C.1}$$

are the **hypergeometric functions**. The solution, which is regular at $x = 0$, is given by the Gauss hypergeometric series

$$F(a, b; c; x) = 1 + \frac{ab}{c}x + \frac{a(a+1)b(b+1)}{c(c+1)}\frac{x^2}{2!} + \ldots$$

$$= \frac{\Gamma(c)}{\Gamma(a)\Gamma(b)} \sum_0^{\infty} \frac{\Gamma(a+n)\Gamma(b+n)}{\Gamma(c+n)} \frac{x^n}{n!} . \tag{C.2}$$

The series terminates, if a or b are a negative integer, it is not defined if $c = -n$, except if a or b is a negative integer $-m$ with $m < n$. The radius of convergence of the series is $|x| = 1$ with the explicit statements ($a,\ b,\ c$ are assumed to be real)

- $(a + b - c) < 0$: absolute convergence on the complete unit circle.
- $0 \le (a+b-c) < 1$: convergence on the unit circle with the exception of $x = 1$.
- $1 \le (a + b - c)$: divergence of the complete unit circle.

The formulae for the derivatives of the hypergeometric series

$$\frac{\mathrm{d}F(a, b; c; x)}{\mathrm{d}x} = \frac{ab}{c}F(a+1, b+1; c+1; x)$$

$$\frac{\mathrm{d}^2F(a, b; c; x)}{\mathrm{d}x^2} = \frac{a(a+1)b(b+1)}{c(c+1)}F(a+2, b+2; c+2; x)$$

etc.

can be derived directly on the basis of the definition, as well as a substantial number of recursion relations as e.g.

$$cF(a, b; c; x) - (c-b)F(a, b; c+1; x) - bF(a, b+1; c+1; x) = 0$$

$$cF(a, b; c; x) - (c-a)F(a, b; c+1; x) - aF(a+1, b; c+1; x) = 0 .$$

The integral representation

$$F(a, b; c; x) = \frac{\Gamma(c)}{\Gamma(b)\Gamma(c-b)} \int_0^1 \mathrm{d}t\, t^{b-1}(1-t)^{c-b-1}(1-xt)^{-a}$$

is useful to derive several formulae for transformations of the function as e.g.

$$F(a,\, b;\; c;\; x) = (1-x)^{-a} F\left(a,\, c-b;\; c;\; \frac{x}{x-1}\right)$$

$$F(a,\, b;\; c;\; x) = (1-x)^{c-a-b} F(c-a,\, c-b;\; c;\; x)\,.$$

The list of functions, which can be represented by the hypergeometric series, contains among others

- the elementary functions

$$\ln(1+x) = xF(1,\, 1;\; 2;\; -x)$$

$$\arcsin x = xF\left(\frac{1}{2}, \frac{1}{2};\; \frac{3}{2};\; x^2\right),$$

- the complete elliptic integrals

$$K(k) = \int_0^{\pi/2} \mathrm{d}\varphi\,(1-k^2\sin^2\varphi)^{-1/2} = \frac{\pi}{2} F\left(\frac{1}{2}, \frac{1}{2};\; 1;\; k^2\right)$$

$$E(k) = \int_0^{\pi/2} \mathrm{d}\varphi\,(1-k^2\sin^2\varphi)^{1/2} = \frac{\pi}{2} F\left(-\frac{1}{2}, \frac{1}{2};\; 1;\; k^2\right)$$

  and the Legendre polynomials

$$P_l(x) = F(-l,\, l+1;\; 1;\; \frac{(1-x)}{2})\,.$$

C.7.2 The Confluent Hypergeometric Functions $F(a, c\,; x)$

The differential equation of the **confluent hypergeometric functions**

$$x\frac{\mathrm{d}^2F(x)}{\mathrm{d}x^2} + [c-x]\frac{\mathrm{d}F(x)}{\mathrm{d}x} - aF(x) = 0 \tag{C.3}$$

has a regular singularity at $x = 0$ and an irregular singularity at $x = \infty$. The similarity with the differential equation of the hypergeometric functions shows, that both differential equations are special cases of a higher level differential equation, Riemann's differential equation.

The solution, which is regular at the origin $x = 0$

$$F(a,\, c\,;\, x) = 1 + \frac{a}{c}x + \frac{a(a+1)}{c(c+1)}\frac{x^2}{2!} + \ldots \tag{C.4}$$

$$= \frac{\Gamma(c)}{\Gamma(a)}\sum_0^{\infty}\frac{\Gamma(a+n)}{\Gamma(c+n)}\frac{x^n}{n!}$$

is called the **Kummer's function** or confluent hypergeometric series. The behaviour of the convergence depends on the properties of the parameters a and c, that is (m and n are positive integers)

- $c \neq -m,\ a \neq -n$: The series converges for all values of x .
- $c \neq -m,\ a = -n$: The solution is a polynomial of degree m .
- $c = -m,\ a \neq -n$ or $c = -m,\ a \neq -n$ with $n \leq m$: There is a simple pole for the parameter value $c = -m$.

For this function exist formulae for the derivatives, recursion relations and integral representations. Examples are

$$\frac{\mathrm{d}F(a,\, c\,;\, x)}{\mathrm{d}x} = \frac{a}{c}F(a+1,\, c+1\,;\, x)$$

and

$$cF(a,\, c\,;\, x) - (c-a)F(a,\, c+1\,;\, x) - aF(a+1,\, c+1\,;\, x) = 0\,,$$

as well as the integral representation

$$F(a,\, c\,;\, x) = \frac{\Gamma(c)}{\Gamma(a)\Gamma(c-a)}\int_0^1 \mathrm{d}t\, \mathrm{e}^{xt} t^{a-1}(1-t)^{c-b-1}\,.$$

The asymptotic behaviour of the function is

$$\lim_{x\to+\infty} F(a,\, c\,;\, x) = \frac{\Gamma(c)}{\Gamma(a)}\mathrm{e}^x x^{a-c}$$

and

$$\lim_{x\to-\infty} F(a,\, c\,;\, x) = \frac{\Gamma(c)}{\Gamma(c-a)}(-x)^{-a}\,.$$

The confluent hypergeometric series can also represent a good number of functions,as for instance the exponential function

$$e^x = F(a,\, a\,;\, x)\,,$$

the trigonometric functions as

$$\sin x = x\, e^{-ix} F(1,\, 2;\, -2ix)$$

and the Bessel functions

$$J_\nu(x) = \frac{x^\nu\, e^{-ix}}{2^\nu \Gamma(\nu+1)} F\left(\nu + \frac{1}{2},\, 2\nu+1\,;\, 2ix\right)$$

$$j_l(x) = \frac{x^{l-1}\, e^{-ix}}{2^l \Gamma(l+3/2)} F(l+1,\, 2l+2\,;\, 2ix)\,.$$

Index

A
Addition theorem
  spherical harmonics, 352
Ampère's
  definition, 8, 333
  differential equation, 250
  law, 249
Azimuthal symmetry
  Dirichlet problems, 148
  examples, 104

B
Bessel function
  differential equation, 120
Biot-Savart's law, 254
Bohr magneton, 273
Boundary value problem
  Dirichlet, 144
  simple, 98
  Sturm's, 220
  uniqueness of solution, 146
  von Neumann, 145

C
Capacitor, 32
  capacity, 175
    partial, 178
  coefficients of capacity, 177
  cylindrical, 177
  dielectric materials, 180
  ideal spherical, 32, 175
  plate, 176
Cauchy-Riemann
  differential equation, 199
CGS-system, 9, 248, 280, 331
Charge
  conservation, 6
  elementary, 2
  free, 182
  mirror, 147
  polarisation, 182
  unit, 8
Charge density, 18
  linear, 20
  space, 18, 24
  surface, 19
Charge distribution, 28, 53, 63
  dipole moment, 113
  polarisation, 194
  surface, 148
Coefficient
  of capacity, 177
Coercive force, 284
Coercivity, *see* Coercive force
Condensor, *see* Capacitor
Condition
  boundary
    Dirichlet, 144
    von Neumann, 145
Conductor, 5
  electric field, 58
Confluent hypergeometric functions, 354
Conform mapping, 207
Constant
  dielectric, 180, 334
Continuity conditions
  magnetic fields, 282
Coulomb
  definition, 8
  force law, 7
  gauge, 262
Current density, 241
  magnetisation, 279
Cylinder function, 120

© The Author(s), under exclusive license to Springer-Verlag GmbH, DE, part of Springer Nature 2024
R. M. Dreizler, C. S. Lüdde, *Electrostatics and Magnetostatics*,
https://doi.org/10.1007/978-3-662-69933-1

D
Delta-function, 63
Diamagnetism, 275
Dielectric
constant, 180, 334
vacuum, 9
displacement, 183
material, 5
normal, 182
polar, 181
polarisation, 180
tensor, 187
Differential equation
Ampère, 250
associated Legendre, 117
Bessel function, 120
Cauchy-Riemann, 199
Green's function, 162
Laplace, 54, 162
Legendre, 109
Poisson, 54, 78, 97
separation constant, 108
separation of variables, 106
Diffraction
law of, 190
Dipole
electric field, 13, 39
field lines, 17
insulator, 182
macroscopic, 31
magnetic, 246
potential, 81
multipole expansion, 124
Dipole moment, 16, 128
charge distribution, 113
magnetic, 247
current loop, 270
permanent, 181
Dirichlet
boundary value problem, 144
mirror charge, 147

E
Electric constant, 9
Electric field
complex potential, 199
conductors, 58
constant, 9
dipole, 13
disk, 22, 43
energy storage, 90
flux, 27
linear charge density, 20
normal component, 57
point charge, 11, 49
quadrupole, 17
ring, 20
sphere, 24, 44
hollow, 30
stationary, 49
tangential component, 57
visualisation, 12
vortex free, 50
Electric susceptibility, 185
Elliptic integrals, 86, 95, 264, 354
Energy density
dielectric material, 191
electric, 92
Equation of continuity, 242
Experiment
Millikan, 3
Plimpton and Lawton, 10

F
Faraday's
cage, 61
Ferromagnetism, 275, 291
Field lines
dipole, 17
law of diffraction, 190
magnetic, 247, 253
mirror charge, 152
point charge, 12
quadrupole, 17
Flux
electric, 27, 200
magnetic, 274
Force
coercive, 284
magnetic
conductors, 295
Lorentz, 293
Function
Bessel, 120
confluent hypergeometric, 354
cylinder, 120
delta, 63
Green's, 162
Heaviside, 285
hypergeometric, 88, 352
Kummer, 355
Legendre, 109
associated, 349
spherical harmonics, 350

G
Gauge
  Coulomb, 262
  transformation, 260
Gauss'
  integral theorem, 25
  law, 49
Gaussian
  box, 55, 188, 284
Gauss-system, *see* CGS-system
Gauss' theorem, 25
  application, 229
  differential form, 52
  integral form, 51
Green's
  first integral theorem, 164
  second integral theorem, 164
  theorem, 164
Green's function, 162
  Dirichlet boundary condition, 166
  simple boundary condition, 166
  symmetry, 163
  von Neumann boundary condition, 167

H
Harmonic analysis, 201
Heaviside function, 285
Hypergeometric function, 88, 352
Hysteresis loop, 283

I
Induction
  electrostatic, 59
  magnetic, 280
  surface charge, 148
Integrals
  elliptic, 86, 95, 264, 354

K
Kirchhoff's rule, 243

L
Laplace
  differential equation, 162
  equation, 54
    cylindrical coordinates, 120
    spherical coordinates, 106, 114
  operator, 54
Law
  Ampère, 249
  Biot-Savart, 254
  Coulomb, 7
  of diffraction, 190
  Gauss', 49
  Lorentz, 293
  Ohm, 244
Legendre
  associated functions, 349
  differential equation, 109
  function, 109
  polynomials, 109, 346
    associated, 117
  series, 348
Light
  velocity, 12
Lorentz's law, 293

M
Magnetic
  dipole, 246
  dipole moment, 247
    current loop, 270
  permeability, 274
  susceptibility, 283
  vector potential, 259
Magnetic field, 246
  continuity conditions, 282
  current ring, 256, 262
  multipole expansion, 267
  solenoid, 257
  sphere, 285
  strength, 280
  thin, straight conductor, 254
Magnetisation, 278
Magnetisation current density, 279
Magneton
  Bohr, 273
  nuclear, 273
Material equations, 187, 282
Millikan experiment, 3
Mirror
  charge, 147
Multipole expansion, 112, 123
  dipole potential, 124
  magnetic field, 267
  quadrupole, 125
  spherical components, 130

O
Ohm's law, 244
Operator
  Laplace, 54

P
Paramagnetism, 275
Permeability

magnetic, 274
vacuum, 247
Plimpton and Lawton experiment, 10
Poisson
differential equation, 97
equation, 54, 78, 97
Polarisation, 61, 182
Potential
complex, 197
dipole
multipole expansion, 130
electric, 53, 78
dielectric cylinder, 202
dielectric sphere, 192
dipole, 81
ellipsoid, 128
Green's function, 165
metal sphere and point charge, 148, 154
metal sphere in electric field, 158
point charge, 124
sphere, 84
spherical charge distribution, 101
superposition principle, 80, 103
thin ring, 84
two charged spheres, 103
units, 79
magnetic vector, 259
quadrupole, 125
Principle
superposition, 12, 27, 80, 103
Problem
boundary value
Dirichlet, 144
simple, 98
Sturm's, 220
von Neumann, 145

R
Retardation, 12
Rule
Kirchhoff's, 243

S
SI-system, 8, 247, 283, 331
Specific electric conductivity, 244
Spherical harmonics, 350
addition theorem, 352
Stokes curve, 57
Stokes' theorem, 50
Sturm's boundary value problem, 220
Superposition principle, 12, 27, 80, 103
Susceptibility
electric, 185
magnetic, 283
System of units
CGS, 9, 248, 280, 331
Gauss, 9
SI, 8, 247, 283, 331

T
Tensor
dielectric, 187
quadrupole, 131
Theorem
divergence, 52
Gauss', 25
Green's, 164
first integral, 164
second integral, 164
Stokes, 50

V
Vector potential
current loop, 268
current ring, 262
magnetic, 258
Velocity
light, 12
von Neumann
boundary value problem, 145